Praise for

A Mood Apart

"A reader-friendly compendium of the most significant research and the most profound thoughts devoted to mood and mood disorders."
—*Philadelphia* magazine

"Seldom has the inner emotional landscape of melancholic depression, mania and manic-depressive illness been mapped with so much clarity, empathy and sensitivity." —*Publishers Weekly*

"A most graceful introduction to the science of mood, by one of the world's experts." —Peter D. Kramer, author of *Listening to Prozac*

"Dr. Peter Whybrow explains everything you ever wanted to know about moods and their vicissitudes, and he does so in wonderfully elegant, highly readable prose."
—Maggie Scarf, author of *Unfinished Business* and *Intimate Worlds*

"Dr. Whybrow has written a wise and graceful book that should be helpful and consoling to patients and their loved ones."
—Irvin D. Yalom, author of *Love's Executioner* and *Lying on the Couch*

"*A Mood Apart* is a beautifully crafted volume that probes into the recesses of the brain and mind to reveal the secrets of depression. Dr. Whybrow distills a lifetime of experience as a master clinician and scientist. . . . I highly recommend this definitive work to the nonprofessional as well as the professional."—Aaron T. Beck, M.D., university professor emeritus, University of Pennsylvania, and author of *Cognitive Therapy of Depression*

"This compassionate book will console through its effective teaching. Anyone who has dealt with depression will be drawn to it."—Judith Rapoport, M.D., chief, child psychiatry, National Institute of Mental Health, and author of *The Boy Who Couldn't Stop Washing*

"*A Mood Apart* is a tour de force, a new standard in writing for the public. . . . A masterpiece. . . . His topic is complex, his explanation scientific and yet deeply rooted in compassion, understanding, perception, and respect for his patients."—Frank Burgmann, president, National Depressive and Manic-Depressive Association

A
Mood
Apart

The Thinker's Guide to Emotion
and Its Disorders

PETER C. WHYBROW, M.D.

■ HarperPerennial
A Division of HarperCollins Publishers

A hardcover edition of this book was published in 1997 by BasicBooks.

First HarperPerennial edition published 1998.

Designed by Elliott Beard

The Library of Congress has catalogued the hardcover edition as follows:

Whybrow, Peter C.
    A mood apart : depression, mania, and other afflictions of the self / by Peter C. Whybrow. — 1st ed.
        p.   cm.
    Includes index.
    ISBN 0-465-04725-4
    1. Affective disorders.   2. Self.   3. Personality.   I. Title.
    RC537.W487   1997
    616.89'5—dc21   96-47974

ISBN 0-06-097740-x (pbk.)

98 99 00 01 02 ❖/RRD 10 9 8 7 6 5 4 3 2 1

For my father,
Charles Ernest James Whybrow

A Mood Apart

Once down on my knees to growing plants
I prodded the earth with a lazy tool
In time with a medley of sotto chants;
But becoming aware of some boys from school
Who had stopped outside the fence to spy,
I stopped my song and almost heart,
For any eye is an evil eye
That looks in onto a mood apart.

Robert Frost
Steeple Bush (1947)

Contents

Acknowledgments

The experience that I write about in this book has been accumulated over a professional lifetime. Many people have contributed to my development as a psychiatrist, and it is impossible to thank each individually. Principal among those who have guided my thinking about these illnesses are those men and women who have suffered them, and who have shared their experience with me. Some of these individuals I have had the privilege of serving as a physician. Many colleagues and friends have also shaped my ideas. Some will find themselves mentioned in these pages; others, equally valued, know my affection for them. I thank them all.

I also wish to thank several individuals who played critical roles in the development and preparation of the manuscript: Suzanne Gluck, of International Creative Management, for her wisdom, commitment, enthusiasm, and also the title of *A Mood Apart;* Sally Arteseros for her support and gentle, but firm, editorial guidance; Helen Whybrow, my daughter and an editor, for her support, clarity of vision, and sound advice in crafting complex ideas; Karen Levinson for her creative and diligent pursuit of the science that is referenced in the book; Jo Ann Miller at Basic Books, who believed in such a book at the beginning, and John Donatich for his enthusiastic leadership, gracious tolerance, and guiding hand during its final months of development; Bonnie Clause and Rosellen Taraborrelli for their gift of time; and Eva Redei, my wife, who has ridden with me the waves of optimism and despair and reviewed many parts of the manuscript, for her valuable insights.

Many individuals and institutions have invested in my research and academic development. For particular contribution to this book I thank: The University of Pennsylvania, The National Institute of Mental Health, The Center for Advanced Study in the Behavioral Sciences at Stanford University, John D. and Catherine T. MacArthur Foundation, The National Alliance for Research in Schizophrenia and Depression, Ethel B. Wister, Joan Piane Fowler, and the M. A. Berman Foundation.

Prologue

The Masks of
the Human Carnival

Late in September 1995, when this book was already well advanced, the European College of Neuropsychopharmacology met in Venice, Italy. One afternoon, our scientific contribution behind us and seeking relief from pharmaceutical exhibits extolling the latest antidepressant, my wife and I left the lush greenery of the Lido (where the conference was being held) and caught an early *vaporetto* back to our hotel. For the rest of the day Eva and I planned to lose ourselves in the dusty labyrinth of the Dorsoduro. We understood from trusted Italian friends that there, within the maze of narrow lanes and minor canals, we would find artists who still created masks in the tradition of the Venetian carnival.

Venice is a city of unique organic form, a place where one is "taught . . . life as a great splendor," as Diego Valeri has described it. Over the centuries, countless writers, painters, and poets have attempted to capture the essence of the creative human spirit in this city of special mood, for Venice is a tribute to both human ingenuity and human emotion. The Venetian carnival, like carnivals throughout the world, is the quintessential celebration of that emotional experience. The creativity, excitement, music, feasting, and sheer self-indulgence of the carnival magnify the emotional energy of everyday life; it is a maniacal springtime catharsis before the atonement and withdrawal of Lent.

The mask has a particular place in carnival revelry. Most obviously, and also most importantly, it covers the reveler's face, hiding both identity and the expression of true emotional feeling. The mask itself usually caricatures feeling, commonly stereotyping deep sadness or great joy, moods which have universal meaning and stand at the poles of human emotion. It is the mask, through what it hides and what it proclaims, that lies at the center of the carnival ritual. Without the mask there is no masquerade.

This book, *A Mood Apart*, is about mood and human emotion, about their function and purpose in our lives. Particularly it is about emotional disorder, about mania and melancholic depression, states of mind that, like masks and carnivals, caricature the emotions of everyday life. In contrast to the masquerade of the carnival, however, mania and melancholia are extremes of mood that touch us all; in their presence rarely can we choose to stand aside as nonparticipant observers. In fact, the chances are overwhelming that most of us, during the span of a normal lifetime, will come face to face with some manifestation of mania or melancholic depression, seeing it in ourselves or in somebody close to us.

So for Eva and me, our quest that afternoon had a particular symbolic purpose. In Valeri's "city of generous madness," at a conference dedicated to the science of the anomalous mind, we hoped to find an ancient and universal link between the carnival of human emotion and the experience of aberrant mood. We sought for ourselves Venetian masks of mania and melancholia.

After an hour or so of confusion, following complex geometric paths, we came upon a workshop that caught our eye. MondoNovo Maschere sits in a little alley off the Ponte dei Pungi, a bright blue door puncturing the exposed anatomy of an ancient wall. Standing sentinel, eyes narrowed, is a camel's head, while a cradled and smiling moon, flanked by modern gargoyles, beckons the visitor inside. After the sunlight of the street, the room is dim. It takes a moment to define the multitude of faces, of every expression and form, that stare from tables, walls, and ceiling. But, as the eyes adjust, the impish humor of the artist-owners, Lovato Guerrino and Spillar Giorgio, becomes rapidly apparent.

The first response to my request, made in mixed and gestured language, produced a mask of alabaster white and sinister black—three masks, in fact, each fused to the other. "These are the faces of life," explained Mariangèla Convento, the young assistant, as she held it before her smiling eyes. "Sometimes we are sad, sometimes happy, and then, of course, there's the other face, the one we keep for ourselves." We smiled back, uncertain that this was the object of our search. "But perhaps this is the one," she added, sensing our reserve. Reaching up, she took from behind a beam a mask with a cherubic face, bedecked in jester's garb, with a coxcomb hood of red and green and bells that jingled from each ear. Turned back and forth in Mariangèla's expert hands, the jester's changing moods were real. In one profile the mask displayed the tears and down-turned mouth of a melancholy clown, while in the other, the jester stood transformed with a vibrant eye and infectious grin. Here was the carnival mask of manic flight and melancholic despair.

In their extreme form, mania and melancholic depression are illnesses. Mania is to depression as fire is to the ashes of the fire. Mania, in Latin, literally means madness, while melancholy is seen as something more noble—even precious. In their disruptive, psychotic forms, mania and melancholic depression can ruin lives, and even end life through suicide. But in their development, and in their muted forms—chronic smoldering sadness, emotional instability, or infectious energy—these illnesses also include behaviors that shade into the daily emotional pleasures, hopes, and disappointments that all of us associate with being alive. These are remarkable and very human disorders, for mania and melancholia exaggerate life.

As is emphasized by the universal meaning of the jester's mask, mania and melancholia are "illnesses of mood" that occupy a special place in the experience of being human. Changing moods, and the emotions that generate them, are central to being alive. Depression and manic depression are thus more than illnesses in the everyday sense of the term. They cannot be understood, as are disorders of other body organs, merely as an aberrant biology that has invaded the brain, for by disturbing the function of the brain—the emotional brain—the illness enters and disturbs the *person,* that collection of feelings, behaviors, and beliefs that uniquely identify the individual self. These afflictions invade and change the very core of being.

As we have come to understand more about the brain, its emotional chemistry and the drugs that change it, these mood states have assumed for many a new significance, even a fascination. Once hidden as a mark of personal weakness, melancholic depression and manic depression (a common form of illness in which mania and depression occur together) have become the object of curiosity and are openly discussed at home, in magazine columns, and in the workplace. After centuries of silence, questions are being asked. When are sadness and swings of mood—those changes that seem so appropriate and universal in the struggle of daily existence—the warning signs of a painful illness? Are these only family illnesses, passed down through the generations, or can anybody acquire them through bad experience? Do bodily changes occur in these illnesses, and if so, is the anatomy of the brain involved? What are these powerful "antidepressant" drugs that some claim can change personality? How do they alter the brain chemistry, and how does that chemistry change something as complicated and personal as our private emotions? Is it true that many successful and famous people have suffered these illnesses or experienced them in their families? If it is true, why, with present knowledge, does stigma still remain? These are just a few of the questions which I shall explore in this book.

My approach to mood disorders is that of the physiologist. That is the way I was trained as a physician in London three decades ago, and it is the approach that I continue to follow. Physiology is the science of the great French physician Claude Bernard; it is a dynamic approach to understanding the integrity of the human body, including the human brain, which accommodates the larger environment in which we live and interact. A central postulate is that the individual elements of the body work together in consort with the whole, and thus, only through consideration of the elements *and* the whole can we hope to understand the living person. Thus the student of physiology must pay attention to both molecular detail and to the behavior of the entire animal. In my science and in my clinical responsibilities I have tried to adhere to such principles.

I saw my first victim of mania as a high school student when one summer in England I worked as an attendant in a mental hospital, located just a bicycle ride away from the small country town where I grew up. Beyond the sufferer's obvious loss of reason and his pain from woefully inadequate treatment, I did not know what I was witnessing. I was frightened and fascinated at the same time. As I look back across those few decades, I find it remarkable how much the therapies in psychiatry have changed and how I have changed in thinking about mood and its disorder. This book is my attempt to document what I have learned and, through the exploration of the experience of men and women I personally have known, to offer some insights about these illnesses, placing them within an appropriate context of humanism and medical science.

During my professional lifetime there has been a revolution. We have learned more about the molecular structure of the brain—the organ that defines us as individuals—in the past thirty years than we did throughout the previous two thousand. Major advances in new drugs and in the brain sciences have given us effective treatments for mood disorders, plus a much better understanding of how the brain chemistry is altered in these disturbed emotional states. The advances have also brought recognition of the enormous complexity of the human brain; how it has developed from a rudimentary collection of nerve cells in simple creatures to the extraordinary intercommunication of billions of such cells in complicated animals such as ourselves.

With this extraordinary explosion of knowledge has come the capacity to better assist those who suffer most. The treatments now available can alleviate and prevent recurring mood disorder and offer a new dimension in self-care. They are interventions as powerful and effective as any in medicine. But all this molecular insight does not replace the knowledge to be gained from understanding the personal experience. Rather, it is complementary.

Mania and melancholic depression are intensely personal illnesses. Although they stand as true aberrations of thought and feeling, extending the range of what we consider "normal" emotional experience, they remain accessible to all of us through empathic understanding and thoughtful dissection of such common mood states as profound grief and great joy. It is within this spirit of introspection and purposeful analysis that I begin *A Mood Apart* with the personal memory of my father's funeral. Then, despite the scientific abstraction that must follow, I will have the comfort of emphasizing, through the personal grief and disorganization of that day, how firmly I believe our advancing knowledge of these illnesses must be rooted in the common denominator of mutual experience. The subjective experience of melancholic depression is an extension of common sadness and grief, in the same way that some elements of mania are an extension of the compelling energy and happiness experienced when we first fall in love. Indeed, they are similar enough that both mania and depression are frequently dismissed or treated with blithe indifference (by the public and the professional alike) in the early stages of their development.

It is true that only a very small percentage among us develop the severe psychoses of mood that threaten life itself. For such individuals there is a special vulnerability. How that happens, and why, is an important part of this book. But the posture of blithe indifference to such suffering is unfounded, even dangerous, for the states of mania and depression each coexist, sometimes in the same individual, with behaviors that we all value as vital to human culture. Many individuals among us who are energetic, creative, and become great leaders, thinkers, and artists have been touched by these disorders of mood. Some such famous men and women have struggled with moments of profound anguish and melancholy, or have experienced periods of disorganization and poor judgment that characterize mania. Others have close relatives who have been severely ill with mood disorder. Many are increasingly willing to publicly acknowledge their struggles. Thus while the extreme moods of mania and melancholia may seem separate and apart from common experience, in fact the spectrum of human emotion is broad. Mood states that have great benefit in some walks of life coexist with others that can do great harm. It is one of the great ironies of Western culture that we revere and socially applaud sustained creativity, drive, and infectious enthusiasm but fail to recognize these qualities as close cousins of the disorganization and suspicion that accompany them in mania and are stigmatized as maniacal madness.

Human behavior is not, as some modern prophets would have us believe, biologically programmed and determined, as pigeon behavior is programmed in the brain of a pigeon. It is a mistake to equate the brain with something entirely organic and structural, a piece of computer hardware from which the programmed person springs automatically to life. Similarly, but at the other end of the philosophical continuum, I am wary of those who claim, from some inspired pulpit, that human destiny is determined entirely by experience. These two views reflect the shrill voices of reductionism that cry for some simple explanation of the way we are. In my experience, the path to understanding mood and its disorder follows not from some partisan theory of mind or matter, but from an intelligent integration of the two.

So in the consideration of moods—including the extremes of mania and melancholia—I favor a dynamic and integrated approach, one that combines objective science with an understanding of the personal experience. To highlight the importance of this integration, I present a series of personal stories, woven together with the clinical and basic science that is now the backbone of our rapidly advancing knowledge. The stories are real, and frequently include verbatim descriptions of illness and recovery, but the characters are composites of the life experiences of more than one person, transported in time and place to ensure anonymity. But I must emphasize that the human stories are as valid, in their contribution to our objective understanding of these moods that stand apart, as are the scientific studies with which they are interwoven.

This is not a conventional self-help book; rather, it is a compendium in the classical sense. It is a digest of what I know and what intrigues me as a practicing physician, about emotion, mood, and the emotional brain, about how the emotional self can become disordered, and about the treatment of those disordered states. The book is intended to be accessible to those who may know little about the brain and behavioral sciences, or about pharmacology, relying heavily upon the clinical stories to place the science in context. Hence the narrative is not exhaustive in its detail, as would be a text on the subject. For those so inclined, therefore, I have appended a series of notes, detailed in their reference, that expand on specific topics and will be of assistance to readers undertaking additional reading or personal research. The book also reflects something of my professional history, for woven through it are vignettes from my travels over the past twenty-five years as an academic psychiatrist. Throughout that education, which is a continuing adventure, I have learned many lessons, but one of particular importance has been that these illnesses of mood distort and magnify what is human; they do not destroy it. In sickness and in health we all share the same fundamental hopes and passions. Indeed there is much that these moods apart, and the emotional twists of life, teach us again and again as we are constantly reminded of their presence—through the personal experience of friendship and through the achievements of those engaged in the visual arts, in literature, in the opera, and in the leadership of nations.

It is my hope that *A Mood Apart* will be particularly helpful to those who themselves suffer from depression or manic depression, or who have experienced the illnesses through someone close to them. It is from such generous individuals that I have learned most about these extreme states of emotional experience. But I believe the subject will also be of interest to those who have struggled honestly with the "ups and downs" of human emotion, and to those who simply have a healthy curiosity about mood, wondering how it is regulated, the purpose that it serves, and why emotion occupies a central place in human affairs—why it is that the moody mask of the jester still survives among the workshops and alleys of the Dorsoduro, and in the celebrations of the human carnival.

One

A Glimpse of Melancholy

Grief, Emotional Homeostasis,
and Disorders of Mood

No one ever told me that grief felt so like fear.

> C. S. LEWIS
> *A Grief Observed* (1961)

After the mind has suffered from an acute paroxysm of grief, and the cause still continues, we fall into a state of low spirits.

> CHARLES DARWIN
> *The Expression of the Emotions*
> *in Man and Animals* (1872)

We buried my father according to his instructions, under the tree he had chosen in the little country churchyard. The weather that day was unsettled, in step, it seemed, with my inner feeling. All morning storm clouds had threatened, with gray hammerheads rising in the warm summer air. Then at noon, just as we gathered together, the heavens parted with a thunderous crack, and rain cascaded down in soaking sheets upon our bowed heads and the open ground.

My father had suffered a long and painful illness, so there had been ample time to come to terms with his death. The long transatlantic flights on the many visits back to England from my new home in America offered time for reflection during those final months. The visible shrinking of his person and his world, our discussions about life in the face of death, all gave me time to prepare. After all, as a doctor I had seen death creep up on people before. And as a psychiatrist I knew something about grief; I had counseled those for whom grief lingered, those for whom it had lapsed into a melancholic mood.

But this time it was different.

Being the elder son, I had fashioned a testimonial of sorts, something that reflected my love and esteem for the man. But at the funeral, when it came time for me to speak, my words dissolved in a flood of tears. Language was abandoned. I learned, as had generations of men and women before me, that for some losses there is no preparation. Faced with the stark reality not of any death, but of my father's death, I was once again a primal emotional being. For a moment, amidst the storm and turmoil of that day in June, I caught a glimpse of melancholy.

Over the natural life span grief touches everyone. It is a profound and numbing experience, and none of us will escape its twisting pain. When it strikes, the raw intensity of the feeling comes as a surprise. Life is rolled on its head, and we find ourselves off balance. Routine patterns and familiar assumptions are called into question. Social attachments of love and friendship that gave meaning and purpose are fundamentally changed. Inevitably we are confronted with the challenge of finding for ourselves a new fit with the world, for that which was once a stable and accustomed part of life's routine has been irretrievably lost. The external world has changed and with it the inner world of personal meaning.

The experience of grief is intrusive and primitive. "No one told me that grief felt so like fear," wrote C. S. Lewis in *A Grief Observed,* after the death of his wife. "I am not afraid but the sensation is like being afraid. The same fluttering in the stomach, the same restlessness, the yawning. I keep on swallowing." In grief there is an uneasiness, an urge to pace, an inability to concentrate that is difficult to explain. That summer of my father's death I felt sad but also strangely tense and anxious, as if in the presence of some primal danger. As a doctor I knew of that feeling, from the many vivid descriptions of my patients. It is the protest stage, the earliest and most obvious phase of grieving.

In protest's wake comes emotional withdrawal. Slowly the sense of disbelief—the denial that the world has really changed—gives way to a mental numbness and profound sadness. In breathing, the chest feels heavy; it takes a conscious effort to fill it with air. To this simple, fundamental act of living, there is added a sense of burden. Favorite interests and familiar appetites are lost. There is a distancing from the daily commerce of the world; a preoccupation with one's own thoughts and a preference for being alone. "I find it hard to take in what anyone says," wrote Lewis. "It is so uninteresting. If only they would talk to one another and not to me." This was the man who had enchanted both children and adults with *The Lion, the Witch, and the Wardrobe,* and other wonderful stories from the mythical kingdom of Narnia—a man who loved storytelling, teaching, students, argument, and the world of ideas—and yet his publisher insisted that *A Grief Observed* be released under an assumed name. "It is a fragmentary work," noted his biographer A. N. Wilson, "lacking any cohesion compared to his other books, it's an example of just what people speak of in bereavement, a shattering of his usual style. Its shooting stabs of pain, its yelps of despair, its tears, its emotional zig-zagging, all bear testimony to such a shattering."

"Shattering" catches the essence of grief. For me the word conjures up a broken mirror, a useful metaphor. In a cracked looking glass it is often difficult to recognize one's self because of the distortion of the image. So it is with grief. Sadness is just a small part of what happens, for there are other distorted reflections in the mirror. We find ourselves stumbling in the management of a familiar existence. Nothing works as it should. Insomnia drains the day's supply of energy. Routine habits become a burden, future planning is neglected, thinking is slowed, and concentration scattered by intrusive memories. Through the looking glass of grief one is reintroduced to oneself as a disorganized stranger, a person apart from the accustomed self.

These fragmented reflections provide a glimpse of what "disordered mood" really means—of what can happen to human beings who suffer depression and melancholic despair, a severe and crippling extension of depression. Although we call depression and melancholia *mood* disorders, they actually involve mental processes that extend well beyond mood and also beyond the emotional awareness that sustains social communication. The essential elements of thought—memory, concentration, and decision-making—plus many of what I call the housekeeping functions of the brain—the cycle of sleeping and waking, sexual behavior, eating and elimination—are also disturbed. Various hormones, the chemical messengers that flow in the bloodstream and through which the brain helps orchestrate the body, lose their daily rhythm. These disturbed patterns of behavior, briefly apparent in normal grief and exaggerated in depression and melancholia, all reflect a progressive upheaval of brain regulation. There is an impairment of the core experience of being—the familiar pattern of thinking, feeling, and behavior that we loosely identify as the "self." In mood disorders, life's usual balance is lost.

Emotional communication, memory and decision-making, and the systematic housekeeping through which the brain orchestrates the body's function—the triad of normal behaviors that are disturbed in the mood disorders—are all intertwined and regulated by a collection of ancient brain centers that, for now, I shall call the "emotional brain." To understand the nature of mood disorders it is important to know something about the function and purpose of these emotional brain centers, how they are organized, what we experience when they are disrupted, and a little about the brain itself. This is what the first part of this book is about.

Disturbance of *emotion* lies at the center of depression and mood disorder. Emotion—so obviously unleashed in acute grief—is an ancient signaling system that we share in common with many other mammalian species living in social groups. Emotion and emotional expression evolved as a preverbal system of communication millions of years ago, when our ancestors first began to seek stable sexual partners, herding together for safety and the protection of the young. It is a system wired deeply into our heads. At once preprogrammed and shaped by experience, it is this system of social communication that is most obviously disturbed in mood disorder, and transiently so in grief.

From the first moments of life outside the womb, long before we can talk, we employ emotional signals—of joy and despair, anger and disgust—to let parents know about our needs and desires, when we want be fed or are frightened and need protection. And right from the beginning we are sensitive to the emotional signaling of others. Each shapes the other in a continuous interaction. Our earliest infant protests upon being separated from those who nurture us can be thought of as mini-episodes of acute grief—the same yelps of pain, the streaming tears—designed to bring back the one who has disappeared. Those primitive outbursts of rage have the same root as the emotions I experienced upon the death of my father. The outpourings at his funeral came from deep inside me, casting aside all verbal facility and my best-laid plans. I was confronting some ancient fear of abandonment that intellectually, and as a grown man, I understood as inevitable, but that for the creature within me—for my emotional brain—was a threat to my safety and stability. My father's death signaled a change in the social order. For the people assembled that day I did not need the language of words; my feelings were communicated in my tears and facial expression, in the preverbal language of emotion.

In Old English the word *emotion* was used to describe a public commotion, an event that disturbed the peace. As it is with words, the meaning has slowly shifted. Emotion came to mean the physical agitation or behavioral disturbance of an individual. The word was introduced as a psychological term in the late 1800s, to distinguish the "natural instinctive affectations of the mind" from those behaviors reflective of "knowledge and reason." So *to emote* became to "dramatize emotion, to act emotionally or theatrically"; to cross the line that in civilized society divides the private person from the public persona.

Charles Darwin, the father of evolutionary theory, was very interested in emotion and its expression. He was one of the first to recognize its importance as a link to understanding the common roots of behavior among animal species. In his very popular book first published in 1872, *The Expression of the Emotions in Man and Animals*, Darwin provided lavish illustration that the emotions, rather than being private and hidden as Victorian society would have preferred, are in daily use and a cornerstone of a stable social hierarchy. He collected extensive evidence that emotional signaling among animals is primarily coded in the activity of the muscles of the face. Darwin also established that many of the fundamental emotions human beings communicate by changes in facial expression are similar to those employed by other social animals, particularly the primates, and other mammals like ourselves who care intimately for their young. Displaying his characteristic obsession with detail, he spent years sending questionnaires across the globe to missionaries and colonial governors, seeking details of how different racial groups expressed the fundamental emotions of joy, grief, and pleasure. At home, Darwin thrust his gray-bearded head into the cribs of unsuspecting infants, took detailed notes about the emotional development of his own children, scrutinized people on trains, and made frequent visits to the London Zoological Gardens to study the facial expressions of Jenny Orang, then the only great ape in captivity.

Darwin's conclusions were a popular scandal in Victorian England. Human emotions, he stated, were not only remarkably uniform across cultures, but similar to the communication of other social mammals—even to that of the household dog. Across the many cultures of man the facial expressions characteristic of anger and fear, disgust and surprise, joy and sadness—a list similar to that which early philosophers considered the primary passions—transcended verbal language as a primitive means of communication. In his book, complete with detailed drawings, Darwin specified the facial muscles involved in the expression of these behaviors: the pulling back of the corners of the mouth and the crinkling of the eyes in laughter and joy; the down-turned mouth and knitted forehead lines in grief and sorrow. He was the first to emphasize that the emotions are a preverbal system of social signaling that, because of our extraordinary facility with language, we tend to forget is still in constant use. For each of us, the recognition and expression of the primary emotions of pleasure, grief, and fear is a fundamental behavior of the brain as distinct and universal as the ability to identify the primary colors of red, yellow, and blue.

Over two thousand years ago, when we in the West began painting a portrait of our own behavior, we called these primary emotions *af-*

fects, from *affectus*, the Latin word for the mental qualities considered opposite to reason—"the natural tendencies of the creature." These were equated with the "animal" passions of rage, fear, and sexual pursuit (we like to think of ourselves as being above the animal herd), behaviors that humans have poorly under control at times of stress and ardent romance. While we still speak of affection—meaning fondness or love—"affect" is not a word we use now in everyday speech. However, it does appear in the professional vocabulary of psychiatrists and behavioral scientists, where it describes states of general physiological arousal associated with "inner" feelings and emotion. Thus, in psychiatry's diagnostic classification, mood disorders, where the most obvious disturbance is of emotion and emotional communication, are clustered together as a family of illnesses known as the *affective disorders*. Depression, and its close cousin mania, are the principal members of this family of mood disorders.

Eleven to fifteen million people in the United States are afflicted with mood disorders of some description, and of these, over two million suffer the severe form of manic-depressive illness. However, fewer than one-third of these millions ever receive treatment or even recognize that their misery could be relieved. This is a remarkable fact. Another intriguing statistic is that for those who do seek professional care, it may take up to ten years and three doctors to make the correct diagnosis. The usual justification for this extraordinary situation—so distinct from other serious medical illnesses—is a lack of public and professional education. But in this era when in medical school the affective disorders take their place with other diseases as part of the standard curriculum, and information about mood disorder is increasingly available in newspapers, magazines, and on talk shows, I find a lack of education difficult to believe in as the *total* explanation. More likely, in my opinion, a major determinant of this collective blind spot lies in the nature of mood disorder itself.

Moods develop from our emotions, and because emotional life lies at the very core of being a person, to accept that emotion and mood can be "dis-ordered" calls into question the very experience that most of us take for granted—the presence of a defined, predictable, and unique subjective entity that we fondly refer to as the intuitive "self." The familiarity and stability of this personal being are threatened when emotion and mood are disturbed. I have found in my professional practice that depression and manic depression initially distort emotional behavior in such subtle ways that most individuals incorporate the changes into their daily lives, and into the image they have of themselves. Often the first reaction is denial that any significant change has taken place in thinking and feeling. "Well, wouldn't *you* be depressed under these circumstances?" is a common challenge I hear when probing a patient's sad mood. The changes experienced in depression are frequently embraced and even defended as an expected dimension of the self, as merely an unusual emotion driven by unusual experience. Depression and manic depression are thus very special diseases of the brain; they are afflictions of the private person—of the emotional self. It is not easy to recognize or accept their intruding presence, for in disturbing the neurobiological systems that regulate the *emotional* brain, they distort the *person*ality. This difficulty in separating self from illness is a recurring confusion in mood disorder and lies at the root of much of the public stigma and misunderstanding.

The changing behavior we call emotion reflects a homeostatic system of brain activity that has been shaped by evolution to increase successful adaptation, especially in complicated social groups. We rely on this system of emotional intuition for basic survival. C. S. Lewis was right: Grief does feel like fear, because it grows from the same root. The loss of somebody important, especially where there is strong emotional attachment, evokes a primitive fear—of being alone and vulnerable to danger, or of a changing social order where one's own position may become insecure. The protest of grief grows out of that fear; it is a cry for assistance, an attempt to retrieve personal balance and accord with others. In the infant it entreats the parent to return while in the adult, grief and sadness provoke sympathy and support. In both instances the goal of the behavior is the retrieval of harmony—to correct adverse social change, solicit aid, and balance changing circumstance.

If the parent does not return—or when death denies reunion—the protest slowly disappears and a phase of withdrawal takes its place. This too is adaptive, conserving energy and resources in a hopeless situation. Volatile, energy-consuming "emoting" is replaced by a prevailing mood of sadness. The emotional regulator has adopted a new set point, one that fosters reflection, regrouping, and solitude for planning and survival. Seen from a Darwinian perspective, both behaviors—protest and withdrawal—serve the evolutionary goal of finding the best fit for the preservation of life and species when faced with difficult or unanticipated challenge. Social experience is thus a powerful regulator of mood and has the capacity to shape behavior over time, for better or for worse. Situations that promote intimacy, attachment, and safety evoke expressions of emotional pleasure and moods of happiness; loss and threat engender the opposite. Thus one way of understanding the function of emotion is to think of it as the brain's continuous search for harmony with the environment.

The scientific word for harmony is *homeostasis*. In literal translation this means to keep something standing still in the same place—in a state of balance. In scientific language a *homeostatic system* is one that "maintains a dynamically stable state by internal regulation, thus balancing those external forces that would disturb the vital equilibrium which the system is set to maintain." Emotional behavior is a homeostatic system; it balances our interaction with the world, especially the complex social world in which most of us live. Such homeostatic systems are essential to life. Probing the natural order of living creatures, we find them everywhere, from the smallest cell to the most complex society. In fact, much of science, from molecular biology to sociology, is concerned with the study of such systems and their regulation.

It is an essential characteristic of all homeostatic systems that they continuously strive to find the best fit with the environment within which they operate. An everyday example in the mechanical world is the thermostat, which governs the temperature of a room. As the room temperature changes, the thermostat makes the necessary correction by switching the furnace on and off, thus maintaining equilibrium around the temperature to which it has been set—technically known as the *set point*. The same general principles apply in biology, and in the human brain. Homeostatic systems are self-correcting in pursuit of an optimum balance.

From the evolutionary perspective, the brain centers responsible for our emotional homeostasis are both ancient and new. Attachment to others and the care of young, which require the elements of emotional communication I have been discussing and which are so characteristic of ourselves and other mammals, are relatively new in evolutionary time. In contrast, the origins of the core brain regulators that maintain the housekeeping of the body stretch back to our reptilian ancestry and beyond. In health, most of the body's physiology is so smoothly harmonized with the rhythms of the world that this ancient and industrious organization by the brain goes unrecognized. But in grief and during severe depression, as the brain's regulators become stressed beyond their normal limits, this natural harmony is lost and the rhythmic patterns of sleeping, eating, and energy are unmasked as they lose their synchrony.

The "housekeeping" responsibilities of the emotional brain constitute the second area of behavior that becomes profoundly disturbed in mood disorder. Here, the experience of jet lag is helpful in illustrating how this unmasking proceeds. Anybody who has taken an airplane flight from New York to London—as I did many times during my father's final illness—remembers the sense of mental distancing and physical discomfort experienced in the first few days after arrival in the new time zone. This occurs because jet travel, like grief and depression, disrupts the brain's regulators of body homeostasis. Over evolutionary time the brain has developed a master clock to help regulate those behaviors best served by being in harmony with the daily and seasonal changes of our spinning planet. But our bodies were not designed for the perturbing effects of rapid flight across the earth's surface, and the brain clock finds it hard to keep up with jet travel. Reprogramming basic bodily functions is just not as simple as resetting a wristwatch. Having got things nicely organized to accommodate the daily cycle of light and darkness at one spot on the globe, to suddenly find the sun rising five hours ahead of schedule is a challenge to the brain's preference for precise regulation.

At first, despite being on London time, the brain continues to conduct business as if still in New York and gives instructions accordingly. This "lag" between brain time and sun time throws the body's rhythmic patterns into sharp relief. There is a disturbance of the subjective sense of a smoothly operating self. Hunger and sleep, for example, seem suddenly to be beasts of independent mind. Then, registering that the environment has changed, the brain resets the clock, accommodates the challenge, and reorganizes the essential rhythms of the body to find a better fit with the new environment. A week later all is well; harmony with the planet is restored—until the return to New York, when the whole process of accommodation must begin again.

This experience of flight also highlights the interesting paradox that harmony in nature—the apparent stability of life—is achieved, in truth, through constant flux. Physical harmony and mental balance are based on resilience to change—moment-to-moment regulation in the service of adaptation—be that to the continuous challenge of the planetary environment, or to the social order that is the fabric of daily existence. In health, we take this resilience for granted; in grief and depression, and in jet lag, it is transiently lost. I know from personal experience that the sense of being "off kilter" and apart from one's world that is felt in jet lag is not dissimilar from the "numbness" of grief, and it has much in common with the loss of social interest that I have frequently heard my patients describe in the early stages of depression. It is not by chance that both jet lag and grief, in vulnerable individuals, can precipitate depression and mania. In each instance the brain's usual ability to maintain equilibrium in the face of environmental demand is momentarily compromised. But there is a vital difference: For most of us, jet lag is resolved in a few days and grief in a few weeks, whereas the disorganization of depression continues on, sometimes for months.

The third area of behavior disturbed in mood disorder, along with emotional expression and the body's housekeeping, is *thinking*. The key elements of thinking—concentration, memory, and decision-making—are all disorganized, but in particular there is an interesting distortion of memory. During depression, memories recalled are predominantly sad or associated with guilt, whereas in a mood of elation past memories are selectively positive.

As we become adult, emotion—or, more precisely, feeling—becomes increasingly entwined with thinking and memory. Beyond infancy, where the primary emotions dominate, the triggering of emotion is highly selective and idiosyncratic. Events we remember with particular clarity are those about which we have strong feeling, something we call meaning, and it is apparent that in adult life emotion, memory, and feeling are somehow bound together. To better understand this interrelationship, and the disturbed emotional experience occurring in mood disorders, it is important to consider what is meant by "feelings," and how they are generated.

An important task of the emotional brain is to continuously integrate the environmental and bodily information gathered from our senses. An awareness of something interesting, novel, or threatening triggers a subjectively heightened state that we describe as *feeling*. (Actually, psychiatrists call this physiological arousal of the brain's emotional circuitry *affective arousal*, and it is the *awareness* of the arousal—our perception of it—that should more accurately be described as feeling.) Most of us are remarkably inept when it comes to characterizing these states of inner feeling. Our verbal descriptions are colorful but imprecise, which perhaps is something to be expected, given that human beings had feelings long before the invention of language.

The words we use to communicate feelings are often descriptions of physical sensations. We speak of the *thrill* of surprise, the tingling sensation that goes down the back of the neck when something extraordinary happens; *pangs* of sadness and of hunger; *twinges* of guilt when an obligation is suddenly remembered, similar to a twinge of pain. In describing the *throb* of passion we are comparing the experience with a wildly beating heart. We speak of the *gnawing* of grief, *hankerings, sinkings, chills, qualms,* and so on. Our language suggests that feelings are tied closely to an awareness of the body's changing physiology. But the recognition of physiological "feeling" alone rarely has meaning. Personal interpretation is required, and that is where memory comes in.

In 1924, in France, Professor Gregorio Marañon reported an interesting observation that was to help clarify this relationship between memory and feeling, and how they are tied together in the development of emotion. He was interested in the effects on behavior of a newly discovered hormone called epinephrine, which we know now is one of the brain's important chemical messengers. It is produced in large quantities in the body when we are frightened, and this is what interested Professor Marañon. He wanted to know whether an injection of epinephrine could precipitate a feeling of fear.

After they were injected, about a third of the professor's subjects reported feeling *as if* they were experiencing heightened emotion, but they could not describe or give it meaning. Others felt nothing "emotional" but recognized a sense of arousal. Some subjects, however, with whom the professor had been in casual conversation during the experiment—discussing personal memories, hopes or fears for the future, et cetera—described the vague nonspecific feeling being replaced by an emotion "appropriate" to the content of the discussion. This suggested to the professor that the "emotion" that the students were experiencing was actually the labeling of a nonspecific feeling by the memory or thoughts that had been triggered during conversation. It was not the feeling of arousal itself that generated specific emotion but the interpretation of that feeling.

Over the next decades, a series of experiments, principally by Stanley Schachter and his colleagues at Columbia University in the United States, confirmed what Professor Marañón had found. It became clear that what we describe as emotion is actually memory and feeling intertwined. You can easily confirm this association for yourself. Moreover, you will discover that the connection works in both directions. Memories of past experience and feeling are so intimately woven that each can access the other. A mood of sadness, as it occurs in depression, thus selectively recruits memories of other sad events, explaining, in part, how a melancholic mood can reinforce itself and escalate into a negative view of the world.

Conversely, the vivid memory of a particular experience draws into consciousness the feelings originally associated with it. For me—and for many others, I have learned—music has a unique ability to open this library of feeling and personal history. Reflect for a moment on some favorite musical theme, one for which you have an emotional attachment. Bring it back from memory, or, better still, listen to a recording of it. The chances are that special memories of friends and places will come marching back. Sensual experience, particularly taste and smell, for some individuals also has this facility to link feeling and memory.

In addition to explaining the distortions of memory occurring during depression, this interesting connection between feeling states and what is remembered has important implications for how mood disorders are precipitated—why the emotional impact of an experience usually has more to do with its abstract meaning and personal significance than with the event itself. As we grow and our experience of the world increases, emotions are increasingly driven by the extraordinary human ability to find meaning in abstract thought. We string our memories and feelings together as personal stories, as emotional tales, that catalogue the significant moments that make each of us a unique person; these are the stories retold when we describe ourselves to others. Brought together as *emotional experience,* memory and feeling sustain individual identity, building for each one of us unique strengths and unique vulnerabilities.

I did not become severely depressed that summer of my father's death. Despite my grief and a deeply reflective mood, I did not develop the other profound changes in thinking and body housekeeping that are associated with melancholia. While the withdrawal phase of grief may throw a shadow of depression in our paths—with sadness, tricks of memory, appetite and sleep disturbance—it is not melancholic depression. In common with depression, grief provokes a heightened introspection, and perhaps some self-blame that all was not said in those last few days before the final parting, but there is no distortion of the way we value ourselves. A passive acceptance of misfortune can develop, but generally in mourning our self-perception changes appropriately with the lessons learned. Grief alone does not breed hopelessness; melancholia does. For the overwhelming majority, grief runs its course. After a month or so, as the regulators of the emotional brain reset themselves, flashes of the old resilience begin to illuminate the sadness: a meal tastes good, curiosity develops about a piece of news, the wish to be alone lessens and we are pleased to see an old friend. A sense of meaning returns. In most individuals, the shadows of normal human sorrow do not lengthen into the anger, self-hatred, and urge for self-destruction that are the hallmarks of melancholia.

Studies suggest that only about five of every hundred people who grieve go on to experience a melancholic depression. It is frequently obscure, to both the victim and the observer, just when a state of mourning makes the transition to one of melancholia. But once the new set point is established, sadness and disturbed emotional diligence are only the most visible core of the disruptions that the illness can provoke. The functions of daily living are rapidly engulfed, and become seriously disorganized. Changes in the regulation of energy, sleep, appetite, and sexual behavior are profound. The restoring comfort of sleep may disappear, or sleep may be broken in the early morning hours by a tortured wakefulness, the prelude to another exhausting day. Attachments to family and friends become meaningless; one no longer eats with appetite, if one eats at all; sex is a forgotten pleasure. Memory and decision-making become impaired. A preoccupation with minutiae and past failure steals away normal concentration, driving out any sense of joy. The skills to adapt to changing circumstance are lost. In their place exist painfully diminished energy and negative self-perception. Social responsibilities are forgotten or excused. Life becomes a burden, and for some, with deepening social withdrawal, a numbing preference for death emerges, born from the dark hopelessness that seeps into every facet of life.

At some point along the path just described, it becomes obvious that life has dramatically changed. From the roots of sadness and disappointment familiar to us all has grown a mood that stands apart from common experience; a new set point in the brain's neurobiological regulation has visibly disturbed the victim's accustomed emotional balance. The illness of melancholia distorts the person. The experience that the sufferer describes has only distant relevance to the real events of the world. What may have started as an appropriate response to a tragic moment extends into some sort of behavioral cancer, a malignant mood that invades and distorts the very nature of the self.

The experience of sadness is common to all; extending from this experience, melancholic depression stands at one extreme of a spectrum of mood. It is the emotional pole we glimpse while in the throes of grief. There exists a similar pole at the opposite extreme of human feeling, but it is one less encountered. About 20 percent of those who suffer debilitating depression also experience something called mania. Whereas in depression energy and positive thinking shrink and contract, in mania they accelerate and expand, magnifying and ultimately distorting the normal experience of happiness. However, the exhilaration and drive that mark the early phases of mania, which is appropriately named *hypomania* by psychiatrists (the word literally means "under"-mania), can be easily confused with a joyous celebration of life.

As in sadness we seek to withdraw, in a happy mood we are expansive and reach out. So it is in hypomania. There is an eagerness to tell others about the good fortune of being alive. Happy people infuse a certain intimacy into their relationships that draws others out of themselves. Happiness is infectious. Its subjective experience is one of harmony, of being in balance and at one with the world. The emotion of joy, frequently a component of happiness, is a vivid sense of pleasure, an exaltation of spirit that brings a feeling of attachment, of belonging. Joy is the emotion of falling in love, with all the promise and sense of security that closeness to a lover can bring. It is this feeling of exhilaration (from the Latin root *hilarus,* meaning cheerful) and the extraordinary energy that is so abundant in hypomania that probably explain its addictive qualities.

While for most of us such feelings are transient, in mania exhilaration is just the beginning. The curiosity, novel insights, infectious gaiety, and love for life may last for days or even weeks. Those experiencing early mania are compelling, amusing companions. Friendships develop, productive partnerships are formed and new sexual liaisons are sought. With the world as one's oyster, sleep becomes unimportant and housekeeping rhythms are forgotten. In contrast to those with melancholia, those who suffer mania do not complain but enjoy their experience. The purpose of life is to travel, to explore, to build; nothing is too great an effort. Thinking changes, too; thought flows fast, in a creative stream, with novel insights and bold plans. Here is the stuff of leadership and in those fortunate few, where confusion never comes, leaders do emerge. But in the usual course of events, as the crescendo builds, this expansive behavior begins to embarrass or concern colleagues and family. Strangers, once drawn in fascination to the energy and promise of hypomania, now find questionable judgment, brassy self-promotion, and excess in every action. As emotional judgment fails, increasing commitments are made, often beyond resources and reality, but the cautioning concern of friends and family is swept aside or met with anger, sometimes even violence. The engaging, infectious humor is replaced by irritability and suspicion as the mania enters full flower. The once compelling companion, now overcome by a profound disorganization, flails dangerously at those around him, courting social alienation and even self-destruction.

In the family of affective disorders, *manic depression* is an illness of opposites; it is the mood disorder that combines the polar opposites of depression and mania. Hence it also called *bipolar* illness in distinction to the *unipolar* disorder of depression alone. For most sufferers of bipolar illness, the periods of high energy, rapid thinking, reduced sleep, and euphoric or irritable mood are each followed by an episode of withdrawal, depression, and ideas of suicide. During these bipolar depressions the time spent sleeping often lengthens, becoming an avenue of social escape—in contrast to unipolar depression, where sleep is frequently reduced. Many bipolar sufferers also develop a compulsive appetite and gain weight, in contrast to those with unipolar depression, where interest in food is strikingly reduced. However, in both forms of severe depression—aptly described by the ancient name of melancholia, meaning "black bile"—distortion of thinking and an urge to commit suicide are a common development. Many plan secretly to kill themselves and, sadly, some succeed.

These are the classical patterns of depressive and manic-depressive illness. As many as 25 percent of us may suffer a mood disorder at some point during the life-span—given the right set of stressful circumstances—but for a vulnerable few, perhaps 4 to 6 percent of the population in the United States, the episodes of unipolar and/or bipolar illness repeat themselves, returning again and again without apparent cause. Frequently in such cases, careful inquiry reveals that other family members suffer too. There is strong evidence that manic depression is particularly likely to cluster in families, suggesting that the specific vulnerability to bipolar illness is passed down through the genes. Often the episodes of illness begin in late adolescence or early adulthood and increase in frequency with passing years. Although each cycle is self-limiting, with usually a return of normal mood and behavior in between, the crippling episodes may span months and sometimes years, severely straining or destroying work and family. We know from careful studies, undertaken late in the nineteenth century before our modern treatments were available, that if the illness is left unchecked, the periods of disorder lengthen over a lifetime until finally the individual is held hostage to the illness.

It is also increasingly recognized that a large number of people suffer variations of these disorders in the form of repetitive behavioral patterns that are variously mislabeled or unrecognized by many physicians, including psychiatrists, as neuroses or "illnesses of character." Followed over a long period of time, however, it is clear that these disabling disturbances of temperament (the Oxford Dictionary defines "temperament" as "constitution or habit of mind") are cousins of the more obvious disorders of mania and depression. There is growing evidence that such individuals may receive significant benefit from medicines that are helpful in the more obvious mood disorders. For example, some find their predisposition to depression is triggered by the seasons, with reduced energy, downcast mood, excessive sleep and weight gain occurring during the dark winter months, to be followed by rapid improvement and euphoria in the spring, a condition that has been named seasonal affective disorder, or winter depression. Other people suffer repetitive behavioral and social patterns, such as the mercurial changes of mood known as cyclothymia, which appear to be driven by an acute sensitivity to emotional attachment, especially its loss. Dysthymia, a chronic smoldering sadness where the world is perceived as alien and hostile and life is a suffering to be endured, is a common cause of isolation and withdrawal. These patterns of emotional behavior, like mania and melancholic depression, may also have an inherited basis, existing as a temperamental encumbrance passed down from generation to generation. But while such patterns are frequently acknowledged within families, rarely are they recognized as the seeds of vulnerability.

Compared to many other human ailments, mood disorders are common afflictions. Just how prevalent we consider them to be depends partly on how we define them, but by the diagnostic criteria of the American Psychiatric Association, they are common. By reasonable estimates, 12 to 15 percent of women and 8 to 10 percent of men in America will struggle with a serious mood disorder during their lifetime. Their illness will impinge painfully upon other family members and the many colleagues and friends with whom they share their daily lives. The economic impact on society is enormous. The cost has been estimated to exceed forty billion dollars each year in the United States alone—a social burden greater than heart disease. Adding to this, in the absence of treatment, is the common complication of addiction to drugs, alcohol, or both, which worsens the disorders and compounds their social consequences. Of even greater significance, with a suicide rate thirty-five times greater than that of the general population, is the extraordinarily high mortality of people who suffer mood disorders.

Thus, while in profound grief and moments of unusual joy we may catch a glimpse of these states of disordered and dysregulated mood, mania and melancholia are moods that truly stand apart from common experience. They are serious medical illnesses that reflect a dysregulation of the core behaviors of the emotional brain.

As I move on to consider these extraordinary disorders in detail, there is one point of semantic clarification to make. While in everyday speech—and at times in this book—the words *mood* and *emotion* are used interchangeably, it is also important to distinguish them. Strictly speaking, mood is the consistent extension of emotion in time. An emotion is usually transient and responsive to the thoughts, activities, and social situations of the day. Moods, in contrast, may last for hours, days, or even months in the case of some depressions. Moods also have their own halo effect: They recruit memory and any ongoing experience and color these with the prevailing mood state. To return to my earlier analogy, expressed emotion is evidence of the homeostatic regulator at work, while mood is the set point around which it oscillates—a set point that for most people is fairly neutral and stable, much as an active thermostat maintains the steady temperature of a room. Thus it is our mood, the state of our emotional balance, that powerfully influences the way we interact and perceive the world. This is critical to remember as we consider the experience of mania and of melancholic depression.

Two

Darkness Visible

The Experience of Depression

Depression is a disorder of mood, so mysteriously painful and elusive in the way it becomes known to the self—to the mediating intellect—as to verge close to being beyond description. It thus remains nearly incomprehensible to those who have not experienced it in its extreme mode . . .

WILLIAM STYRON
Darkness Visible (1990)

Try, for a moment, to imagine a personal world drained of emotion, a world where perspective disappears. Where strangers, friends, family, and lovers are all held in similar affection, where the events of the day have no obvious priority. There is no guide to deciding which task is most important, which dress to wear, what food to eat. Life is without meaning and with meaning has gone motivation. This color-less state of being—the very antithesis of the emotional outpouring experienced in grief—is exactly what happens to some victims of severe melancholic depression. Emotion drains away to be replaced by a visceral void.

Claire Dubois was such a victim. Indeed, her experience was so debilitating that she had come to question the worth of life itself.

Claire's first visit to my office was in the dead of winter—during that week between Christmas and the New Year, which is such a bad time for depressed people—in the middle of a particularly bleak and snowy period. It was the 1970s, when I was Professor of Psychiatry at Dartmouth Medical School. Elliot Parker, Claire's husband, had telephoned the hospital desperately worried about his wife, who he suspected had tried to kill herself with an overdose of sleeping pills. The family lived in Montreal, but were in Maine for the holidays, and Mr. Parker knew he needed some immediate help. I agreed to see them that same afternoon.

Before me was a handsome woman approaching fifty years of age. She sat mute, with brows knitted and eyes cast down, holding her husband's hand, but without apparent anxiety or even interest in what was going on. In response to my questioning she said very quietly that it was not her intention to kill herself but merely to sleep. Everything was hopeless. She could not cope with daily existence; she felt occasionally disoriented and life didn't mean anything. There was nothing to look forward to and she was of no further value to her family. She wondered whether she might have a brain tumor because she felt so strange. She could no longer concentrate sufficiently to read, which had been her greatest passion. Before, she had always been able to escape into her books. Her voice had trailed off; I could hardly hear her. I remember distinctly how she was staring past me, through the long office windows, apparently focusing on a small stand of maple trees, silhouetted black against the snow in the failing light of late afternoon. I asked her for her thoughts. "I feel no more than those trees, frozen there in winter. 'Things have dropped from me. I have outlived certain desires; I have lost friends, some by death . . . others through the sheer inability to cross the street.' " She looked at me for the first time. "The last part's not original; it's from Virginia Woolf, from her book *The Waves,* I think. But it does sum up the way I am. Feeling has drained from every crevice of my life, even from my favorite literature."

In her severe depression, Claire was describing what psychiatrists call anhedonia—from *hedone,* the Greek word for pleasure. Anhedonia means in literal translation "the absence of pleasure," but in fact in its most severe form anhedonia becomes an absence of feeling, a profound blunting of emotion such that life itself loses meaning. Anhedonia can accompany morbid grief, as when Hamlet, stricken by the brutal murder of his father, questions the meaning of life: "How weary, stale, flat, and unprofitable seem to me all the uses of this world!" But this lack of feeling, especially of pleasure, is most frequently present in melancholia. Melancholia lies on a continuum with depression, extending the illness to its most disabling and frightening form. It is a depression that has taken root and grown independent, distorting and choking the accustomed feeling of being alive.

In a letter she wrote me later, Claire Dubois described the experience very well: "It is like falling into a deep black pit; or being drawn down into a dark vortex led by only a pinpoint of light, which growing smaller and smaller, finally flickers and goes out. With it goes all feeling. There is no despair for there is no meaning; all is as white as the absence of color, as black as all color. It is a state of nonbeing; there is no cure, there is no illness. I was convinced that I was dead, emotionally dead. I have no words to describe this thing that was totally alien to my life experience. I see it now as comparable to the astronomer's discovery of black holes. My compulsion to give it description and a name is very strong but the closest I can come is that of a living void; of being condemned to life. And as the ability to live recedes, the most terrifying part of all is that it leaves a certain serenity. At that point only the idea of death itself gives hope."

For most of us, as social animals, a meaningful existence is based on our interaction with other people, with ideas, and with the physical world in which together we live and work. For Claire Dubois in her melancholic state, this normal interaction had been lost. Deprived by the severe depression both of feeling for her family and of the intellect required to retreat into her abstract world of literature, her existence had become meaningless—a "living void." Claire had become preoccupied with this shutdown of feeling. Closer questioning, however, revealed that she had also been experiencing changes, so characteristic of melancholia, in the body's housekeeping rhythms—disturbances of her appetite, sleep, and energy. This lost harmony had crept upon her slowly, indeed so slowly that she had hardly registered the change until upon looking back she was forced to confront the wreckage of her daily existence.

Claire's experience is not uncommon. Even the most dramatic forms of manic-depressive illness and melancholic depression, in their early stages of development, can be notoriously difficult to pin down and are frequently ignored or mistaken for something else. This holds true for professionals—who should know better—and for family members as well. The clues are there but they are inconsistent, varying in number and in severity over time. Those who suffer, and their friends and relatives, and the physicians and mental health professionals involved, all experience the same confusion in the early stages of mood disorder. It is very difficult to differentiate the changes in the person, seen later as illness, from those naturally present in the emotional ebb and flow of daily life. This is a major reason why it sometimes takes so long to identify depression as something alien and pathological.

So it was for Claire Dubois and her husband, Elliot Parker. In her own mind and in his, the whole thing had begun after a serious automobile accident which had occurred the winter before. On a snowy, icy evening in Montreal, while on her way to pick up her teenage children from choir practice, her car had slid off the road and down the embankment. The injuries she sustained were miraculously few but included a concussion from her head hitting the windshield. She was admitted to the Royal Victoria Hospital, and after a careful evaluation by the neurosurgeons she was declared very fortunate and in good health. However, despite this good fortune, she began to experience headaches in the weeks following her discharge. Her sleep became fragmented, with frequent awakenings especially in the early morning hours. With this insomnia came an increasing fatigue. The family doctor prescribed sleeping medications but despite increasing doses they had only marginal effect. Eating held little attraction and she began to lose weight. She was irritable and inattentive, even to her children. Completely out of character, she began fighting with her husband.

Elliot Parker was bewildered by the changes he was witnessing. Claire had become a different person. As far as he could tell, his wife had not been herself since the accident. Some days she failed to get out of her dressing gown, and was in bed when the children returned home from school. This annoyed him as unnecessary and a very poor model for their daughters, especially when she became argumentative and angry in front of them. He employed somebody to help in the house, hoping to relieve her burden of daily chores. Now there wasn't a great deal for Claire to do, but to Elliot's surprise this seemed only to make things worse. He was perplexed. How could Claire want for a better life?

Claire spent most of the day reading. Her favorite author was Virginia Woolf, with whom she expressed a kinship—an affinity with "a similar tormented soul." Through the day she began sipping steadily from a glass of wine, a habit she had acquired during her solitary teenage years in France. By the spring, Claire was complaining of "dizzy spells." She was seen by the best specialists in Montreal, including Wilder Penfield, the famous neurosurgeon and a friend of the family, but no explanation could be found. The summer months, when she was alone in Maine with her children, brought minor improvement, but with the onset of winter, the disabling fatigue and insomnia returned. Elliot found himself distraught and angry, wrestling inside with a rising sense of impotence and hopelessness. The medical profession, and his own wits, had failed them. In the words of the family doctor, Claire had been reduced to "a diagnostic puzzle." Then, at the very worst time, in the middle of the Christmas holidays, came the suicide attempt.

Why should sliding off a stretch of familiar, but icy, road have precipitated Claire Dubois into this black void of despair? Many things can trigger depression. In a sense, it is the common cold of emotional and mental life. In fact, depression can literally follow in the wake of the flu, it can herald the beginning of a serious physical illness like cancer, be a sign of a hormonal disturbance such as thyroid disease, or be an early symptom of Alzheimer's dementia. Just about any physical trauma or debilitating illness, especially if it lasts a long time and limits physical activity and social interaction, increases our vulnerability to depression. As I have already indicated, loss, the reversal of social fortune, or damage to cherished hopes can also usher us along the path. But only a minority among us, perhaps 8 or 10 percent, travel far enough to experience the torments described by Claire Dubois. What determines this vulnerability to severe depression is complex. It is not a single gene that we inherit from our parents, although indeed, as I shall explore later, genetic inheritance can increase the risk. Similarly, isolated childhood experiences are rarely capable of scaring us so deeply that we hover forever on the edge of despair. Rather, what makes us vulnerable over time is the subtle interplay of both these elements: what we inherit, and what we experience. And particularly, in what we experience, it is the meaning of individual events and the control we have over them that appears to be most important.

To be trapped in uncertainty, alone, and without purpose or control, is the nightmare of our species. Throughout life, human beings strive for meaning and for a stable future in which change can be predicted and controlled to serve a personal destiny. Transient episodes of sadness and depression are ubiquitous, and are important points of emotional reference as we navigate the complexities of everyday life. These moments of common experience are entirely distinct, however, from the illness of melancholia. The roots of serious melancholic depression grow slowly over years and are usually shaped by many separate events, each of which combines in a way unique to the individual. In some, a predisposing shyness is amplified and shaped by adverse circumstance, such as childhood neglect, trauma, physical illness, aging, and so on. In others who experience mania and depression, there are specific genetic factors that determine the shape and course of the mood disturbance, once triggered. But even in manic depression the environment plays a major role in determining the timing and frequency of illness.

The resilience and freedoms we have at our disposal to adapt to everyday challenge are continuously changing, and highly dependent upon such mundane matters as physical health, social circumstance, and whether we are male or female. Events too vary in their impact, depending upon the threat or opportunity they present, the control we have over them, and their personal meaning. The death of Ginger, an old and favored family cat, had a very different meaning for me—and therefore a different emotional impact—than it did for my daughters who had grown up with him. Significant also is whether there is anybody around to share the burden—to provide comfort, support, and offer explanation. In short, the only way to understand what has the power to kindle and precipitate a depression is to know the personal story that lies behind the experience and the illness.

Claire Dubois had been born in Paris. Her father, a politician and businessman, was much older than her mother, and had died of a heart attack shortly after her birth. She never knew him except through the stories that were told by her mother and other family members. A stepbrother—from her father's previous marriage—insisted that Claire's mother, an actress, had been a long-standing mistress of Monsieur Dubois, and that his father had only agreed to marry her when she found herself pregnant with Claire. He also suggested, as Claire entered adolescence, that she might consider performing the same services for him. Claire considered and refused; she preferred books. Claire was by necessity a solitary child, and no stranger to grief and loss. She had discovered literature at an early age. Books offered a fairy-tale adaptation to the reality of her daily life. Indeed, one of her fondest memories of adolescence was of lying on the floor of her stepfather's study, sipping a glass of watered wine and reading *Madame Bovary*. Her mother had remarried when Claire was eight, but drank heavily and was in and out of hospital with various ailments until she died in her late forties, probably of cirrhosis of the liver (although in the family story it was pneumonia). The other good thing about adolescence was Paris. Her stepfather was a lecturer in art history at the Sorbonne and the family lived in the respectable quarter of Montparnasse, near the Luxembourg Gardens. Within reasonable walking distance were all the bookstores and cafes that any aspiring young woman of letters could desire, and Claire made full use of them. These few blocks of the city became her personal world. Everybody, be they bookseller, street vendor, concierge, or gendarme, knew the blue-eyed Claire Dubois; the people of the neighborhood became her extended family.

However, with the increasing threat of war in Europe, this life that

Claire had created so carefully collapsed. Just before the outbreak of hostilities she was loaded onto a boat bound for Canada, to live with a stepsister in Montreal—a woman some twenty years older than herself—while she attended McGill University. There she spent the war years, lonely and unhappy, avidly consuming every book she could lay hands on. Flaubert, Proust, Chekhov, and Virginia Woolf became her closest friends—perhaps her only friends. Her mother died in Paris, but she hardly noticed. She majored in English and became a freelance editor to supply the little money she required. Then, shortly after the war ended, on impulse she returned to Paris at the invitation of a young man whom she had met when he was studying in Montreal. He proposed marriage and Claire accepted. She had sorely missed France; living in Canada was not the same. Because of the war, her departure had been precipitous and painful and she yearned for the familiar Paris of her youth. Her new husband offered her a sophisticated life among the city's intellectual elite, but after only ten months he declared that he wanted a separation. Claire was never able to fathom the reasons for his decision, neither then nor subsequently. Her only explanation to herself was that he had discovered some deep and fatal flaw in her that he would not reveal. To complicate matters she was pregnant, but after six months of further turmoil she agreed to a divorce. She returned to Montreal, to the family of her stepsister and to her old job as an editor. There her first child, a daughter, was born.

Much saddened by her experience and considering herself a failure, she entered psychoanalysis and her life stabilized. She began writing the first of her several destined-to-be-unpublished plays. For the next five years she lived in her stepsister's household with her growing daughter. Then, at the age of thirty-three, "in a desperate effort to create some sort of future," Claire Dubois married Elliot Parker, a wealthy business associate of her brother-in-law. Approaching fifty, and in his own mind a confirmed bachelor, Mr. Parker was somewhat surprised but delighted with the match. Claire rapidly had two more daughters.

Claire had initially valued the marriage. Her husband came from a Scots-English background, and Claire felt that his lack of demonstrative interest in her was probably a cultural thing, reinforced by his many years alone. His work kept him busy and frequently out of the country, but she made a life for herself with her growing daughters.

The sadness of her earlier years did not return, although at times she drank rather heavily, much to the consternation of her husband, who felt that her behavior when drunk was socially inappropriate. This led to a cooling of the already tepid relationship. With her daughters now growing rapidly, Claire proposed that the family should live in Paris for a year. The children could be enrolled in French schools and learn something of the true French culture that Claire valued as part of herself, but which she felt her daughters were missing by growing up in Montreal. Her husband grudgingly agreed. With the rekindled enthusiasm of youth she eagerly planned the year in every detail. "The children were signed up for school; I had rented houses and cars; we had paid the deposits. We had our ship cabins booked from Montreal; we were going to Florence and to the Alps for a couple of months during the winter. I had maps of routes we would take, pensions that we would stay in. I had budgets the details of which would surprise you. And then one month before it all was to begin Elliot came home one evening to say that money was tight and it couldn't be done.

"I remember crying for three days, but after that it was just another loss like the many I had experienced in childhood—like the promises mother had made that she never followed through on—and I just dismissed it. I felt angry with Elliot but totally impotent. I had no allowance, no money of my own, and absolutely no flexibility." Four months later Claire slid off the road and into a snowbank.

Placed in the context of her life story, the roots of Claire's melancholic depression become clearer. We begin to understand, through this familiar construction, something about the person—the private self—of Claire and the personal world in which she really lives. Stories and storytelling are integral to being human. Within the privacy of our own heads we experience life as a series of interrelated stories, which when collected together become the anthology of our personal experience. These are genuine narratives, constructions that link factual events and episodes of emotional significance. As we tell ourselves these stories we give important people, places, and events their own special meaning. Paris becomes the magic city of our dreams, embodying hope. That tragic events have occurred there is forgotten, for the mind is a tidy place; when facts don't quite fit with our hopes and dreams, we distort them a bit. Each day we struggle to build a future that brings us closer to the dream we seek, for ourselves and our children, attempting to shape the reality of what we experience by what we do. It doesn't always work. Under such circumstances hope can be crushed, and the grief that follows is every bit as profound as in mourning the death of a person.

Although Claire's story on the surface was one of pain and loss, of loneliness and coping with being alone, nonetheless for many years she had managed artfully to find meaning in her life and accommodate to difficult circumstances. That was one of the reasons why, in her private world, she preferred Paris to Montreal, and books to people. In the stories of her favorite literature she always knew the ending. As the three of us—she and her husband and I—explored her life story together on that snowy winter day, it was clear to all that the event that kindled her melancholia was not her automobile accident but the devastating disappointment of the canceled return to France. That was where her hope, energy, and true emotional investment had been placed; that was the next chapter of her life story that Claire had been writing when Elliot withdrew his support, and her future and private hopes collapsed. She was grieving the loss of the personal dream of introducing her adolescent daughters to what she herself had known and loved as an adolescent, of taking them to the streets and bookshops of Paris where she had crafted a life for herself out of the loneliness of her childhood. Her three days of tears we may see as the protest phase of that grief, and then the inevitable return to her literature, which had always worked for her in the past, as the phase of withdrawal. It was in this state of grieving—perhaps through inattention or preoccupation, it is difficult to reconstruct—that she had driven off the road that icy afternoon.

Depression, as William Styron describes in *Darkness Visible*, the masterful memoir of his own experience of the illness, is a disorder "so mysteriously painful and elusive in the way it becomes known to the self—to the mediating intellect—as to verge close to being beyond description." The "manageable doldrums" of everyday life only give "a hint of the illness in its catastrophic form." For Claire, the accident was little more than what Styron describes as an "ominous way station" between the initial grief of disappointment and the melancholic descent that was to follow. In addition to the physical trauma, the driving accident was another challenge to her already weakened sense of competence and value. Now she was not even trustworthy in the care of her own children; the accident corroded further a self-esteem already weakened by the emerging depression. But, ironically, with the accident came suddenly and unexpectedly Elliot's undivided attention, something which was destined to further obscure the true nature of Claire's illness.

Elliot Parker loved his wife and worried deeply about her, but he did not understand her. While he could readily comprehend the physical trauma of having a car roll down an embankment, he did not truly understand the emotional trauma of the cancellation of a year in Paris. And it was not in Claire's nature to explain how important it was to her or even to request an explanation of Elliot's decision; after all, she had never received one from her first husband when he had abandoned her. There were some questions that she had learned long ago it was better not to ask. And besides, her husband now seemed concerned and attentive. Thus the accident itself further confused and obscured the true nature of her disability. Her restlessness, fatigue, and pessimistic preoccupation were taken as the residue of a nasty physical encounter. A flurry of medical consultations ensued. She began taking sleeping pills for her growing insomnia and returned to her old habit of sipping wine as she buried herself once more in the stories of her favorite authors. And so, the lethargy of the emerging depression was exacerbated by a growing addiction to barbiturates and alcohol. What echoes of her own mother's alcoholism, genetic or otherwise, the drinking had for Claire, is hard to fathom. Certainly as an access to the muse it had magical power; it became her traveling companion as she wandered alone through a storybook world. But it also obscured the nature of the physiological changes that she was beginning to experience as her melancholy deepened.

Many people who have mood disorders become dependent upon alcohol or stimulants. In fact, studies estimate that over 50 percent, a majority, will be caught up in such dependence at some point in their illness. For Claire, as for many, the alcohol confounded her illness by disrupting further the daily housekeeping functions of appetite and sleep, dulling her senses, and, together with the rising dose of sleeping tablets, increasing her loss of emotional control. Arguments with her husband escalated; in all probability the dizziness that first took her to a physician was related also. This peculiar collection of disturbed behaviors, plus the memory of the accident, convinced Elliot Parker that his wife had sustained some sinister progressive injury to the brain. The details of her own emotional anguish were overtaken by the quest for a "medical explanation."

Again, such presentations are all too common in advancing depression. It is especially confusing—to the sufferer and to the family—when the mood goes beyond the experience of sadness to that of anhedonia and then to no feeling at all. After all, the very word "depression" suggests that sadness should be at its core, but the name is inadequate to the breadth of disturbance that occurs in the disease. "A true wimp of a word for such a major illness," complains William Styron, and I agree. Melancholia has always been my preference, better capturing the "veritable tempest in the brain" that marks the experience of inner turmoil and confused thinking as harmony and emotion drain away, often to be replaced—as for Claire—by a withered imitation of life. In reality, the portrait of severe depression is complex and varied and many have been misled by its common name.

Looking carefully into the events that precede depressions, some precipitating social happenstance can be found in most cases. Changes that remove people from social interaction with others seem to be the most malignant and are commonly associated with the onset of depression and melancholia. Eugene Paykel, Professor of Psychiatry at Cambridge University and an expert on depression and grief, has called these "exit events," in distinction to "entrance events" where social opportunity—and demand—are increased. While desirable life changes—a new job, public acclaim, or increased responsibility—can tip some vulnerable souls into depression, undesirable events such as a lost love or death are by far the most frequent. For Claire, the loss of the year in France was a seminal "exit" event. It came at a time when her children were growing rapidly and were soon to leave home; this was in her mind the last opportunity to introduce them to French culture. She had planned the trip in meticulous detail, emphasizing the extraordinary emotional investment she had made. Most important of all in the genesis of her melancholia, that investment was tied closely to the fragile foundations of her own life history, and to her brittle self-esteem.

Claire Dubois had had a remarkably lonely childhood. She had never known her father; her mother was preoccupied with a whirl-wind social life and her own tragedies. Claire had been given little con-sistent nurturing, and if it hadn't been for the stability of the stepfather's household and extended family she would have faced even greater privations because of the war and her own mother's death. Apart from her stepfather, whom she genuinely thanked for introduc-ing her to books, most of the men in her life—and particularly her first husband—gave Claire the profound sense that there was something about her that was unworthy, even undesirable. Through experience, Claire had become wary of attachment. In her marriage to Elliot it was her children who were the focus of her existence, and when well she made every effort to provide for them the mothering she had missed. But Claire was not without her triumphs. In fact, the life she had con-structed for herself in Paris as an adolescent, she remembered with warmth and fondness. She was extraordinarily well read and a fine ed-itor. She had loyal friends and acquaintances through her work, and loved the energy she discovered in the world of books. But her true sense of self was tied to those formative adolescent years in Paris, and she wanted desperately to pass this experience on to her children. Liv-ing in the English culture of her husband's family in Canada, even in French-speaking Canada, she felt trapped and not herself. These were natural barriers that a return to Paris might overcome. The year in France she had constructed for her daughters had been constructed also for herself.

While unique in detail, Claire's story has a familiar outline. Many stumble along life's road, burdened by hardships such as hers, but do not fall victim to melancholy. Why was Claire so vulnerable? Unfortu-nately I cannot give an absolute answer, and indeed that luxury we may never have, considering that much of what we each experience is driven by chance. What we do know, however, from various studies, is that when children lose a parent, or parents, in childhood they are at increased risk for depression, and also for suicide. There appear to be two periods of particular vulnerability: losing a parent before the age of five, and also during adolescence. Of course it is not just the loss of the parent alone—as if that were not trauma enough—but the social con-sequences of the loss that make the tragic differences. Before the age of five, the death of either the mother or father emerges as of special significance in most studies, in all probability because of the social, and sometimes economic, chaos that frequently follows such a loss. Divorce can have the same impact. In early adolescence it is most frequently the absence of the father that seems to tip the scales.

Thus, through the early death of her father and her sense of neglect in childhood, there is little doubt that Claire was sensitized to loss and to the disappointments that were to come. But, as do many, she also learned from these privations: that to commit herself was to be vulnerable, and that grief is often the price of attachment. Thus Claire Dubois learned to keep her distance and her intimate attachments to a minimum—a lesson that was reinforced by the experience of her first marriage. Apart from her daughters and her books, she was committed to very little; that was how she had sustained herself. She had learned the value of conserving herself through emotional withdrawal. This was not her preference, but a rescue strategy from an early age. Impulsively, she had attempted to break out of this loneliness in both of her marriages. And while in the bond with Elliot Parker she had found material stability, there was little emotional commitment. Elliot, too, through his natural temperament, reinforced this distance between them.

During the early winter months, as Claire Dubois slipped deeper into the pit of melancholy, she withdrew to the world of books and her favorite authors. Claire turned once more to Virginia Woolf's complex novel *The Waves,* a story of six intertwined characters, for which Claire had a particular affection. But in the realm of books the "luxury of concentration," as William Styron has described it, is the essential coinage. As the shroud of illness fell upon her, Claire found sustaining such attention increasingly difficult. The draughts of wine and sleeping medicine multiplied, and a critical moment arrived when Virginia Woolf's woven prose could no longer occupy Claire's befuddled mind. During that darkest week that separates the Christmas festival from the New Year the coinage finally was spent; Woolf's characters slipped beyond Claire's grasp. Access to her favorite author was now denied. Deprived of her last refuge, there was only one thought in Claire's mind, drawn possibly from her identification with Virginia Woolf's own suicide: that the next chapter in Claire's life should be to fall asleep forever, to hope for death itself. This stream of thought, almost incomprehensible to those who have never experienced the dark vortex of melancholy, is what preoccupied Claire Dubois in the hours before she took the overdose of sleeping pills that, during the next afternoon, first brought her to my attention.

Those bleak midwinter days would mark the nadir of Claire's melancholia. That was my promise to her, and to her husband, as I explained the nature of the illness and how it could be brought under control. But there was one stipulation. The first step on the road to recovery would necessitate a stay in the hospital. I was worried about Claire's drinking and the number of sleeping pills she had been taking—quite apart from the overdose itself. I outlined for them how the brain can get accustomed to the presence of such powerful substances, and how when they are suddenly removed, seizures can occur. Better that we taper slowly Claire's dependence, freeing the brain from its addiction, before beginning actively to treat the depression with a more appropriate pharmacology.

As the old year turned, and for most of January, Claire was in the hospital. Her confinement was welcomed, she said, although she soon missed her daughters—which I took as a reassuring sign that the anhedonia was beginning to crack. What she found difficult was our insistence on a routine—on social reengagement. Getting out of bed in the morning and taking a shower, eating breakfast in the company of others, choosing from a menu for lunch, taking a walk—simple things which in health we do every day—were for Claire giant steps, comparable to walking on the moon.

During normal times, emotion swings like the pendulum of an old Regulator clock, back and forth at a steady pace, marking the experiences of the day. In melancholia, that comfortable excursion slows and falters until, with anhedonia, emotional regulation is lost. The pendulum sticks. Simply conceived, the goal of treatment in melancholic depression is to release and re-regulate the pendulum to swing freely and steadily again.

The psychiatric units at the Dartmouth-Hitchcock Mental Health Center had been developed for that purpose. Light and airy with vaulted ceilings, bright colors, and comfortable furnishings, they broke the mental asylum stereotype of battleship gray and bedpans. Local people—both my colleagues working in other specialties at the hospital and those living in the surrounding community—had named us the Hitchcock

Hilton, and secretly we were pleased. These were intensive-care units for the emotional brain, meticulously designed for behavioral rehabilitation. A regular routine, as we insisted upon for Claire, and social interaction are essential emotional exercise in any recovery program. They are calisthenics for the emotional brain. The action of antidepressant drugs can be similarly understood. They prod the pendulum to assume once more its regular excursion.

Other drugs will do the same, although not always with a beneficial, stabilizing effect. Those who suffer mood disorders, together with many who don't, have discovered that stimulants—alcohol, coffee, and particularly cocaine—can all profoundly alter mood. That is what Claire had been doing since childhood, first with her watered wine and later with the heavy drinking that had so troubled her husband. Where individual attempts have been made to treat melancholy mood, alcohol has a very long history.

Such "self-medication" has only a short-term gain in depression. The disturbed communication among the brain's neurons—the billions of cells in the emotional brain that in health ensure the pendulum's steady swing—is ultimately made worse by alcohol, cocaine, and other stimulants. Antidepressants are not perfect either, although, as we have learned more about the brain chemistry of depression, the precision of the drugs available in the market has improved. In the late 1970s, when Claire Dubois was my patient, the most commonly used class of antidepressants was the "tricyclics"—so named because of their chemical structure. Tricyclic antidepressants, and essentially all other antidepressant medications, alter the pattern of communication among neurons in the emotional brain by changing the activity of the chemical messengers carrying information. The major difference between tricyclics and drugs developed in the late 1980s, such as Prozac™ (the brand name for fluoxetine), is that the tricyclic antidepressants influence several different messenger systems, including some best left alone, whereas Prozac affects only serotonin, the messenger thought to be a key player in the emotional brain. As a result, patients find Prozac more friendly, with fewer disturbing side effects, such as dry mouth and tremulous hands.

But the tricyclics are comparable to the newer drugs in lifting depression, if such side effects can be tolerated. Studies show that approximately two-thirds of those suffering depression experience improvement in their symptoms with both classes of antidepressant. Sometimes it is necessary to try several before the right one is found—similar in some ways to finding a comfortable suit of clothes—but in Claire's case I was lucky. The first one off the peg—amitriptyline, a drug named Elavil™ by its manufacturer—seemed to fit. We had substituted for the alcohol and sleeping pills with diazepam (Valium™, a drug with addictive properties of its own), which initially satisfied the brain's craving, subsequently permitting a controlled withdrawal from the addiction by slowly lowering the daily dose. Toward the third week of her hospital stay, as this integrated pharmacological and behavioral treatment took hold, Claire's emotional self began showing distinct signs of a reawakening.

Recovery from severe depression is often erratic; a little advance is made—a meal tastes good, conversation is easier—then comes the backslide of a wakeful night. It has much in common with climbing a steep and slippery hill. Emerging from the pit of despair, as Claire observed, is not for the fainthearted. But given time, as the emotional brain's regulatory precision returns, the body's housekeeping improves; the daily contrast between bleak mornings and the relative calm of evening—what psychiatrists call diurnal variation—decreases and the pleasure of sleep is regained. For Claire, the greatest gift of all was the return of concentration, opening her mind once more to literature. Telephone calls were made, books arrived, and Claire's life again assumed an outward calm.

Beneath this tranquillity, however, troubling thoughts persisted. Our conversations were making clear to both of us that, although her deep melancholy was in retreat, Claire carried burdens unlikely to be relieved by 200 milligrams of Elavil and a short stay in the hospital. There was a need to "rethink" the way she felt about herself and conducted her relationships with others, particularly with men.

While depression is an illness named for disordered mood, it is also a thinking disorder. Claire's view of herself during the period of melancholia had been dark and negative: She was a useless soul from whom friends had fled, a burden to her husband and family. But even as these dark shadows receded, there remained a distortion in her thinking, especially in regard to Elliot and her sense of self-worth as his partner. That individuals predisposed to depression are frequently pessimistic about themselves and their place in the social order is something many clinicians and researchers have witnessed and written about over the years. It is a thinking pattern both subtle and pervasive that disturbs important relationships and distorts decision making by explaining the world in negative terms. This "explanatory style" in many instances seems to be a learned behavior—something that has grown from adverse experience—while in others it is a subtle shaping of behavioral patterns—particularly shyness and fear of novelty—that are evident in early infancy.

To Claire, at the center of her negative self-perception lay her relationship with her husband, Elliot Parker. She felt she needed him but at the same time was angry, considering herself trapped by the marriage. Quoting Virginia Woolf, Claire described to me how she had been "slowly sinking, waterlogged, her will into his" for more than a decade. It was true that Elliot was a willful man and by nature undemonstrative in his emotional expression, but Claire had brought her own difficulties to the marriage. She had long believed that something about her was damaged, making her fundamentally unattractive to others and particularly to men. Thus she assumed the worst regarding Elliot's remote behavior. I had little doubt that the roots of this self-appraisal lay in the capricious nurturing she had suffered as a young child. It was not difficult to imagine how her mother's whirlwind social life and repeated illnesses, plus the early death of her father, had made Claire's young life a chaotic experience. Circumstances had combined to deprive her of the stable attachments from which most of us securely explore the world, and hone our social experience, in childhood and adolescence. In the absence of any consistent support, Claire had mounted her own rescue operation. Armed with a fine native intelligence, she had discovered a love of intellectual pursuits and in the solitude of books a brittle independence that avoided emotional closeness. But secretly she longed for intimacy and considered her isolation another mark of her unworthiness. Her fragile self-esteem stemmed from these roots and had been further weakened by the experience of her first marriage. The same roots dictated her "explanatory style"—the assumptions upon which she habitually explained her daily experience. Faced with social circumstances where she believed she was being rejected or denied some opportunity, rather than logically exploring an explanation she responded by withdrawing, feeling a mixture of guilt (that she had sought special favor) and anger (that she was being deprived because others considered her unworthy).

Such patterns of thinking drive behaviors that are socially maladaptive, even self-damaging, because they destroy opportunity for new relationships. But they are patterns commonly found in those who suffer depression. With motivation to examine them in the context of the true social circumstances, however, it is possible to shed this depressive cast of mind through psychotherapy. While psychotherapy comes in many forms—some of which are better suited and more focused upon treating depression than are others—psychotherapeutic intervention and education are an essential part of the recovery from any depression. Indeed, there is now strong evidence that for those who suffer repeated illness, both psychotherapy and antidepressants are important in preventing further episodes of depression.

Claire and I began work on the psychotherapeutic reorganization of her thinking while she was still in the hospital and continued after she had returned home to Montreal. I was initially concerned that the three-hour drive from her home to the hospital was an unnecessary hardship but she insisted that starting over again with another physician was of even greater concern to her. Claire was now committed to change and each week she creatively employed the commuting time to review the tape of our psychotherapeutic session, making detailed notes after she had returned home. Playfully, Claire referred to this activity as her "homework." While rather unorthodox, in practice it served very well. Any ambiguity and misunderstanding between us was rapidly addressed, for the tape remembered everything, and Claire's notes highlighted her central concerns for our next discussion. In fact, looking back I realize that Claire had invented for herself something akin to the process of the "structured" psychotherapies—particularly cognitive therapy—that were to gain ascendancy in the 1980s. (It is not only antidepressant drugs that have evolved but also the psychotherapy of depression.)

A preoccupying theme of our discussions during the time that Claire and I worked together was the tension she had always experienced between her need for closeness and a fear of rejection. This personal dynamic was rapidly apparent in her relationship with me and was one of the reasons why she insisted upon remaining a patient of mine. Having decided to trust herself to my care—in itself a major step toward health—the threat of separation provoked considerable anxiety. Attachment is a primary drive for human beings; in infancy and childhood it is through attachment that we each develop a sense of integrity and autonomy. For individuals such as Claire, and many others who suffer depression, it is frequently the inconsistency of early attachments that have impaired the smooth evolution of a comfortable autonomy, damaging self-esteem and distorting intimate relationships later in life. Such impairments of psychological maturation are often at the center of the exaggerated dependence and acute sensitivity to disappointment that many depressive individuals experience.

Claire described it well in one of the notes she wrote to me some months after she began psychotherapy.

I have been thinking about my pathological dependence on Elliot —reflecting upon it now that we have started the untwisting. When he is away I feel that I want to go to sleep, to go to bed until he returns, and yet I am afraid of him and will agree to almost anything when he's home, to keep the peace. I need him but I hate myself for it. Switching back and forth requires tremendous adroitness and sometimes with the tension I can feel the ulcer forming—except in my case it is stalagmites in my head. I want his approval, but it never comes; nor does his disapproval for that matter. It's a non-relationship between peers. If I'm still looking for a father and someone to talk to, to care for me as a child (as it sometimes seems in my behavior towards you), then the lack of give-and-take as adults seems reasonable. At least it's an explanation, because I'm not behaving as an adult. As an adult I would have to make decisions about my daily life or even argue my own point of view if I didn't agree—that's what really frightens me. I would have to be a real person. Am I afraid of physical wounding of some sort? I don't think so. So it must be that I'm frightened that he will leave me, and so it goes round and round, that's the twisting part. It seems that what I really want I've missed by about fifty years—a comfortable attachment in which I can learn to be myself. Now it's like learning to walk for the first time as an adult, slow, awkward and seemingly impossible at times. I don't really feel at the turning point, yet I must believe that there will be one, that I will break the angry dependence I have on Elliot.

Slowly, for it took the best part of a year, Claire did break that dependency. All together, after recovering from her melancholia, Claire Dubois and I worked intensively together for almost two years. It was not all smooth sailing. On more than one occasion there were panicked midnight calls when in the face of uncertainty and squabbles with Elliot a sense of hopelessness returned. And occasionally, at some social gathering, she would succumb to the anesthetic beckoning of too many glasses of wine. But the curve was upward; slowly Claire was able to put aside old patterns of behavior—the rescue strategies developed in her youth—and address the specifics of managing the realities of her adult world. While it is not the case for all, the experience of depression for Claire Dubois was ultimately one of renewal.

The experience of melancholic depression varies among individuals, depending upon the nature of their special vulnerability. and the details of the life story. The emotional content of the suffering is driven by what has personal meaning, the aspirations and the people to whom attachments have been made, and the control that can be exerted over the events that challenge those attachments. Despite the enormous diversity of these circumstances, once the illness is established, in almost every individual a similar core of disabling symptoms emerges. Objectively, as I have said, the behaviors that change in melancholia can be clustered into three groups. These are the aberrations of emotion and mood, changes in the housekeeping functions of the brain that regulate sleep, appetite, energy, and sexual function, and disturbances of thinking and concentration. Each of these areas represents a part of the functional system of the emotional brain which, with a rapidly advancing neuroscience, we are increasingly able to define anatomically and physiologically.

That the changes in behavior are so consistent in established melancholia, despite the extraordinary diversity of events and experiences that can trigger the illness, suggests that there is a *final common pathway* of dysregulation in the emotional brain that determines the core dimensions of what is experienced. While in the beginning these

changes are subtle enough to be dismissed as the emotional consequences of the buffeting of daily life, as the symptoms progress they become incapacitating. Then, with this final common pathway of
disability established, the loss of emotional and bodily harmony in
melancholia stands clear for all to see. It is not only mood that has
changed but the ability to thoughtfully negotiate the social complexity
of everyday life—the essential key to adaptation and survival—and to
orchestrate the fundamental rhythms of the body. One of the major
reasons that we do not diagnose depression earlier is that it takes time
for this core of the characteristic profile to emerge, and frequently—as
in the case of Claire Dubois—the right questions are not asked. Unfortunately, we find that this state of ignorance is often present as well in
the lives of those who experience mania—the expansive, colorful, and
deadly cousin of melancholia.

Three

A Different Drummer

The Experience of Mania

If a man does not keep pace with his companions, perhaps it is because he hears a different drummer.

HENRY DAVID THOREAU
Conclusion, *Walden* (1854)

There is a particular kind of pain, elation, loneliness, and terror involved in this kind of madness. When you're high it's tremendous. The ideas and feelings are fast and frequent like shooting stars. . . . But, somewhere, this changes. The fast ideas are far too fast, and there are far too many; overwhelming confusion replaces clarity. Everything previously moving with the grain is now against—you are irritable, angry, frightened, uncontrollable. . . . It will never end, for madness carves its own reality.

KAY REDFIELD JAMISON
An Unquiet Mind (1995)

"In the early stages of mania I feel good—about the world and everybody in it. There's a faster beat; a sense of expectation that my life will be full and exciting." Stephan Szabo, elbows on the bar, leaned closer as voices rose from the crush of people around us. We had met years earlier in medical school and on one of my visits to London, when this book was just an infant, he had agreed to a few beers at the Lamb and Flag, an old pub in Covent Garden district. With the noise and jostle of the evening crowd, however, I was beginning to question whether we had chosen the best rendezvous. Stephan seemed unperturbed. He was warming to his topic, and it was one he knew well. Stephan was describing his own experience with manic-depressive illness.

"It's a very infectious kind of thing. We all have an appreciation for somebody who's positive and upbeat. Others respond to the energy. People I don't know very well—even people I don't know at all—seem happy around me. It's just very easy to make friends. There's also the personal sense—and it's an accurate one, I believe—of an enhanced ability to act and reason. I become particularly aware of women, although it's not necessarily a sexual thing at first, and, interestingly, women seem very aware of me.

"But the most extraordinary thing is how the quality of thinking begins to change. Usually I'm a good planner. I think about what I'm doing with the future in mind; in private I'm almost a worrier. But in the hypomanic periods everything focuses down upon the present. I like it. All of a sudden I have the confidence that I can do what I had set out to do. I take on more projects, largely because I'm not worried about running out of energy. I'm not grandiose, but I feel vigorous and active; accelerated, willing to take more risk. People give me compliments about my vision, my insight. Suddenly I seem to fit some stereotype of the successful, highly intelligent male. It's a feeling that can last for days, sometimes weeks, and it's wonderful. There's no other way of describing it."

I felt fortunate that Stephan was willing to talk so openly about his experience of mania. During the planning of this book I had written to him, and found that he was enthusiastic about helping me. As a medical student Stephan had been ahead of the rest of us. A Hungarian refugee, he had begun his medical studies in Budapest before the Russian occupation of 1956 (his father, a prominent physicist, had been active in the uprising) and in London we had studied anatomy together. Stephan had an uncanny facial resemblance to the man who played the medieval knight in Ingmar Bergman's *The Seventh Seal,* a popular film at the time, and we all teased him unmercifully—the knight of The Seventh Seal, fresh from the Crusades in Eastern Europe. Given to humorous allegory, Stephan was a wry political commentator, linguist, storyteller, extraordinary chess player, avowed optimist, and a good friend to all of us. Everything he did was energetic and purposeful. Secretly I had envied him as a natural leader. Then two years after graduation from medical school came his first episode of mania, and during the depression that followed he had tried to hang himself. Those of us who thought we knew him were dumbfounded. In recovery, Stephan had been quick to blame unfortunate circumstances. Two terrible things had happened: He had been denied entry to the Oxford graduate program to study physics and, the worst imaginable, his father had committed suicide. Stephan, insisting that he was not ill, refused any long-term treatment and over the subsequent decade—before he finally agreed to take lithium carbonate, a proven mood stabilizer—he suffered through several further bouts of illness. There was little doubt that when it came to describing mania from the inside, Stephan Szabo knew what he was talking about.

He lowered his voice, confiding in me. "Then, as time rolls on, my head really speeds up; ideas are moving so fast they're stumbling over each other and I begin to get this sense of power—power over other people. I begin to feel that what I think and do is of significance to those around me, even to the universe at large. I think of myself as having special insight, as understanding things that others do not, and with a special capacity to lead. I recognize now that these are warning signs. But typically, at this stage people still seem to enjoy listening to me, especially if the subject is scientific or about ideas that they don't easily grasp, as if I have some special wisdom.

"Also I become very self-centered. It's much easier for me to insist on what I want for myself and I get irritated if I don't get my own way. I become demanding, expansive, even exaltative about things I like. Then at some point in the process I start to believe that because I feel special, maybe I am special. Now, having been through it several times, I recognize that feeling as the beginning of genuine madness; the early distortion of reality. I have never actually thought I was God, but a prophet, yes, that has occurred to me. Inside my own head I'm a reluctant prophet. I get the sense that I am at the center of some kind of plan, a script. Then, later—probably as I cross into psychosis—I sense that I am losing my own will, that others are trying to control me. It's at this stage that I first feel agitation and the early twinges of fear. I become wary and suspicious; there's a vague feeling that I am the victim of some outside force and then after that everything becomes a terrifying, confusing slide which is impossible to describe. On the two occasions it has gone that far I ended up in hospital; it's a crescendo—a terrible tornado—that I wish never to experience again.

"You can see the way it unfolds, although in the living reality it is not as organized as my description. That's important to remember. And it's one of the reasons why it's not easy to blow the whistle and get help in the middle of it. You move seamlessly through something wonderful to the plausible, although far-fetched, to ideas and thoughts that are completely implausible, before sliding into a self-deluded confusion."

I asked at what point in the process he really considered himself ill.

Stephan smiled. "It's a tough question to answer, especially in the beginning. My experience in observing people who are very successful in life in the conventional sense, like my father in his earlier years in Hungary, suggests that it is difficult to tell the difference between the way, in their success, they appear to feel and some of what I experience in the early stages of mania. They just don't seem to have the normal reaction to dangerous situations. They seem to feel they are blessed. I remember the student leaders in Budapest, jumping onto tanks as if they were immortal. They didn't have mania, but they were experiencing something similar that lifted them.

"I'll bet you that many successful businessmen, who have taken risks and almost lost their company—and many good businessmen have—can describe something similar to my experiences in early mania. But they edit them out; they decide that such feelings have no relevance to anything but competition and risk, and they put them aside. It's the same with depression. If you're not depressed for too long you can dismiss it. And, of course, there are always loving friends, and others if one is fortunate, who cover these things up. Obviously I cannot prove any of this, but I think there are certain experiences in life—very competitive athletic experiences for example, war, severe stress—that put people into extreme states, similar to what I have experienced, but only for brief periods.

"If there is any validity in comparing how I feel in the beginning of mania with the way these people appear to feel, then I have to say that there must be some value beyond the medical label of mania that doctors give to this condition; even some survival value perhaps. I think the 'illness' is there, in muted form, in some of the most successful among us—those leaders and captains of industry who sleep only four hours a night. My father was like that, and so was I in medical school. It is a feeling that you have the ability to live life fully in the present. What's different about florid mania is that it goes higher until it blows away your judgment, causing terrible pain and disorganization. So it is not a simple situation to determine when I go from being categorically normal to being categorically abnormal. Indeed, I'm not sure that I know what a "normal" mood is. My moods have a broader range than most people's and in the extreme I am definitely ill but I have now accepted that as part of who I am. Over the years, now that the highs are pretty well controlled by lithium, I have come to see it rather like having bad asthma. After an especially severe bout the patient can remember the details and recognizes that he has suffered another attack. But looking back, while battling *in extremis* to draw the next breath is a terrifying experience, no obvious line was crossed in getting there. There is no Rubicon on the road to mania. To draw a line, at least in a subjective sense, I find impossible."

I believe there is much truth in Stephan's musing. The experience of hypomania—of *early* mania—is described by many as a preferred state of being, comparable to the sustained exhilaration of falling in love. There is no rush to label it as illness. It is the antithesis of depression, which in withdrawal excludes the world. Hypomania, with exuberant expansion, embraces everything within reach. Those with early mania do not complain; rather, they live their experience. When the extraordinary energy, enthusiasm, and self-confidence of the condition are found harnessed with a natural talent—for leadership or the creative arts—such states can become the engine of achievement, driving accomplishments much revered in human culture. Eloquence and compelling argument, as Stephan discovered, spawn leadership. Cromwell, Napoleon, Lincoln, Churchill, perhaps Joan of Arc, to name but a few, appear to have experienced periods of hypomania and discovered the ability to lead in times when lesser mortals failed.

Many of these leaders suffered periods of depression too, for the acceleration of mania rarely occurs alone. But that is less discussed. As a culture we are more willing to accept melancholy in the artist than in the political leader. However, many artists—Edgar Allan Poe, Byron, Vincent van Gogh, Sylvia Plath, Anne Sexton, Robert Schumann, and others—who suffered depression also had extraordinary periods of productivity during hypomania. Handel, for example, is said to have written "The Messiah" in just three weeks, during an episode of exhilaration and inspiration.

Virginia Woolf—the favored author of Claire Dubois—was one artist whose insightful struggle with mania and depression helped transform the modern English novel. In London in the 1920s, Virginia and her husband Leonard were at the center of a circle of artists and writers known as the Bloomsbury Group, so named for the neighborhood in which most of them lived. Her extraordinary talents, and her periods of illness, were recorded in elegant style by Virginia's many literary friends and diligent husband. This unusual collection of information has been carefully explored in *The Flight of the Mind*, a biography by Thomas Caramagno of Virginia's art and manic-depressive illness. Virginia Woolf suffered through several malignant depressions, in the last of which she killed herself by walking into the River Ouse with her pockets filled with stones. But there was another Virginia, given to flights of hypomanic fancy (of which Claire Dubois had no experience). What Virginia's friends particularly remembered were her fantastic stories, the "gift for sudden intimacy" and "conversational extravagances," her "mercurial public manner." "Her imagination was furnished with an accelerator and no brakes," wrote Nigel Nicolson. "It flew rapidly ahead, parting company with reality." She had a way of "magnifying simple words and experiences. One would hand her a bit of information, as dull as a lump of lead, and she would hand it back glittering with diamonds. I always felt on leaving that I had drunk two glasses of excellent champagne. Virginia was a life enhancer."

Nigel Nicolson's description is of mania with a social face—a seductive hypomanic prelude to the destructive, suspicious rage that Virginia and her husband knew could sometimes follow. There is often an urge to romantically embellish or to stigmatize madness as distinct from normal experience—perhaps to remove ourselves from risk—whereas in fact for those who suffer there is an extraordinary and terrifying continuity between the normal self and the madness of manic depression. Kay Jamison has exquisitely captured the seamless quality of this personal experience in her compelling memoir, *An Unquiet Mind*. For Virginia Woolf, who wove the insights gained from her own encounters with mania and depression into her fictional characters, there was a similar disturbing continuity. Writing about her experience was, in her own words, "a very intense and ticklish business." "It was a subject I kept cooling in my mind until I felt I could touch it without bursting into flame all over," she wrote in one of her letters. "You can't think what a raging furnace it is still to me."

Manic-depressive illness is both familiar and strange. Where early mania may be exhilarating and within the reach of common understanding—even desired as a state of being—mania in full flower is confusing and dangerous, seeding social violence and sometimes self-destruction. The same core regulatory systems of the emotional brain disordered in depression are disturbed in mania, but the set point of oscillation is at the other behavioral pole. As in melancholia, there is a clustering of behavioral disturbance—a final common pathway—as the dysregulation progresses. But in contrast to melancholia, in mania thoughts and decisions flow rapidly while attention skips to keep pace. Sleep shortens as energy grows. Mood is expansive and emotion mercurial and euphoric, turning easily to irritability. And contrary to the personal withdrawal and quiet pain of melancholia, mania in its expansion is a public disease, incorporating all those caught in its turbulent path, sometimes with tragic consequences.

The crush of revelers in the Lamb and Flag had diminished. Clutches of people were drifting away. The bar was quieter and emptier, giving me room to think. Stephan seemed to have changed little with the years. True, he had less hair, and what was left was flecked with gray, but there before me was the same nodding head, the long neck and square shoulders, the same mobile face and dissecting intellect.

Stephan had been lucky. Over the past decade or so, since he had decided to take care of himself and accept his manic depression as an illness—something he had to control lest it control him—he had done well. The social scars inflicted by his early turbulence had healed. Lithium had smoothed his path, reducing the malignant manias to manageable form. The rest he had achieved for himself.

It was my meeting with Stephan that evening in London that had reminded me about Emmett Jones. As somebody who had been less fortunate, Emmett's story needs to be told, providing contrast to Stephan's success. Emmett had been briefly a patient of mine in the late 1960s, shortly after I arrived at the University of North Carolina. It was early in my career as a psychiatrist and before I had much knowledge or experience about mania. Quite frankly, and tragically, Emmett's illness had slipped through our fingers.

Emmett was about my own age—in his late twenties when I met him. He lived with his wife and young son in a small town in the western part of North Carolina, where the hills rise to meet the Blue Ridge Mountains. In those days it was a growing region, the economy driven by an expanding tourist trade, some light industry moving up from Charlotte, and a significant migration of older couples seeking a quiet country retirement. However, despite these changes, Hamilton was still a sleepy place. The Jones family, into which Emmett had been adopted when he was just a few months old, had lived there for generations and had recently turned to real estate development with great success. Emmett, despite being adopted—or perhaps because of it—was accepted by the community as the family's star performer. Having graduated from the University of North Carolina with honors in business, he had returned to Hamilton to marry his high school sweetheart, and they were now the proud parents of three-year-old Emmett, Jr. For all these good reasons, to his family and to the citizens of Hamilton, the events of that day in late summer when Emmett was brought to the North Carolina Memorial Hospital in Chapel Hill just didn't make sense.

From the information available on the evening of his admission, Emmett's disturbed behavior seemed to have developed suddenly, that same afternoon. At around two o'clock Emmett had come out of the Beechwood Inn, a downtown restaurant and motel, in the company of a young woman I learned later was Beth Parsons, the new junior associate in the family real estate firm. There was nothing peculiar about that; the Jones family owned the establishment and the couple had been having lunch. Nor was it surprising that the local police chief, who knew Emmett well, was dozing quietly in the motel parking lot, behind the wheel of the town's only official cruiser. In fact, Chief McBride frequently stopped by, in the early afternoon, for coffee at the Beechwood Inn. He had been enjoying his break when Emmett burst from the restaurant, apparently in a state of great excitement. The Chief payed little attention, although he realized the woman was not Emmett's wife, until he noticed how Emmett was pulling the woman by the arm, and saw him run to his bright new sedan and accelerate rapidly out of the parking lot. Sideswiping a stationary vehicle, he roared away in a cloud of dust and flying stones, heading out of town. The police chief, fearing some dreadful emergency, followed in fast pursuit.

An extraordinary chase ensued at high speed through back country roads, until Emmett lost control of his car and headed off through some trees and into a creek. Fortunately, he wasn't injured. The young woman, who had climbed into the back seat for safety, was terrified and a little bruised, but otherwise unhurt. Scrambling down the bank, perplexed and angry, Chief McBride wondered if Emmett had been drinking. But before he could demand an explanation, Emmett, shouting as he struggled out of the car, accused the police chief of spying upon him and prying into his affairs. Agitated and combative, Emmett proclaimed that people working in league with his father had been tailing him for weeks.

The commitment paper said that as Chief McBride had moved forward to help him, Emmett had attempted to strike the police chief, claiming that McBride was a leader in the group working to destroy him as an adopted bastard and a womanizer. By this time the chief, although confused about details, had recognized that something was seriously amiss. He knew the Jones family well and this was not Emmett's normal behavior. However, neither he nor Beth could persuade Emmett to come with them to the family home. After a scuffle, Emmett took off on foot through the woods, threatening anybody who might dare to follow.

Beth's story, as it was patched together by Chief McBride while he drove her back to town and from telephone calls that I made to her later in the week, was somewhat different. The couple had been having an affair, but only for a very short time. On that particular day, after lunch in the family motel, Emmett wanted to find an empty room and go to bed, but she had refused. He had seemed very irritable at lunch and not his usual funny self and her refusal seemed to really upset him; he became angry, and they were arguing when they left the motel.

The affair had begun in Charlotte, just three weeks earlier, when she had been traveling with Emmett, visiting banks and businesses to develop contracts for Hamilton's first major shopping center, which the Jones family proposed building on the outskirts of the town. It was the first venture for which Emmett had been given primary responsiblity since he graduated from the university five years before, and Beth, as the new associate, had been assigned to help him. The initial plan had been to entice local merchants to relocate or extend their businesses into a convenient area, within easy driving distance for retired individuals and those coming from outside town. But Emmett soon developed a grander scheme: He suggested that investment from larger retail businesses in the Charlotte area would put Hamilton "on the map." The first visit to Charlotte had been gratifying. Emmett was a charming and articulate spokesman and Beth found his extraordinary energy exciting. Emmett's ideas had expanded as the summer progressed. Soon he was arguing forcibly with his father about missed opportunities. The family should invest in attracting some of the big-name national stores. Hamilton was growing, and this was an easy way to dominate the local market. His father was not convinced. Nonetheless, taking Beth with him, Emmett returned to Charlotte to contact some of the major banks. "After a whirlwind of meetings and intense dinners," said Beth, "we found ourselves in bed."

However, as Emmett's father had predicted, the Charlotte banks had little interest in the Hamilton development. In their opinion the project was already too big and the town too small. They refused to advance Emmett any money. He was at first angry, but then came ideas of seeking support from venture capitalists in New York and other big eastern cities. In the week before being hospitalized, without telling his father, he engaged an architect to draw up plans for a new, larger, complex. According to Beth's account, Emmett was now "in a very high gear," sleeping little, working constantly, and telling all who would listen about his bold new project to put Hamilton and the Jones family on the map.

Emmett was also becoming sexually demanding and publicly flirtatious, which made Beth increasingly uncomfortable. But only during the argument after she refused to go to bed with him at the family motel did she become aware of his anger. This was also the first time she had heard him talk about a family conspiracy. He had spoken frequently of being adopted and the need he had to be successful, but this was different. Coming out of the motel to find the police cruiser in the parking lot seemed to trigger something in Emmett's mind, precipitating a rapid stream of tangled fears—about adoption, being disowned by the family, and disliked by everybody, including Beth. As the chase began and they raced through the back roads, Beth's preoccupying thought was that they both would be killed. Tearing herself from the grip he had placed on her arm, she climbed into the back seat and prayed. She thought that perhaps he'd been drinking before she met him but he didn't seem to smell of alcohol; perhaps he had been taking drugs. Although she didn't know much about how people behaved on drugs she understood they could be energized and violent. Emmett certainly had been staying up very late. She knew that because on several occasions he had called her in the early hours of the morning wanting her to meet him at the office, which she had always refused.

I met Emmett for the first time in the locked unit at the Memorial Hospital. He had been found at the family's hunting-cabin in the woods, where his father suspected he had been heading during the chase, and by the time he arrived at the hospital it was close to midnight. Upon the insistence of Chief McBride, his father and the local doctor had signed legal papers to commit him to the psychiatric unit for observation. This had confirmed in Emmett's mind that his family was against him. "Why else would they put me in here?" he demanded, but he seemed less agitated than I had anticipated, given the story I had heard over the telephone. He did not want to be in the hospital, but preferred it to the local jailhouse. As I conducted the physical examination, he pressed upon me that his confinement was a serious mistake, one that eventually somebody would need to explain. He looked exhausted, and I decided to delay a detailed history of the episode until the next morning. His doctor had given him chlorpro-mazine (a tranquilizing drug) before his journey and, with the addition of a barbiturate for the insomnia he had described, Emmett was soon asleep.

To the uninitiated observer—a category which in sad hindsight I know included me at the time—and to innocent family members, the flowering of manic excitement is bewildering and frightening. For the Jones family, in the space of a few hours, the life of an energetic, engag-ing young man had dissolved into violence, flashing lights, sirens, newspaper headlines, and shame. To those closest to the episode—to the key witnesses, Beth and Chief McBride—Emmett's bizarre out-burst had occurred so suddenly, and was so "crazy," that they were convinced Emmett must have been taking "speed" or some other pow-erful street drug.

It is not uncommon in the bursting of florid mania that a series of chance events—in this case Beth's refusal to have sex, the police chief in the parking lot, Emmett's rapid exit, and the subsequent chase—con-spire with the nascent distortions of manic thinking to trigger behavior that catapults the illness into the public arena. Then, with events out of control, anything can happen. Frequently as the behavior escalates the police are called (if they don't happen to be in the parking lot) and tragedy may not be far away. It becomes a game of chance. Chief McBride knew Emmett and his family and realized that something odd was afoot. However, if Emmett had been in Charlotte or some other strange city when the public outburst had occurred, he might not have been so fortunate. I have seen manic individuals come into the hospital handcuffed or, worse, with knife or gunshot wounds from brawls and altercations before the police arrived. Some police officers are wise about the different presentations of violence but too frequently individuals with acute mania find themselves in jail as their first stop on the way to the hospital. Mania can be a socially violent and danger-ous illness, as those who have experienced it or have been close to its turmoil will quickly attest.

The next day, somewhat slowed down from the tranquilizers, Em-mett told me his version of the events. The facts were similar, but his interpretation was remarkably different. He was not sure what had triggered his panic when he saw the police chief; probably the guilt over his affair with Beth, which had not been a smart idea given his social position in the small town. He knew he would eventually be discovered. Ironically, Beth had little interest in sex, which was all he had wanted. His wife was heavily pregnant with their second child and sex had become awkward and infrequent. Over the summer months he had become preoccupied with sexual thoughts, eyeing women on the street, secretly buying "girlie magazines" and masturbating increasingly. He had found Beth attractive but, usually shy with women, had been surprised when she thought him clever and funny.

Emmett was more eager to talk about his family and his job. He claimed that he singlehandedly had put Jones Real Estate on the map. It was obvious to him—and to a lot of other unspecified people—that members of his family, his brothers in particular, were jealous of his accomplishments. Both brothers had been born after he was adopted; by his own account they were rather dull and self-centered. Without his intelligent leadership of the family business, he asserted, they would not even have a job. Yet, because of the adoption, Emmett feared his brothers would be favored and placed in authority over him. He firmly maintained that only a major investment in the shopping center development could save the family from financial ruin. When I questioned why under such circumstances he was personally spending large amounts of money—on a new car and plans for a new house—his answer was that he deserved them and anyway it was his money, generated from his own real estate investments.

Emmett was a handsome fellow who spoke rapidly with great animation and an amazing flow of ideas. He would switch frequently from one topic to the other, and I found his Southern accent, new to my ears, hard to follow at times. But he had great personal charm and an engaging smile. I could easily imagine his success as a salesman. He seemed born to it. In response to my direct questions he denied taking street drugs; he wouldn't even know where to buy them. But he did drink coffee, and alcohol, although only socially and not excessively. In short, he denied that he was ill—"just too much darn pressure with the baby, business, and Beth and all"—and he certainly hadn't been ill before. What he needed was to be out of the hospital and back with his wife, to explain and straighten things out in Hamilton.

I found myself going through a mental inventory of the individuals with mania, even schizophrenia, I had seen in London. Emmett was different. He seemed so put together and self-confident, almost aggressive. At moments I had felt I was the one under scrutiny. He wanted to know what I was doing in America, where I was from, was I married, did I have children, did I understand his situation. In London I had been the resident physician for a large psychiatric "receiving" unit, a converted wing of an old tropical disease hospital, situated just behind the three major railway stations of Saint Pancras, King's Cross, and Euston. Those with mental illness, economic privation, addiction, and disability poured through our doors day and night in a continuous stream. They came from all over England with florid illness and often bizarre stories—stories not dissimilar from that told about Emmett. But now, in the flesh, as he looked me straight in the eye and explained his complicated circumstances, he seemed so normal. Was there indeed some sort of mistake here?

Emmett was committed to the psychiatric unit for three days. He spent most of them in the hospital gym sparring with the punching bag, shooting baskets, and running on the small track in the exercise area. A drug screen that had been run on the urine sample taken the night before came back negative. The physical exam had indicated he was in excellent shape and the results of the routine blood tests fell within normal limits. The night nurses reported that he slept two or three hours a night and that throughout his brief stay he was charming but intrusive. They saw nothing of the paranoid ideas about his father that had been reported on admission. One aide speculated that he would make a great host on the late show—that he was just a narcissistic guy from a small town who didn't like to lose and now found himself in a tight spot. His father was eager to agree.

I spent considerable time with Emmett's father. Mr. Jones did not think that his son was mentally ill. The affair was the problem, that and working too hard. Beth was from New York and they should never have hired her. She had led Emmett astray and given him big ideas. That had affected his judgment, especially at a time when he was stressed at home with a new baby on the way. Yes, Emmett had been foolish to get involved, but in a way that was part of growing up. The one thing Emmett's father did fault him for was spending too much money. Emmett seemed to have forgotten the value of a dollar. But even there, Mr. Jones confessed to me that he felt partly to blame. The shopping center was Emmett's first big project. Emmett had always been sensitive about being adopted and in trying to build his son's self-esteem he had intentionally given him a very free hand. He now realized that was a mistake; he should have been looking over Emmett's shoulder. It was not in his son's character to ask for help. He was a proud boy who just "marched to a different drummer." He wanted Emmett back home as soon as possible; being with all those "crazy" people was not good for him. And if, back in Hamilton, it got out that Emmett had been in the "psych unit" at Memorial Hospital, then the whole family was in trouble. Mr. Jones confessed that the committment had been a terrible mistake on his part. He should never have listened to McBride.

Looking back—and knowing what I know now about the illness—there is not a shadow of doubt in my mind that Emmett had suffered an episode of mania. But his story illustrates how illusive manic illness can be, especially during a first episode. Despite the drama and social chaos mania produces, it has a nasty way of slipping through the medical grasp. Emmett's extraordinary energy, which had built over the summer months (mania is most common in the late summer and early fall), his grand illusions for the future of his family, business, and hometown, plus his increased sexual drive, shortened sleep, irritability when crossed, and fearful suspicion, are all typical elements of unfolding manic excitement.

So if everything was so characteristic, one may ask, why was the diagnosis not made? Well, through my naïveté for a start. But I was not entirely to blame. In Emmett's case—and it is frequently so in similar cases—there was a resistance to making *any* diagnosis. Sometimes a social and medical alliance develops that actively denies the presence of psychiatric illness. This was particularly true in America during the 1960s, when almost all behavioral disturbance was viewed through a psychodynamic lens and circumstance was considered the root of most evils, including madness. Not only did Emmett and his family wish to avoid the stigma of psychiatric "labels," but in general, psychiatrists avoided them too, regarding them as a social burden rather than a disciplined exercise that facilitates treatment.

At that time the treatment of psychotic illness—which can roughly be defined as behavior and thinking divorced from reality—regardless of diagnostic classification, was hospitalization and tranquilizers. Lithium, now the most common drug for the specific treatment of acute mania and the prevention of manic-depressive attacks, had been introduced into Europe but was still being tested for use in the United States. In the face of overwhelming social stigma—which has lessened but is still with us—and with the lack of any specific medical intervention readily available, the diagnosis of manic depression was rarely made. This was especially so when, as in Emmett's case, the period of "psychotic thinking" was brief, and evidence of previous illness absent. Thus Emmett, in the official psychiatric classification of the day, was a man with a "narcissistic personality" who had suffered an "adult situational reaction with psychotic features." When the three-day commitment period was up, Emmett was released from the hospital with a modest supply of tranquilizers into the care of his family. They did not wish to return for any follow-up appointments. Should another crisis emerge, they would contact their local doctor.

The diagnostic slipperiness of mania is abetted by the masks that it can adopt. One, which Emmett displayed, is that of "plausible sanity," where, under the pressure of public scrutiny, self-control reasserts itself along with plausible "explanations" for past excess. Emmett was a different man while in hospital from the one described to us on the day of admission. With confinement, he quickly pulled his thoughts together. Rather than being wild-eyed and crazy he was charming and persuasive, offering reasonable arguments—the stress of work and marriage—for the breakdown of his normal behavior. Although energized and intrusive at times, he no longer spoke about his suspicion and hostility toward his father, or voiced the theories of conspiracy that had preoccupied him while at home in Hamilton. So plausible in fact were his explanations that some of us found ourselves questioning the motives of those who had signed the legal commitment. Certainly from what we had witnessed there were no legal grounds upon which to detain him against his wish, and that of his family. A judge would have released him immediately. This pattern is not uncommon. Except in the most severe manic excitement, under the pressure of necessity a mask of sanity transiently appears, confusing both the uninitiated and the unwary. The result, unfortunately, is that many in the throes of mania have successfully charmed their way out of the care they desperately need.

Suspicious thinking—what the medical profession calls paranoia—is also a mask that can confound the diagnosis of mania. Many physicians, including some psychiatrists, tend to associate paranoia only with schizophrenia, an illness that fragments mental life and usually runs a more malignant and deteriorating course than does mania. But in fact suspicion and fearful behavior can occur in mania too. Virginia Woolf, during her manic flights when she lost track of time and had to be restrained, was convinced of the conspiracy of her nurses and her husband, just as Emmett felt alienated from his family and conspired against by his father. During these suspicious interludes the normal processing of information is distorted. This is similar to what occurs in melancholia, but the ideas are more obvious to the observer—driven by the manic's effusive urge to communicate.

I learned during my evening conversation with Stephan Szabo at the Lamb and Flag that he, too, had suffered such suspicious distortions. Stephan told me how a particularly strange episode had occurred shortly after his father's suicide.

After the death, in a state of shock, Stephan had gone to stay with his mother. As he explained, "We just needed each other." In Hungary, Stephan's father had been a scholar of great prominence, but he was not able to repeat his achievements in England. When he hanged himself at the age of fifty-two, without warning and with no explanation, Stephan felt shattered and abandoned. I knew that in the wake of this tragedy Stephan and his mother had shared the same house for about six months. "She was depressed and I was denying my own grief. I had gone through some rapid job changes, working at various hospitals while trying to get into graduate school, and I was running out of money. Perhaps it was the beginning of my mania—some sort of denial of how I really felt. I was irritable, excited and depressed all at the same time. Then as I became more agitated the strange thought developed that the person in the house was not my mother but an actress *playing* her part. For a few days I was convinced that my mother was an imposter."

Stephan went on. "Later, after I had recovered, I found myself wondering what had seeded the idea, where the lack of certainty about her identity had come from. After all, like most people, I know my own mother pretty well. I concluded that what had triggered the whole thing, in my fevered grief and hypomanic state, was a dramatic change in her physical appearance. While we spoke frequently on the phone, being a newly minted physician and continuously on call I hadn't seen her for a year. After my father's death, in her sadness, she had aged enormously. What I saw was a very thin, approximate rendering of my mother, but not my mother. And there was a reversal of roles. She had begun talking about her own death, the need to tidy up the family affairs in case she died, and so on. Suddenly, from being cared for as a son I felt obligated to pull her out of her misery. In my own grief for my father her depression was more than I could bear. Psychologically, I think it just was easier at the time to believe she wasn't my mother."

Stephan's vignette had triggered in my mind two additional aspects of the manic experience that are important. First, the psychotic tricks of mind that emerge from the turmoil of deepening illness, although bizarre to the observer, usually have roots in deeply personal and rational concerns. In the early phases of mania the feelings of power and special ability are compatible—or congruent, as psychiatrists like to say—with the expansive mood and accelerated thinking. Later, as disorganization clouds normal thinking and memory, suspicion and increased vigilance emerge, perhaps as a last vestige of declining self-awareness and a desperate effort to find explanation. Thus, paranoid images often stem from everyday troubles. Stephan, faced with a crumpled and grieving mother, in his own debilitated state preferred to see her as somebody else, a stranger to whom he had no obligation. Similarly, Emmett's long-standing worry about being the adopted son, and about the ultimate place he would hold in his father's esteem, was expanded and distorted by his illness and impending business failure. Paranoid ideas in mania are not a suit of new clothes, created *de nova*, but the exotic elaboration of, usually, some commonplace concern.

Secondly, I was reminded by Stephan's story that grief can precipitate mania just as it will trigger melancholia. Mania is sometimes a pathological reaction to death or disappointment. Many studies have shown that acute or chronic stress may both precede the onset of mania and help precipitate repeated episodes. For Stephan, the suicide of his father was the event that first pushed him into florid madness. Virginia Woolf, too, suffered major episodes of mania and depression after the death of her mother, around the publication of *The Voyage Out* (one of her early and successful novels; it took her ten years to write, with much vacillation), and after the bombing of London in 1940.

Mania does not happen to everyone; it requires a special vulnerability. The stress associated with grief and loss, while an important contributor, is not a sufficient explanation for the manic experience. Virginia Woolf weathered the deaths in her family and wrote and published many other works beyond *The Voyage Out* without experiencing psychosis. Stephan, likewise, after recovering from the illness precipitated by the suicide of his father, went on to navigate the rigors of a Ph.D. in physics. Those who experience the extraordinary flight of manic moods are not randomly distributed throughout the population; they have relatives who also suffer mania and mood swings. Although the particular patterns vary, the vulnerability for mania is passed down with the genetic legacy. In Virginia Woolf's family, for example, her mother, father, and paternal grandfather, as well as her sister, brother, and two cousins, all suffered mood disorder in one form or another, including severe depression. Mania and melancholia travel together, and because those who suffer mania are rarely visited by it alone, the experience of mania for many families includes melancholic depression—and the tragedy of suicide.

For Emmett, life ended that way, as it had for Virginia Woolf, for Stephan's father, and for far too many others with mood disorder. In the United States, a suicide occurs approximately every twenty minutes—some thirty thousand persons a year. Other nations, particularly the Scandinavian countries and Hungary, in proportion to their population far exceed this rate. Of those who kill themselves, probably two-thirds are depressed at the time and of those, about half will have suffered manic depression. Indeed, it is estimated that of every one hundred people who suffer manic-depressive illness, at least fifteen will eventually take their own life. A large Swedish survey conducted in 1981 found that a history of depression multipled the risk for suicide seventy-eight-fold compared to those with no such history. Such statistics are sobering and a reminder that mood disorders are comparable to many other serious diseases in shortening the natural life span.

Some individuals kill themselves—or are killed—during the flight of mania, but many more die of their own hand in the depressive phase that so often follows. In the mind of the victim, who is witnessing the world through the distorted lens of depression, the desperate purpose of suicide is to resolve intolerable and hopeless circumstance—occasionally some dreadful nightmare spawned by the mania itself. The final act of self-destruction is invariably accompanied by a profound sense of helplessness—that the personal degradation or social tangle that drives the unbearable suffering is beyond repair and control. Suicide becomes a final desperate means of regaining that lost control. We know from those who have survived their attempts at self-destruction that it is this loss of control in managing their social affairs, and the associated experience of helplessness and alienation, which drives them to the final act. Feelings of helplessness and alienation are the most powerful predictors of suicidal behavior.

My thoughts are drawn back again to memories of Emmett Jones and how he must have struggled with such feelings in the days and hours before his death. Emmett died in a hunting accident some two months after his discharge from the hospital, but it was clear to me, and to his father, that his death was an act of suicide. Pursued by the specter of his manic indiscretions, Emmett killed himself during the period of melancholic despair that followed.

It was a week or so before Thanksgiving when his father telephoned me. I had heard nothing of Emmett since his discharge from the hospital. Apparently on the day of his death Emmett had left home before dawn to go hunting—something which he and his father had always done together at that time of the year. But on this occasion he was alone. When Emmett did not return by late afternoon his wife began to worry, for he had been sad and withdrawn after his strange behavior of the summer. She called his father, hoping that Emmett had driven directly to his office, but he had not been seen all day.

Upon putting down the phone, under some strange forewarning that he could not explain, Mr. Jones drove out to the family's hillside cabin—to the place where Emmett had fled after the fateful car chase that had first brought him to my attention. He found the cabin door ajar and from the silence he knew instinctively that Emmett was dead. Emmett lay just beyond the porch steps, crumpled on his side, with a gunshot wound to the head. From the oily cloth and the other paraphernalia about the body, it appeared—for those wishing to believe it—that Emmett had been about to clean his rifle when the weapon had accidentally discharged into his face. Driving slowly back to Hamilton and to the police, Mr. Jones knew that was not the explanation. His own belief, he confided in me, was that his son had *intended* to kill himself.

Two days after the accident, a letter from Emmett had arrived at Jones Real Estate, in a typed envelope and marked confidential. The postmark was the date of Emmett's death. From the snippets that Mr. Jones provided, it was not really a suicide note—Emmett said nothing about any intention to kill himself—but rather a long apology for the shame he believed he had brought upon himself and the family. I asked Mr. Jones about Emmett's mood in the days preceeding the tragedy, but it was difficult through his grief to gather a clear understanding. Emmett apparently had been withdrawn. In the days immediately following his return from the hospital, however, Emmett had remained irritable and argumentative. A confrontation had occurred when Mr. Jones insisted that Emmett work directly under supervision until he could control himself. In tears, Emmett's father described to me how during that argument, in a moment of frustration, he had told his son angrily that his behavior brought shame upon the family. Looking back, Mr. Jones was convinced that it was this argument, and the accusation he had made, that had directly triggered his son's guilt and subsequent death—a death for which he held himself responsible. "I didn't understand; I failed him," said his father. "I think we all failed him: the hospital, the doctors, everybody. Emmett *was* ill. He really *was* marching to a different drummer."

Suicide—what A. Alvarez, the English writer and poet, has aptly called the "Savage God" of mood disorder—cuts down unique individuals but conducts a common inquisition of those who remain. What should we have done differently? How could we have known? If only we had not argued, not taken that high moral tone. How could the suicide have been avoided? About Emmett I have asked myself that question on many occasions. His was an unnecessary death.

Had I been more familiar with mania at the time, had we known something of Emmett's biological parents—of possible illness in other family members and earlier generations—or had his illness declared itself a few years later, the story might have been different. The apparent profile of Emmett's illness—manic excitement followed by depression—is that which is most responsive to lithium carbonate. Lithium can dramatically reduce recurrent episodes of illness and was the first specific treatment for manic excitement. But in the 1960s lithium was not yet commonly available, even in university medical centers. But that is no excuse. Had we looked behind the mask of sanity he displayed in the hospital—had he and his family been warned appropriately—a supporting hand might have guided them through the depression, breaking down the distorted logic that had proved so irresistible to Emmett in his final days.

But powerful social forces were also at work, and many of those same forces remain in place today. Emmett was killed not only by his own hand but by the social stigma that surrounds depression and manic depression. While we may aspire to the energy and vivacity of early mania as a form of life enhancement, at the other end of the continuum depression is still commonly considered evidence of failure and a lack of moral fiber. This will not change until we can speak openly about these illnesses and publicly recognize them for what they are— human suffering driven by dysregulation of the emotional brain.

I had reflected this to Stephan during our evening together, and he had readily agreed, but reminded me that it was past time for our supper. "Look at it this way," he had said, amiably, as we got up from the bar, "things are improving. Twenty years ago neither of us would have dreamed about meeting in a public place to discuss these things. People are interested now because they recognize that mood swings, in one form or another, touch everybody every day. And what's more, we are well on the way to explaining my subjective experience with an objective science. Times really *are* changing."

I had smiled to myself. Here was the Stephan I remembered. Despite an early visitation from the Savage God, the allegorical knight of the Seventh Seal was still in the saddle, still playing chess, and still optimistic. It was a good feeling, and helped a little with my memories of Emmett. It *was* time for supper.

Mania both seduces and destroys. It is an illness of emotional expansion—of extraordinary social and physical appetites—the very antithesis of depression. The romantic descriptions of the early stages of manic excitement, or of budding episodes of hypomania that never reach full flower, can be compelling. But in its fulminating expression mania poisons rational thought and the objectivity of an accustomed self, ultimately endangering both person and property. For those who suffer manic depression, life can be an emotional roller-coaster, the dimensions of which many can only imagine.

Four

A Mind of One's Own

The Development of the Emotional Self

All life depends on a flow of information. Information is carried by physical entities such as books or sound waves or brains but is not itself material. Information in a living system is a feature of the order and arrangement of its parts, which arrangement provides a code or language. When these codes are transmitted, they provide the control that maintains the order that is the essence of life.

J. Z. YOUNG
Programs of the Brain (1978)

Ivan Ilyich saw that he was dying, and he was in continual despair. The syllogism he had learnt from Kiezewetter's Logic: "Caius is a man, men are mortal, therefore Caius must die," had always seemed to him correct as applied to Caius, but certainly not as applied to himself. . . . He had been little Vanya, with a mamma and a papa, with toys, a coachman and a nurse, and with all the joys, griefs, and delights of childhood, boyhood, and youth. What did Caius know of the smell of that striped leather ball Vanya had been so fond of? Had Caius kissed his mother's hand like that, and did the silk of her dress rustle so for Caius? Had Caius been in love like that? "Caius really was mortal, and it was right for him to die; but for me, little Vanya, Ivan Ilyich, with all my thoughts and emotions, it's altogether a different matter. It cannot be that I ought to die."

LEO TOLSTOY
The Death of Ivan Ilyich (1886)

The brain's specialty is the active storage and control of information, just as the heart specializes in pumping and controlling the flow of blood. The vitality of both these organs is essential to life. Thanks to medical science, should the heart fail, we can now transplant it. Not so with the brain. The brain is who we are; when the brain dies we are pronounced dead, even if the heart is still beating. It is the working brain, the subjective activity of which we call "mind," that determines the self.

In illness and in health we are each a singular self. Our coherent image of the world is a unique interactive process between the brain's biology and what we experience. Each can change the other. Brain chemistry and neuronal communication—even the brain's anatomy—are not rigid and predetermined. Although the general structure and organization of our brain is laid down by the genes that we inherit from our parents, each individual nervous system has its own evolution which continues over a lifetime. Through this interaction with experience we each design a brain—and a self—which is uniquely our own.

An understanding of this interaction between biology and experience is essential when thinking about mania and depression, or any other disturbance of behavior. The perennial argument about whether depression is an illness biologically pre-programmed, or psychologically determined and thus acquired through experience, has no place in reality, for behavior is an amalgam of biology *and* experience. Although thinking about the self and mood disorder in this integrated fashion does not come naturally, it is time to set aside debates about mind and body. Interestingly, rapid advances in genetics and neurobiology, and the recognition that biology is mutable and elastic, now make this difficult task increasingly possible. In this chapter I shall explore some of the ways in which learning from experience molds, and is molded by, the genetic instructions that we each carry within us. Specifically, I will discuss how emotion develops, and how variations in emotional temperament may guide what we learn, predisposing to emotional style and sometimes to mood disorder.

I begin with an evolutionary perspective. Staying alive, preserving oneself and the species, depends on accurate information—from the body and the environment—and its meaningful interpretation. The human brain is the body's information processor, and its elasticity is essential to survival. To keep us safe despite changing circumstances, the brain must not only organize a continuous flow of new information, but must constantly compare that with a personal model of the world that has been established from prior experience. In choosing a response to any situation, the brain seeks the best reconciliation of the prior model and the present information, and the experience gained from the outcome of the decision updates the reference files for next time. We call this "learning from experience." Thus, inside our heads, all of us are continuously re-creating ourselves—modeling the world in which we live, and updating that model with each new decision we make. This constant process of evaluation and accommodation we call *adaptation*. Adaptation is the catalyst that binds the amalgam of behavior—an amalgam in which experience modifies biology and biology shapes experience in a continuous cycle.

Nature has been artful in the development of the brain. The brain's operating system does not function like that of a personal computer, in a rigid linear fashion, conducting one task at a time and storing information for later retrieval, but in parallel systems that are capable of modifying behavior depending upon the information received and the output that is required. The brain not only manages to do several things at once, but usually can approach and solve the same problem in several different ways, while constantly upgrading its efficiency. While the genetically programmed biological systems of the brain actively set down the principles of regulation and control, it is experience, and learning from the continuous flow of information that experience produces, that modifies brain biology and determines our adult patterns of behavior.

The brain does not passively accept and store information, but presuming the world to be a coherent and tidy place aggressively sets out to evaluate and understand it, continuously adapting to what it finds. Essential things—the important things to control for immediate survival—are genetically programmed into us and are available for immediate operation as soon as we enter the world. Other behaviors, necessary to manage a complicated and unpredictable environment, are encouraged to develop through trial and error, when learning from experience can help build the brain's basic programs of organization. Still others, like emotion, sit somewhere in between.

During the first weeks of life, babies sleep most of the time and during waking moments seem to indulge in purposeless random movement. But just touch the cheek and the infant's head will turn; the lips will open to grasp and suck your waiting finger. This "rooting response" is an example of a genetically programmed reflex, a complex set of behaviors that can be triggered by a particular stimulus. At birth, the mouth is a far more efficient organ than the hand and also more important, for sucking is the behavior of survival, the only way to draw food from the mother's breast. So most of what we know about sucking is already wired into us at birth.

But within a few weeks things begin to change. The principles of trial and error take center stage as interaction with the surroundings begins to actively shape behavior. When one of my daughters was tiny—perhaps four or five months old—I remember watching as she grasped at a colored ball hanging above her crib. With an expression of intense concentration, legs pedaling furiously, she slowly extended an arm—fingers pointed like pencils—into space. At first, when the small hand closed, it was usually before the ball had been reached; the next time she would overshoot. But soon she had accurately scaled the distance, and the ball was in her grasp. Before long even moving the ball to a new position did not fool her. With gurgles of satisfaction she was learning to organize the essential information required to master coordination of hand and eye in three-dimensional space. Her brain had begun its lifetime task of model building. My daughter had moved from relying solely upon genetically programmed behaviors to learning from purposeful exploration of her environment.

This drive for control, and for mastery of our interaction with the environment, is an essential developmental process present in the first moments of life. Even before the hand becomes a useful tool, infants who are given a device through which, by how actively they suck, they can alter the focus of a brightly colored picture, will work to bring the picture into optimum focus. We seek such self-control throughout our lives; in fact, games are born this way: grasping a ball, throwing it, running, jumping—all are games of mastery that have become institution-alized as sport. The Olympic Games is a festival dedicated entirely to such tasks of agility. The control and integration of eye and limb neces-sary to perform these astonishing feats, coordinated in time and space often at extraordinary speed, involve years of repetitive training. Most of us fall far short of Olympian accomplishments, but all learn at an early age—through the same process of repetition—to coordinate the body with agility, operating the arms and legs in space and time with deft precision and little conscious thought. That is what my daughter was practicing in her crib. As we learn this bodily balance, the brain is modeling a body image which is precisely represented in the memory currents and physical organization of the brain. Thus, even in total darkness, most of us can easily touch the right ear lobe with the left in-dex finger. In the blackest night the brain maintains for safety and guidance a sense of the body's dimensions and positioning; we know exactly where our feet are placed and the balance of weight upon them, whether we're sitting or standing, and so on. This three-dimensional image of the physical self in space, which the brain constructs from the information returning from its many sensory systems, is in constant use, helping to guide every movement we perform. It is part of the sub-jective awareness that we call the body image or physical self.

The models we build of social interaction are similar to those em-ployed by a baby in the development of body image, only they are more flexible to accommodate the wide range of information that must be con-trolled. Social models—again initiated by early experience but updated throughout life by the things that we learn—cluster information into manageable units that facilitate social behavior. These models, or schema as psychologists call them, are the essential road maps of daily living: eat-ing from a plate at a table, greeting others, driving a car, finding routes through a new city, buying something at a store. In addition, beyond in-fancy, schema play an important part in emotional development.

The human capacity for emotion is both innate and molded by what we learn. Infants have a core repertoire of emotional behaviors present at birth, which they put immediately to work letting their mother know the state of their inner feelings. Anger, fear, surprise, and disgust—as in identifying a bad taste for example—are *primary emotions,* obvious in the first weeks of life. These behaviors are compa-rable to the sucking reflex, genetically programmed and expressed in response to the environment. They are part of an innate pre-verbal sig-naling system that we share with other primates and many other social mammals. Human emotion is rooted in this mammalian heritage, but because of language and our extraordinary capacity to learn from expe-rience we also have developed a complex range of *secondary emotions.* As Darwin recognized, primary emotions are expressed through the muscles of the baby's face. Behavioral expression is the "language of emotion" and the movements of our facial muscles provide the vocabu-lary. In fact, the brain exerts such fine control over these muscles that the area of the cortex devoted to managing them is even greater than that devoted to managing the muscles of the hand.

Within weeks of birth, the infant begins to smile and the expression of happiness becomes increasingly obvious; it is clear too that the emo-tions expressed by others are understood by the child. Speaking with an angry voice results in fear and crying, while soothing, calming tones generate wriggles and happy gurgles. This cooing and gurgling is the beginning of what later will be shaped as language, but in the first weeks of life it is facial expression that dominates the developing bond between mother and child. Research has shown that in the early months the mother's face is the interactive key with her child, and her facial expression the infant's guide to managing any uncertain situation and novel experience. In responding emotionally to strangers and strange situations, babies first check mother's face. This is well illus-trated by what has been called the "cliff experiment." In this laboratory test an infant, crawling toward his mother over the surface of a table, comes to a transparent section through which he can see the floor. This visual cliff gets careful attention and the child hesitates, looking up to check his mother's expression. If she is smiling and beckoning, he pro-ceeds; without that reinforcement, however, he hesitates and cries. Thus it is through the facial expression of the mother that the infant first learns to validate his or her experience of the world.

It is the repertoire of primary emotion and the learned social schema of infancy that mold the early image of the self. Our innate programs expressing fear, surprise, and happiness at first shape and then, in turn, are shaped by what we experience. By the first birthday most of us have attached a primitive emotional meaning to much of our immediate environment—both people and things—and have begun to build a set of social schema in which emotion plays the organizing role. It is the attachment between parent and child that provides the safety in which all of this takes place, emphasizing the vital importance of a stable emotional bond for the normal development of the infant. Later, as we grow and learn about the world—and the need to seek reassurance diminishes—language development emerges as a powerful modifier, muting the obvious expression of primary emotion. Emotion becomes tempered by abstract reasoning as we begin to measure ourselves against a set of learned social and internal standards. From these moral schema, fostered by a growing ability to express emotion in words as well as in body language, emerge the self-conscious or secondary emotions—pride, embarrassment, envy, shame, and guilt—and the capacity to modify or postpone emotional expression, depending upon social circumstance.

In adult life, language is so dominant as an instrument of communication—including the communication of emotion—that there is no single word to describe pre-verbal emotional expression. This dedication to language is actively acquired very early in life. We cluster the sounds that we make, first in association with the feelings of pain and pleasure and other significant interactions we observe about us. In listening and imitating "noise" that is associated with our experience, we are learning to talk. Babbling, we string the noises together, rapidly learning that some sounds, rather than others, produce behaviors to our advantage. We become selective in our noise making and soon we are talking, extending the pre-verbal communication of infancy. While it may seem that we are taught language, in fact there is considerable evidence that language is shaped actively by the inherent way in which the brain organizes information. That there is an underlying common structure to the diverse languages that human beings speak across the globe suggests the presence of a genetic template in the brain that guides their structure, and which at a critical time in childhood enables us to learn languages at astonishing speed.

Once language is acquired, not only are communication and social interaction enhanced, but also plans and dreams can be constructed in the "mind's eye"; every facet of life's experience, complete with words and pictures, can be reviewed and reworked. Most young children first talk to themselves as they experiment with this developing ability. Only later, beyond the "innocence" of childhood, do we achieve the sophistication of thinking "in private"—of experiencing "the soul talking to itself," as Plato described it. This we call imagination, but the thoughts and ideas constructed are as enduring as any "real"-world memories. The physical feats of the Olympic gold medalist and the most lithe of animals pale by comparison with this extraordinary ability, facilitated by language, to manipulate images and abstract ideas. However, in the statesman's rhetoric, the scientist's dissection of the brain, the gifted artist's separate vision, even when one is playing a simple game of chess, emotion remains central to judgment and to decision-making. Throughout their development, language and abstract reasoning extend the emotions; they do not replace them.

Even with my broad brush strokes, from the picture I have painted of emotion and its development it is apparent that we cannot meaningfully divide the nature of brain—the genetically prescribed biology of the organ—from experience and the environment that nurtures us. Only by recognizing that these elements are intertwined can we hope to understand emotional behavior. In this age of empirical science probably few would argue with such statements. And yet surveys suggest that over 40 percent of Americans still believe that the cause of depression is apart from brain; a deficiency of willpower, a "mind" problem, or poor nurturing and bad luck that can be overcome, given sufficient fortitude. Many, it would appear, believe depression to be an illness quite different from any other in the body, or even from the "organic" disorders of brain such as epilepsy or Alzheimer's disease.

The results of such surveys about depression are revealing, although not really surprising, for they reflect difficulties we have had in thinking about ourselves that are as ancient as human history. Trying to reconcile the apparent differences between mind and body and the relative importance of nature and nurture in human behavior has preoccupied academics and philosophers for centuries, and even today remains a source of debate. It was once fashionable to blame Descartes, the French philosopher and scientist, for the mind–body split. But Descartes did not invent this dilemma, although his clever rationalization of mind and body as two independent clocks, standing side by side and each telling the same time, did enable the investigation of human anatomy to proceed without religious opposition. (In Europe, at the beginning of the seventeenth century, the "being" or soul of man, thought to be revealed in the "passions" of everyday life—love, hatred, desire, joy, and sadness—belonged to God. Mortals did not tamper with such things.) We now enjoy far more information and sophistication in science than did Descartes, but the existential dilemma of the self remains.

The difficulties we have in thinking about ourselves as integrated beings of flesh and thought stem not from the limitations of empirical science but from the parallel way in which we understand objective and subjective experience. The mind–body problem is bound up with the experience of being alive. Each individual rediscovers this dilemma during the evolution of the unique self. As that self grows, struggling to develop a sense of order, meaning, and purpose in life, we each in our own soul experience something comparable to that which is described by Tolstoy in his story of Ivan Ilyich. As an individual, I know beyond doubt that it is *I* who played as a child, learned to speak, to love, to run, to cherish music and smell the new-mown hay. This subjective emotional self, private and inside my head, is very distinct from my awareness of the physical body, of my mortality, and what others see of me. Although intellectually I can acknowledge the integrated singular self of flesh and thought, I do not naturally think about myself in those terms. Yes, I accept that it is *my* body, but I live in *my* mind, and my mind can view my body, even objectively, as something distinct and separate. This sets the stage for what is commonly called "the mind–body dilemma."

Bodies can be located in space, and the activity of the body can be inspected and verified by those observing us. The body is open to objective scientific assessment. Minds, on the other hand, don't appear to be subject to any obvious scientific or mechanical law; they are private and intangible things governed only by subjective reasoning and somehow separate from the bodily self. What is going on in somebody else's mind is largely unknown and not open to scrutiny by the general public.

Mind and brain may not be divisible in theory, but as a practical matter it does seem so, and most of us think that way—in parallel. After all, we have two essentially parallel lives or histories; one consisting of what happens to our body and our behavior in the social realm, and the other what happens within the boundaries of our own thoughts. The first goes on in the "physical" or objective world; the second is an activity of the "private" or subjective world—the world of the mind— where we live as unique individuals. Like Ivan Ilyich, having struggled with the complexity of developing a unique and private self, most of us do not yield easily to death or illnesses of mind. While acknowledging vulnerability in body, we cling to invincibility in spirit.

But the two are interdependent and indivisible. Gilbert Ryle, a professor of metaphysical philosophy who taught at Oxford in the 1940s— and a man of whimsy—suggested in his book *The Concept of Mind* that the mind was rather like the University of Oxford as it must appear to a foreign visitor. "Those visiting our noble university for the first time," wrote Professor Ryle, "are shown a number of colleges, libraries, playing fields, museums, and administrative offices . . . and then comes the inevitable question: but where is the university?" As Ryle explained, "it is impossible for the visitor to see the university in the sense that he seeks it, for it—rather than a building comparable to a college or a library—is the way in which the colleges and libraries are organized (and the information and ideas that flow within them). The university is the activity of its many (ordered) parts." So it is with mind. What we experience as mind is the orderly activity of the brain. Mind is the subjective awareness of the continuous processing of information among the many neurons that make up the brain's organic parts.

Now it follows, if the experience of mind is the orderly flow of information among the neurons of the brain, then any disorder of that flow will also change mind. And of course there is ample support for such a proposition. Just drinking a glass of wine will immediately prove the point—brain chemistry changes and with it both behavior and our subjective state of mind. But in such circumstances we have an external explanation for the subjective change that we experience—the drinking of the wine. So it is not that we deny the biological underpinning of mind—we even socially sanction its manipulation with coffee and alcohol—but rather that we find it difficult to conceive of a *disability* of mind.

Depression and mania are disabilities of mind. Hence, when a mood disorder first disturbs the way we think, the tendency is to incorporate the changes into our worldview and search for an explanation. We look around and find the equivalent of the glass of wine, and frequently there is some social strain that can be identified. In the absence of such "objective" evidence, however, we search for hidden psychic conflict or, failing that, turn to "moral" issues, noting some blemish in the invincible self, perhaps determined by a loss of "willpower." These explanations are commonly clustered together as "psychosocial" explanations of depression. But for the neurobiologist there is an alternative level of description that emphasizes the role of the brain as the organ of mind. From this perspective depression is understood as a disordered flow of information and loss of behavioral resilience secondary to changes in neuronal regulation and a disturbed brain chemistry. Here we see parallel thinking at work, and the temptation is to choose between the two explanations. But they are equally valid—different rather than divergent.

And yet we do choose between the explanations, and do so again and again. Such a compulsion to choose, it would appear, is another example of the human preference for parallel thinking, a longing for "either–or" answers to the complex questions of mind and body. In some cultures, such as the United States of America, where individual destiny—life, liberty, and the pursuit of happiness—is considered a function of social opportunity and dedication to hard work rather than anything written in the genes, for many decades psychological explanations for depression have been preferred. After all, having spent years laboring to achieve a goal, suggestions that one's intelligence, energy, or mood are influenced by nature removes a certain sense of control and accomplishment. And reciprocally, given such social mores, failure and depression are considered a matter of personal responsibility, not a disability of mind. But the culture may be changing. The considerable interest in *Listening to Prozac,* a book in which a practicing psychiatrist, Peter Kramer, explores the social implications of new antidepressant drug therapies that some claim can change the behavioral dimensions of the "self," suggests that American public opinion is shifting, with greater acceptance of biology's contributions to emotional experience. But even with this greater acceptance, argument remains regarding the relative contributions of nature and nurture in the development of mood disorder.

A hundred years ago, at the end of the nineteenth century, mania and melancholia were considered "organic" diseases of the brain, perhaps akin to the disease of syphilis. Many psychiatrists and neurologists of the time studied the anatomy of the brain, hoping to define a pathology of the disorders. Indeed, Sigmund Freud began his career as a neurologist with an interest in neuropathology, and had some minor success with a gold chloride stain, which he developed to identify brain neurons. It was only after visiting Professor Charcot, the preeminent French neurologist with an interest in hysteria, that Freud turned to a consideration of behavioral illnesses, including melancholia. Even then, Freud initially formulated a biological theory of mind. He considered behavior to be driven by the innate instincts of sex, life, and aggression (which he called the "death instinct"). These biological drives, argued Freud, when in conflict with experience, are the source of neurotic behavior, including disturbances of mood. The conflicts, existing at various levels of consciousness, play themselves out in repetitive patterns of behavior, the origins of which are largely unknown to the sufferer. Through the free association of psychoanalysis, neurotic behavioral patterns (a similar concept to schema) are brought into consciousness and thus can be changed to healthier forms. Thus, although the work of many of his disciples has tended to obscure it, Freud's original intellectual premise was that biology and behavior are inextricably intertwined.

On the other hand, Ivan Pavlov, the great Russian physiologist who demonstrated in the early 1900s that environmental cues could condition the eating behavior of dogs, was a confirmed environmentalist. Pavlov's work spawned the movement of American behaviorism that dominated psychology for several decades. Behaviorism views the brain as essentially a black box, with none of the drive systems postulated by Freud. Hence the behavior expressed by an individual is directly proportional to what has been learned. Happiness and sadness, therefore, are conditioned by experience, and are behavioral responses to reward and punishment.

Certainly, we have all noticed examples of conditioning in our own behavior, as when the mouth begins to water before we succeed in getting the candy out of the machine. But does this confirm that the brain is a passive agent programmed only by experience? In the 1960s, Martin Seligman, then a graduate student in psychology at the University of Pennsylvania involved in classical Pavlovian conditioning experiments on dogs, began to wonder. Seligman's original plan had been to expose the animals to a mild random foot shock and condition them to a warning tone in a situation where they could not escape, then later provide them with the opportunity to avoid the shock by jumping over a low wall upon hearing the warning. However, the experiment did not go as planned. Even when the dogs had the opportunity to do so they did not jump over the wall. Most of them just whimpered, lay down, and accepted the shock with apparently little interest in trying to escape. Seligman, fascinated by this turn of events, reasoned that the explanation must lie in the first phase of the experiment, when the animals had no escape regardless of what they did. In anthropomorphic terms, recognizing the "hopelessness" of avoiding shock the dogs had "learned to be helpless." In the absence of any obvious way of controlling the situation they had given up the struggle and accepted the inevitable. Subsequent experiments in dogs and rats have confirmed that the element of control is a critical ingredient in determining the development of helplessness behavior. When the animal had control over the shock device—being able to turn it off by nosing a switch, for example—learned helplessness did not occur.

These observations marked the beginning of what is now commonly called cognitive psychology; they questioned the strictly behaviorist view that the brain is a passive input–output system, and the classical psychoanalytic formulation of neurotic illness as unconsciously driven conflict. The cognitive psychologists, in contrast, saw the brain as an active, thinking, machine. Learned helplessness, Seligman argued, was probably what happened to people in depression—the world became so oppressive and beyond their thoughtful control that they just gave up. Lack of assertiveness, a negative viewpoint, and the depressed person's grim acceptance of inevitable tragedy, Seligman proposed, were learned "attributional styles," essentially schema of helplessness. Such schema were evidence that, in the past, efforts to satisfactorily control unpleasant events had failed, fostering a predisposition to depression.

Seligman's pragmatic view of depression, as a thinking disorder, is closely allied with that of Aaron T. Beck, a professor of psychiatry at the University of Pennsylvania, and the founder of Cognitive Therapy, which has become a very important treatment for depression. Dr. Beck, himself trained as a psychoanalyst, was intrigued during his treatment of depressed patients with the consistency of their negative thinking, similar in pattern to the thinking I described in the case history of Claire Dubois. Dr. Beck called this peculiar depressive pattern of thinking—a negative conception of self, a negative interpretation of experience, and a negative view of the future—the "cognitive triad." Dr. Beck further postulated that these negative mental schema, similar to Seligman's concept of attributional style, persist beyond the episode of clinical illness and so predispose individuals to repeated depressions.

What Beck and Seligman have in common is their emphasis upon thinking—cognition—as the factor determining mood in depression. The cognitive approach to mood is compelling in its simplicity. It fits with the general experience that when good things happen we feel good, while disagreeable events commonly make us feel sad. Optimistically, Beck and Seligman argue that if negative thinking can precipitate depression, then it should be possible to think your way out of a depressed mood. The research evidence suggests that this is so, in many instances, although rarely when the depression is melancholic. However, even under such severe circumstances, negative thinking improves as the illness improves, regardless of the treatment. This suggests, at least in melancholia, that even if negative thinking precedes the disability it is also a product of it, varying with the biology and mood state of the illness.

The cognitive theories of depression have their roots in behaviorism, but this does not mean that they are strict theories of nurture, denying biological contribution. In fact, some qualities of what Seligman describes as attributional style, and Beck has identified as the pessimistic cognitive triad, are to be found early in life and clustered in families. Such behavioral styles, molded by experience, may have their predisposition in the temperament of the individual.

Temperament is an ancient, but still vital, concept. The word refers to "habit of mind," to the characteristic manner in which we react emotionally to the world, rather than the content of what we say or do. Temperament is a constellation of personal characteristics—from shyness and pessimism through to sociability and optimism—that can be distinguished in infancy and persists into adulthood. The word *temperament* comes to us from the Greeks, who compared emotion with the qualities of the earth's changing environment, namely hot, cold, dry, or humid. Individual temperament was thought to be determined by the timing of birth and the seasons and the stages of the life cycle, and was a mixture of four "humors" expressed in the activity and functions of the body. The blood (hot and moist) represented the sanguine humor of warm air and springtime; yellow bile (hot and dry) was of fire and summer; black bile (dry and cold) the melancholy of earth and autumn; while phlegm (moist and cold) reflected water and winter.

We owe to Galen, a Greek physician who lived in Rome nearly two thousand years ago, the elaboration of the humoral theory of temperament as the "constitutional" explanation of mood. Galen devoted particular attention to melancholia (in Greek meaning "black bile"), a group of diseases where fear and despondency were considered evidence of the humor of black bile casting its shadow over the brain. Today Galen's ideas about black bile and melancholy serve principally as metaphor, but his language of mood is enjoying a renaissance. Melancholia is again the word preferred by many to describe severe depression. And with the scientific advances in molecular genetics defining, in practical terms, what "constitution" really means, the concept of temperament has returned to prominence.

By constitution is meant the composition of the body. Despite evidence to the contrary (for example, comparing the image of one's face seen in the mirror with the photographic image captured during past years, and preserved in the family album), constitution is commonly considered something static and rigid, like the bricks and mortar of one's house. Molecular genetics has taught us that this is not so.

The building blocks of the body, and the enzymes that maintain the chemistry of life, are proteins. Proteins are complex molecules created within the cells of the body, where each is created by stringing together a unique sequence of amino acids. How the instructions required to manufacture the many proteins necessary in the development of a human being, or any other living creature, are transferred from one generation to the next only became clear in 1953, when James Watson and Francis Crick determined the structure of the genetic molecule called deoxyribonucleic acid or DNA. DNA molecules exist in the cell nucleus, as templates known as genes, within strands of material called chromosomes (because they stain easily with colored dyes). Genes, collectively known as the genome, make up the unique library of comprehensive instructions that we each inherit from our parents, and contain all necessary information to manufacture the kidneys, lungs, liver, muscles, bones, brain, and so forth, plus the vital developmental time sequence for doing so and the necessary instructions for lifetime operation.

The "books" in the genetic library are written in a coded language—hence the name "genetic code"—with a short molecular alphabet of adenine (A), guanine (G), thymine (T), and cytosine (C). These simple chemical units have certain physical and chemical properties (and are thus perhaps more akin to a Lego set than an alphabet) that bind them together in long chains, forming the complementary double helix of DNA. There is no substance in the body more important than DNA, for each gene is specifically coded to produce a protein molecule required in the building and operation of a cell's machinery. Nobody is quite sure, at the moment, how many genes are required to build a human being—estimates usually are in the 100,000 to 150,000 range—but discovering which gene is coded for each specific protein will enable us to understand the physical and chemical architecture of cells, including brain cells, and how they function at the molecular level of organization. A massive international project—including the American Human Genome Project—is now underway to map the human genome, and eventually we will have a complete reference library.

It is through the genetic templates of DNA that all cellular activities of the body are precisely controlled, and continuously maintained. While every cell in the body carries a full complement of the books in the genetic library, not all of the information available is used by each cell. Thus the thinking cells of the brain, the neurons, use a different set of "library books" than do the sugar-storing cells of the liver. A general rule is that the more complicated the tasks performed, the greater the information drawn from the genetic library. Thus the complex neuron is thought to use about sixty percent of all the information available in the genome, and uses it continuously, in daily maintenance and in responding to environmental experience. This is important to realize. While the basic architectural plan of the body does not change, the individual building blocks—the protein molecules that make up the body—are continuously being destroyed and reformed. Thus, over a time frame of about two weeks, some ninety percent of the proteins in the brain will be replaced, and the detailed configuration of the neurons will change as we learn about the environment and build the models that I described earlier. Within certain genetic parameters, therefore, the brain is a flexible organ, changing with experience.

This brings me back to constitution and to temperament. The present-day use of Galen's ancient designation refers specifically to the inherited, or genetic, parameters of emotional behavior. The unique library of genetic instruction that each of us inherited was encoded at conception, when the sex cells of our parents came together to create a fertilized egg. That conception requires the genes from two parents ensures a continuous variation in the gene pool, and promotes successful adaptation and survival of the species in a changing environment—the process described as evolution. Parents, children, siblings, and non-identical twins share approximately fifty percent of each others' genes, and grandchildren some twenty-five percent, while identical twins each inherit an identical set of genetic instructions. This explains why members of the same family frequently have similar physical features, something which is most pronounced in identical twins, who frequently cannot be distinguished from each other except by those who know them well.

But families also share common behavioral idiosyncrasies. Professor Thomas Bouchard, who directs the Center for Twin and Adoption Research at the University of Minnesota, from studies of identical twins separated at birth and reared apart in different environments, estimates that the genetic contribution to "personality" is around forty percent. Some of the stories that have emerged from these studies make for popular science journalism, such as twins who after a lifetime spent apart appear for their joint interview wearing identical shirts, and with children having first names in common. But beyond the sensational anecdotes, Professor Bouchard has developed a solid body of evidence confirming that many character traits, including attachment, dependence, tidiness, energy, social dominance, intellectual achievement, *and temperament,* are all partially shaped by genetics.

Temperament does not directly determine the emotional self but helps shape one's emotional attitude toward the world, and thus the way the world behaves toward the individual. Temperament and experience work together in the development of a singular self, each molding the other. Hence for a shy person social involvement may be uncomfortable, especially in novel situations. Shy people are more liable to be frightened by threatening behavior or punishment and, should they be unfortunate enough to be reared in such an environment, they are likely to become further inhibited and thus more cautious in reaching out. Similarly, what the shy person remembers, and the emotional context of the memory, is likely to be different from what an optimistic, outgoing person remembers. The shy person, when something goes wrong, is more likely to blame him- or herself and be pessimistic about the outcome than the optimist, who will frequently see the situation as unfortunate, but not of his or her making. The pessimist takes personal responsibility for a situation; the optimist shrugs it off. Thus the pessimist and optimist, predisposed by temperament, develop different attributional styles.

It is readily apparent to parents observing their infant offspring that such differences in temperamental tone are present at an early age. One child sucks energetically and sleeps well, and thus is described as a "good" baby; another, who seems tentative when feeding and wakes frequently during the night, is thought of as "finicky." And like the delightful anthropomorphic characters Tigger and Piglet, in A. A. Milne's classic book *Winnie the Pooh,* some children are bold from an early age, and some are shy. Furthermore, and this is the essence of temperament, these emotional characteristics commonly persist over a lifetime. Such persistence is even honored in the culture of children's stories.

Tigger, you will recall from A. A. Milne's characterization, is a young tiger, and a bold and bouncy one indeed. So bouncy in fact that, in one story, a tired Rabbit suggests that Tigger should be "unbounced"—taken "for a long explore" and lost to make him an "Humble Tigger . . . a Melancholy Tigger, a Small and Sorry Tigger." This sets little Piglet—the shy and cautious one—worrying about Tigger's future, but Rabbit is reassuring. "Tiggers never go on being Sad. They get over it with Astonishing Rapidity." Tiggers are the optimistic children of the world, the extroverts who seem to bounce back from any adversity, apparently taking everything in their stride. Piglets, by contrast, are the pessimists, the introverts who seek shelter behind their parents on the first day of kindergarten. "It is hard to be brave," as Piglet points out, "when you're only a Very Small Animal."

The contrasting behaviors caricatured in these stories—the willingness to approach an unfamiliar person, or situation, as opposed to withdrawing—provide a parallel to the behaviors we all experience in moods of sadness (emotional withdrawal) and happiness (emotional expansion). These behaviors form the temperamental poles of an emotional continuum that has been described over many centuries. Galen contrasted the temperament of the suffering melancholic—the shy and pessimistic Piglet—with that of the sanguine individual, the affable ruddy-faced optimist, amorous and confident, who is the bouncy Tigger of A. A. Milne's characterization. And, in more recent decades, the writings of the psychologist Carl Jung have given us the popular cultural stereotypes of introversion and extroversion, which emphasize the same polar behaviors.

Jerome Kagan, a professor of psychology at Harvard, describes, in the book *Galen's Prophecy,* his careful study of these variations in temperament, and their relative stability during the development of the emotional self. Screening over 400 children from intact, highly functional families in Boston, Professor Kagan and his team were able to identify approximately fifty young children who were consistently shy—behavioral Piglets, we might say—and a similar number who were uninhibited, committed Tiggers. To maximize his chances of finding persistent differences, Professor Kagan had selected children for his study at the extremes of the shy–uninhibited continuum of behavior. There were equal numbers of boys and girls in each group, and the children were followed over several years to determine whether their apparently polar temperaments persisted as they grew older. The studies were meticulously developed, and included a carefully de-signed, age-appropriate novel situation to draw out the characteristic behavior of the children at each evaluation. When the children were first tested, at approximately two years of age, new toys and a stranger were present in the evaluation room, in the absence of the child's mother. The uninhibited children tended to explore this novel situa-tion with energy and curiosity, in marked contrast to the inhibited in-dividuals who withdrew to the safety of a corner. At five years, unstructured play with unfamiliar children of the same age and sex was the challenge to the two groups, embellished later with competi-tive games as the children grew older.

The two groups of children were also chosen to make possible the study of changes in temperamental style, presumably resulting from experience. (Piglets might learn, as Professor Seligman has suggested, to be optimistic Tiggers.) Indeed, not all the initially shy, introverted, children remained shy. By the age of seven and a half years only about a third of them were painfully so—although three-quarters remained socially inhibited—suggesting that although caution and pessimism lessen with experience, it is a genetically driven temperamental style. However, interestingly, most of those who began life as Tiggers re-mained that way, being much more outgoing and engaging with the examiner at the age of seven than any of the inhibited children (which suggests that Rabbit was right: Tiggers do bounce back with Astonish-ing Rapidity). In fact, despite the trend towards the middle, the ex-tremes of behavior appeared to "breed" true in Kagan's study. Piglets tended to grow up to be shy, introverted youngsters while Tiggers re-mained extroverted.

What we actually inherit in the library of DNA we receive from our parents are developmental processes—the regulatory chemistry and anatomy of the brain that guides the organization of information, memory, and emotional expression. Subtle variations in the coding for the protein molecules, singly or in combination, that sustain these regu-latory processes result in individual differences in behavior—just as vari-ations in our genetic complement determine differences in weight, height, coloring, and other obvious physical characteristics. The genome lays down these parameters but it is interaction with the environment, especially during critical periods of development, that determines how the genetic blueprint reveals itself. This amalgam of genetic blueprint, shaped in its "expression" through continuous interaction with the environment, is called the phenotype. And, just as good physical nutrition during development will increase ultimate body size, changing the phenotypic expression of the genetic inheritance, a nurturing emotional environment of secure parental attachment during childhood will enhance the development of the emotional self.

If at conception mistakes occur in the organization of the genetic code then "genetic illness" may result; with incorrect instructions in the genetic library, the manufacture of the cellular machinery is impaired and the cells do not work properly. In the most severe coding disorders the fetus dies before birth, or the adult is incapable of reproduction, and therefore the illness rarely occurs, or disappears. However, minor genetic variations may persist through generations, especially when those variations require the participation of several genes in combination to have a damaging effect, or actually have a survival advantage under certain environmental conditions. Sickle cell anemia is an example of the latter, where mild forms of the illness appear to protect against malarial infection. Variations in emotional temperament, including mania and melancholia at the extremes of emotional experience, may also fall into this category.

A coding variation that predisposes some families to manic depression—a regulatory enzyme, for example, that promotes an unusually broad oscillation of emotional homeostasis—may in combination with other genes determine melancholic psychosis in one family member, and high energy and short sleep in another, resulting in an optimistic temperament. While both are genetically equally deviant, the phenotype of psychosis marks the first individual as a psychiatric patient and the second as a successful leader. Emotional expression is essential to social interaction, and in adulthood *the way* we interact determines in part our place in the social hierarchy. Most of us like expansive, energetic people who get things done, and we tend to follow them. Emmett Jones rode the rising wave of his hypomania with some success until he overreached and fell into suspicion and grandiosity. But there are many other people who experience the energy and drive without illness. Frequently they are found in the same families where mania and depression are prominent and often they are leaders, as Stephan Szabo observed. On the other side of the coin, the withdrawal and conservation of depression (but *not* the psychosis of melancholy), offers time for reflection, rebuilding, and creative thought. Thus, within the spectrum of mood disorder and emotional temperament may be found behaviors of both positive and negative value to the individual, to society, and in the evolutionary struggle. Some would argue that it is because mania and depression, in their muted forms, have this evolutionary value that the disabling illnesses persist, together with the social advantages.

Temperaments are not moral labels, or preordained pronouncements of emotional destiny, but descriptions of an emotional landscape—genetic templates that in a manner similar to the way language develops help organize the grammar of mood and especially the development of self-conscious emotions. As we grow and develop, each constructing a mind of our own, the emotional self is in continuous evolution, constantly being eroded and rebuilt. How that evolution proceeds determines in large part how we behave toward others and how they respond. Most individuals move back and forth across this emotional landscape, appropriately to the task at hand. While the hills and valleys that must be traveled are greater for some, we each have a personal reference point along the optimistic–pessimistic, sanguine–melancholic continua, from which the journey begins. But are there foothills, within this misty terrain of human temperament, from which the polar peaks of mania and melancholia are more easily encountered? Where an inherited predisposition to process information in some special way, when later played upon by environmental circumstance, gestures toward an unstable emotional path?

Probably 20 to 30 percent of the population meet criteria for a general notion of shyness—reluctance to start a conversation, few spontaneous gestures to strangers, finding new social situations uncomfortable—but even the most expansive of surveys classify no more than 10 to 15 percent of the population as seriously depressed. To be shy does not mean that one is destined for major depression. But do a shy, introverted way of relating to others, sensitivity to criticism, and difficulty in developing the friendships that are so helpful in times of adversity make for vulnerability to melancholy? Is shyness one of the genetic constellations that under the right social circumstances will be expressed as a phenotype of depressive illness? Shyness and instability of emotional life—

the Piglet profile—have been found to be associated with symptoms of depression, especially in adolescent girls. A number of other studies have chronicled shyness and social withdrawal in the preschool years as predisposing factors to loneliness and depression in later life, with an adult emotional style that is ruminative, reflective, and withdrawn.

That some temperaments, like shyness, increase vulnerability to mood dysregulation later in life becomes very important when we consider the medical diagnoses of mania, melancholia, and the family of illnesses with which they coexist. Frequently, a spectrum of mood disorder clusters in an individual family. A knowledge of family history, as an index of genetic predisposition, is thus critical in accurately defining the immediate illness and in helping to determine its likely course and outcome over time.

So it was when I met John Moorehead and his family.

Five

Unique and Similar to Others

Moods, Morality, and Medical Diagnosis

Man desires a world where good and evil can be clearly
distinguished, for he has an innate and irrepressible desire
to judge before he understands. Religions and ideologies
are founded on this desire. . . . .

> MILAN KUNDERA
> *The Depreciated Legacy of
> Cervantes* (1988)

We must first find out what is wrong with you
Before we decide what to do with you.
All cases are unique, and very similar to others.

> T. S. ELIOT
> *The Cocktail Party* (1950)

Probably the best way to describe John Moorehead was "craggy." A tall man in his forties, slightly stooped but lean and tanned, he had the sort of face that would be at home among the presidential carvings of Mount Rushmore. From the testimonial of Angela, his sister, John had been a typical "Tigger" all his life. Inquisitive and bold as a boy, in manhood he was charming and compassionate, with bounding energy and the capacity for inspirational leadership. John Moorehead was a success.

When he first came to see me as a patient, Angela's description of him was hard to believe. His hair was almost white, very full but cropped short and unkempt. Bushy eyebrows flecked with gray, and blue eyes that seemed not to look at me, were the most striking features of his thin, almost emaciated face. On that day of his first visit to my office, the eminent Jesuit scholar and college provost was dressed in a rough workshirt, chino pants, and old loafers. That in itself was unremarkable. Over a typical summer season, on the streets of the small college town where I lived in New Hampshire, I would pass a hundred people like him: New Englanders enjoying their leisure. But when he finally looked at me—and then lowered his eyes as he caught my gaze—I saw on his face what I had seen many times before: the mask of profound depression.

As Angela and John Moorehead unfolded their story, an irony was immediately apparent. Angela had struggled with smoldering depression on and off for most of her life, especially during the winter months. That was one reason why, a decade before, she had fled the East Coast for California—that and a series of disastrous love affairs which had exasperated her devoted but puritanical brother, threatening the security of the love they felt for each other. But that summer the tables had turned. John had stumbled and fallen into a depression of his own. The family's colossus showed signs of crumbling, and Angela feared for what he might do under the circumstances. As his sister knew better than anybody, the Reverend Professor John Moorehead was not one to tolerate distraction and incompetence, especially in himself. Fortunately, as his ruminations about death and inadequacy worsened, John had confided in Angela. Much against his wishes she had immediately canceled her own vacation to join him in New Hampshire. After ten days of watching her brother suffer, Angela, the little sister whom John had counseled all her life, insisted that he seek psychiatric advice.

The first thing John Moorehead wanted me to know was that he had agreed to see me only out of respect for his sister and his persistent, unsettling thoughts of suicide. He could handle the rest. He did not believe in psychiatry; psychiatrists eroded personal responsibility. He'd understood something of depression from observing his mother and sister struggle with it. In fact, during his years in seminary, he himself had known periods of emotional withdrawal, which he had presumed then were the natural course of events. However, he was prepared to admit now, after several months, that something had distinctly changed. He had fallen into chronic indecision and pervasive anxiety, and had great difficulty in concentrating and an appalling lack of energy. Among his colleagues at the Jesuit college where he taught in Massachusetts, his energy was legendary. He had always been a short sleeper. Four or five hours a night were usually enough for him. But he had seen more dawns this summer than he could count. In previous years, when on vacation, his insomnia would have meant some very good fishing, "enjoying the sounds of lapping water and loons calling through the hanging mists," he added, with a sudden poetic flare that gave me a brief glimpse of what others knew of him when he was well. But such pleasures required energy and concentration, and at the moment he had neither. In contrast to his usual outlook, he now found little joy in anything. The thought of getting back to teaching and his new administrative duties as provost in the fall was horrifying. He couldn't imagine how he had once organized himself to teach thirty students; now he was expected to lead one hundred faculty.

In response to my questions, John Moorehead described how he had brought upon himself this disabling state of mind. "I have fallen into bad habits in the past few years," he told me. "I'm just not fit to be a member of the Jesuit community. I've become slothful—it was called acedia in the Middle Ages—a well-known problem for seminarians. If you're spiritually strong you can work through it. If not, you can expect the worst; the Devil *does* take those who are idle. You may laugh and consider such statements some sort of metaphor, but to me sloth is still one of the seven deadly sins. I'm a gaudy-minded man. I deserve this. I've been headstrong, proud, and complacent. You probably remember the old drinking song, *Gaudeamus igitur, juvenes dum sumus* (Let us be merry while we are young). Well, it keeps plaguing me, running through my head. I have misspent my youth and now it has gone, and my energy with it. While it may sound strange to you as a psychiatrist, I see this mental pain as a warning. It's time for me to do something about my spiritual turpitude. I've been arrogant and lazy long enough; this suffering flows from a moral laziness."

I was fascinated with what I was hearing. As do many patients with depression, John Moorehead had adopted a deeply personal meaning for his anguish. But this particular explanation reached beyond the pain of everyday concern to a profound knowledge of medieval scholarship—to John Moorehead's core identity as a devout Christian and a Jesuit scholar. It reached back to an interpretation of human depression that, although we seem to have abandoned it long ago, remains deeply embedded in our Western culture.

An urge to judge and find an order in things is in the human character. It is one of the things which the human brain does best. Milan Kundera, exploring the allegorical tale of Don Quixote, has emphasized how this need to reduce the ambiguity of a complex world has driven human history. Faced with the unknown, we search for a Supreme Judge and the comfort of explaining experience in familiar terms. Ideology, religion, and science all have their roots in this human need. Classification is a large part of the method of science. Through objective scientific observation and classification we seek to discover the true association among events, making past experience more intelligible and the future more predictable, with the general goal of reducing fear and promoting understanding. Thus the theory of evolution, which is fundamental to our understanding of biology, was originally based upon one grand classification that was Charles Darwin's life's work.

But as Kundera also emphasized in his essay, both as individuals and as social beings we have "an irrepressible desire to judge *before* we understand." That is what John Moorehead had done when faced with the ambiguity of his depressive illness, and it is what we have done as a society for over two thousand years when trying to explain disorders of mood. Because, as I have suggested earlier, mood and emotion lie at the center of being human, we have been slow to grasp the true nature of their disorder. It is only recently that the severe mood disorders have been categorically considered as illnesses and as requiring the attendance of a physician. Previously, except in Greek and Roman times, Western society has been more inclined to condemn than to cure when confronted with manic and depressive disability. Hence my dual purpose here is to outline the principles of current medical diagnosis in mood disorders while placing their classification firmly in an historical context. John Moorehead, struggling to explain his changing mental state within the terms of his own experience, had missed the mark; he was suffering not from sloth but the medical illness of severe depression.

Medical diagnosis is a special form of scientific classification, a clustering of information, based upon the observation and experience of physicians over many centuries, that helps judge the nature of the disability suffered by an individual patient. But medical diagnosis is also a classification that carries a special burden, for it must not only describe and characterize the disability, but seek to predict accurately the eventual outcome of events. The necessity for such a burden is obvious. When we suffer we are most interested in relief from our suffering. We seek the advice of a physician not out of intellectual curiosity but to reduce our pain and worry. Therefore, in medical diagnosis we ideally look for a classification that can predict as well as describe. The physician, to achieve this, seeks accurate information upon which to make an informed judgment.

It was Hippocrates, some 400 years before Christ, who first gave us methods for medical diagnosis. He and the many students in the school of thought that bears his name were the first to teach the importance of longitudinal observation, the careful study of a disease over time. Most importantly, Hippocrates recommended separating the illnesses of the body from the "knowledge of the world called theology or philosophy," and thus he is remembered as one of the first "scientific" physicians. Because of his objective, unbiased approach to illness, he is revered as the "father" of Western clinical medicine.

He introduced detailed clinical description—what we now call the case history—to medical science and was meticulously honest in recording what he observed. In the forty-two case histories that have survived for posterity, well over fifty percent of the patients died, but the course of their illness was written down so that others might learn. It was Hippocrates who first noticed that those who suffered head wounds in battle and survived were different in their behavior and would sometimes develop a disease that shook their bodies. This association proved to be so common in his experience that he concluded—and gained public acceptance for the idea—that brain injury, and not divine visitation from the gods, was the explanation for epileptic seizures. Epilepsy had previously been called the sacred disease; after all, who but the gods could pick a man up, throw him to the ground insensible, and later let him recover? Because of such careful observation, Hippocrates became one of the first physicians to associate abnormal behavior with the brain, rather than with some supernatural being.

But, as I have emphasized earlier, most of us find it difficult to be objective about human behavior, especially when those behaviors include a disturbance of the mind—that part of our behavior we consider to be the essence of ourselves. Losing control over the arms and legs is very different from loss of control of thinking, or how we feel emotionally. Objective classification of the body and its organs is a simple business compared to that of classifying mind, mood, and emotion—the complexities of the private self. It is only comparatively recently, in the last century or so, that an objective classification of mood disorders has emerged.

In their descriptions of the temperaments, however, the Greeks, who dominated medical science at the height of the Roman Empire, did make a very thoughtful start. Indeed, had John Moorehead been a Roman—and he probably would have made a very good senator or centurion—one of the Greek physicians of Galen's time would have recognized his ailment. John would have been considered an individual of sanguine temperament; a man of confidence and optimism, of brave and hopeful disposition. Such men are leaders—the Winston Churchills of this world—individuals who seize their time and place and employ themselves well, frequently to the advantage of all of us. But such individuals can get depressed and when they do, they tumble badly. Frequently, unable to tolerate the sudden loss of their familiar, ebullient selves, they take to alcohol (as did Winston Churchill), or worse, to suicide. What Angela Moorehead had intuitively understood is in fact correct: Upbeat, sanguine people, especially men, who become seriously depressed for the first time in midlife are prone to self-destruction.

John Moorehead's opinion of himself as a man of sloth has a complex cultural root, one in which a theology of the soul—the seat of the passions—has been indivisibly entwined with emotional afflictions of the mind. He was a man molded, through family and vocation, by a history that reaches back almost two thousand years to the teaching of the early Roman Church and the origin of the seven deadly sins. In the fourth century, when Rome finally became too difficult to defend against the raiding Scandinavian and German hordes, Constantine had moved the center of his empire east to Constantinople. Europe fell into what historians now call "the Dark Ages" and most of the teaching of the Greek physicians was lost. In the intellectual and religious struggles that followed were sown the seeds of John Moorehead's self-judgment and the pervasive cultural confusion that still haunts us today—whether depression is a medical illness or some sinful sickness, an outcome of laziness and moral turpitude.

Christianity had begun to thrive in the latter days of the Roman Empire. The real key to the religion's popular growth lay in the simple message that there was one God, who had a compassionate, tangible missionary in the person of Jesus Christ. The Church was the infallible custodian of emotional life, and the Pope the keeper of all earthly souls. There were debates among the churchmen of the time, just as there are today, regarding Papal infallibility and the control that God—and therefore the organized Church—exerted over human feelings. Augustine, who lived late in the fourth century, was one of the most influential of the monks who thought about such things. His was a miserable view of the human race—we had crucified the Son of God—and he preached that all humans were born corrupt and helpless sinners. For Augustine, salvation could be gained only through a life of atonement. These early Christians were particularly concerned with the temptations of the soul while in pursuit of spiritual perfection, and their deliberations paved the way for what became known as the seven deadly sins. (There were originally eight, but Pope Gregory reduced them to seven by making pride—as John Moorehead explained to me—the root cause of all the others.)

Acedia, the name given for the sin of slothfulness by which John Moorehead felt consumed, was first given prominence by an influential monk named John Cassian, who lived at the turn of the fifth century and established monasteries for both men and women in what is now southern France. Later included under the broader sin of dejection, acedia was recognized as a common problem in monastic life. From contemporary descriptions that have survived, individuals suffering acedia expressed "impatience in work and devotion, wrath against the brethren, and despair at ever reaching their spiritual goals." Carelessness, weariness, exhaustion, apathy, anguish, sadness, plus sometimes a yearning for family and their former lives, were frequent complaints—ones that have echoed across the centuries, and which we now characterize as the symptoms of serious depression. The remedy prescribed for these troubles was manual labor. In the early monastic world of John Cassian, and as John Moorehead expressed to me centuries later, "laboring in the vineyard" was imperative to guard the soul against sloth.

Confession and penance became early forms of psychotherapy. They were obligatory in atoning for the seven deadly sins of vainglory, anger, envy, dejection (acedia or sloth), covetousness, gluttony, and fornication. Initially these "afflictions of the emotional spirit" were the concern of the monasteries and monastic life. But over the next several centuries, as the Church's control expanded through the influence of the clergy to the "flock of common souls," stern general warnings were issued against idleness as the root of mischief and evil. Perhaps you recall, as do I growing up, the familiar admonition—a clarion call of conscience from a thousand years ago—that "idle hands make devil's work." So, increasingly, as the Church extended its control over emotional and spiritual life through the Middle Ages, penance became established as the medicine of the soul. The act of confession and catharsis was established as healing and therapeutic, and the father confessor—the priest—became the physician of souls. Indeed, even as late as the eighteenth century, it was necessary in England for any physician wishing to attend the "mentally deranged" to first be licensed by the local bishop.

So in my office that summer afternoon sat a man not only confronting the demons of the day, but also the demons of the past. In the private explanation of his experience as evidence of a slothful life, John Moorehead drew deeply upon a cultural heritage that many of us share. It is a commonly held sentiment in our society that depression is moral weakness. It is in our belief system and at the root of the shame and stigma that many who suffer depression still bear. There are many idle people in this world, and doubtless some of them are slothful, but not all of them suffer as did John Moorehead. By all objective criteria, the eminent professor was neither an idle nor a slothful man. His pride had been built not on complacency, but on diligence and hard work. The meaning and justification that he offered for his plight was an explanation culturally driven. But it was insufficient as an objective explanation of the dramatic change in his energy, behavior, and emotional life. Like many others, John Moorehead had judged himself before he understood the true nature of his sickness. The explanation he gave of himself did not withstand objective medical and scientific scrutiny.

Today we tend to associate science with extraordinary technology, electronic machines, complex formulae, and molecular structures invisible to the naked eye. But these are the tools and products of science, not science itself. Science is a method, a simple way of asking questions about ourselves and the world, that fosters the development of agreed-upon, objective opinion designed to guide present judgment and future investigation. The scientific method is just as reasonably applied to medicine and to human behavior as it is to physics and chemistry. Some have argued that it is impossible to find common symptoms in something as complex as disturbed emotional behavior. There is no truth in such statements. As the uninvited guest in T. S. Eliot's play *The Cocktail Party* wryly observed, we are each "unique, *and* very similar to others [emphasis added]." Darwin demonstrated a century or more ago that there are many behavioral denominators, including emotional expression, that bind human beings together as social mammals.

Similarly, denominators of common symptoms, across individuals and generations, are found in those extremes of emotional experience that we identify as the illnesses of mania and melancholia. Standing back from the private meaning and preconceived judgment that John Moorehead had adopted to explain his symptoms, and simply describing the changes in behavior that he had experienced (just as Hippocrates might have done), clustered patterns emerge from behind the cultural stigma. That these patterns can be traced back consistently over many centuries, and through varied cultural and social circumstance, suggests that mood disorders are of the flesh, not the spirit. It is through careful attention to this consistent clustering of the symptoms and signs that an accurate diagnosis of mood disorder is made and appropriate treatments are decided upon.

So during my examination of John Moorehead in my office that afternoon, as I listened carefully to his complaints I was also busy cataloging them in my mind. At the core of John's concern was a disturbance of his accustomed emotional balance, a change in the usual regulation of his mood. Rather than his usual optimism and pleasure in daily tasks, he complained of anxiety and an uncomfortable, irritable mood, a combination that psychiatrists classify as dysphoria. Even his most leisured pastimes, such as fishing, failed to give him pleasure, a condition called anhedonia. You will remember that this also lay at the center of Claire Dubois's experience of melancholia. Clustered about these core disturbances of emotion, John also experienced a profound lack of energy, making it difficult to complete, or sometimes even contemplate, his usual daily tasks. In addition there was a change in sleep patterns, a loss of appetite, and an inability to concentrate and make decisions. In his appearance he was disheveled, suggesting a change in concern about his personal grooming. Furthermore his sad, heavily lined face and stooping, burdened shoulders, plus his reluctance to meet my eye, did not fit the social mantle described by his sister—that of an exceptional leader and revered scholar. The description he gave of himself was peppered with guilt and self-blame; he saw himself as old and incompetent. Finally, and most terrifying to him because of their strength and alien nature, he described intrusive, ruminative thoughts including those of suicide, thoughts he had never previously experienced.

Being by now increasingly accustomed to the behavioral disturbances of depression, probably you have already sorted John Moorehead's complaints into the familiar triad of mood (anxiety and irritability), thinking (poor concentration, indecision, and rumination), and housekeeping functions (difficulty sleeping, lack of appetite, and lack of energy), just as I did. If so, then you will also have begun to appreciate the clustering of the symptoms common to this illness. As I shall explain when we explore the anatomy of the emotional brain, there are reasons why the abnormal behaviors of depression cluster together in this way. The clustering reflects a disturbance of a number of centers in the emotional brain, which usually begins with a disturbance of mood and recruits other important regulatory centers as the depression develops.

In medical diagnosis this clustering together of particular symptoms and signs is called a syndrome. Symptoms are what the patient describes or complains of, and signs are those things which the physician observes. Thus the complaint of loss of pleasure and the intrusion of suicidal ideas described by John Moorehead are both symptoms, whereas in my observation that he was gaunt and unkempt, I was identifying a sign. A syndrome is really just a medical name for a category, a cluster of symptoms or pathological signs which consistently occur together. Recognizing syndromes is not the same as making a diagnosis, although it is frequently the first step. A diagnosis usually implies that we know something beyond the mere description of symptoms— something about the natural course and outcome of the illness, or the mechanism through which the disease develops in the body.

In identifying a syndrome, it is the *consistency* of the clustering, across individuals and over time, that is critical. Take breathlessness, for example. In itself, just like sadness it is unremarkable. We all become breathless occasionally, but we dismiss it—and rightly so—as normal experience. However, if episodes of breathlessness become common when that was not so previously—when one is walking quickly in the street, or in the course of daily chores—they take on a different meaning and are cause for alarm. So what breathlessness means depends upon a particular context, and the other symptoms and signs that cluster with it in time. In association with a wheezing chest it may represent asthma; with swelling ankles, possibly a failing heart; or with a loss of healthy color in the skin, tiredness and a loss of energy, perhaps anemia. Each of these clusters, all of which include the symptom of breathlessness, represents a different syndrome that leads in turn to a different diagnosis and the identification of a distinct illness. Thus the meaning of a symptom, or a sign, changes with the association that it keeps.

So it is with sadness and joy. Alone, they are the extremes of a familiar dimension of behavior we accept as normal emotion. But when sadness or joy extend relentlessly for days or weeks, driving a mood that colors everything we do, and are clustered together with disturbed sleep, a change in energy and appetite, and even distortion of thought itself, then the meaning changes. These clusters are the patterns of a gathering illness which reflects a dysregulation of the emotional brain. These are the syndromes of mania or melancholia.

The first steps in developing a diagnosis from any syndrome—and it is no different in mood disorder—are to accurately describe the individual symptoms, how they cluster together, and how that clustering evolves over time. Only then will it be possible to objectively distinguish the significant changes from among the many complaints and interpretations that usually surround illness, and classify them with sufficient precision to arrive at a true diagnosis. Doctors call this objective description and classification *phenomenology*. Thus the phenomenology of the American Psychiatric Association's diagnostic classification of melancholic depression requires the presence of a sad and diminished mood, or anhedonia, that has persisted for more than two weeks. For a reliable diagnosis of melancholic depression to be made, at least five other symptoms from those commonly occurring—sleep disturbance, fatigue, body restlessness or slowing sufficient to be observed by others, feelings of worthlessness and guilt, difficulties concentrating, and sometimes ideas of suicide—must be present in addition. The length of time that these behavioral changes have persisted is also important. We all have days when we feel tired or sad, even sometimes without being able to identify a satisfactory explanation, but rarely do such "blue" days come back to back, disrupting sleep, thought, and appetite for days on end.

Mania, at the opposite pole to depression, is defined as a distinct period of abnormally expansive or irritable mood, lasting for a week or more, which is associated with three or four of the following behaviors: grandiosity, a decreased need for sleep, sometimes increased sexual interest, great energy, and excessive indulgence—with little regard to the risks being taken—in pursuits that most of us would normally consider dangerous to person or property. Another frequent and striking feature of mania is rapid speech and a pressured flow of ideas, with many changes of subject and sometimes joking, rhyming, and punning, something which Stephan Szabo had described to me as characteristic of his manic periods.

These "diagnostic syndromes" of mania and melancholia have been given the general name of mood disorder because disturbances of mood are commonly the core symptoms. When mania and melancholic depression occur (usually following each other) in the same individual, the syndrome is called manic depression. Because mania and melancholia represent the extreme poles of mood change, manic depression is also called bipolar illness and is distinguished from unipolar illness, where one mood state (usually depression) occurs alone. Unipolar mania, which is exceedingly rare, describes the syndrome of mania without episodes of depression. In the American system of classification, when an individual suffers predominantly repeated depressions and only occasional mild hypomania, the diagnosis of bipolar II disease is made to distinguish the illness from the severe manic form of bipolar I.

While it is not necessary to know the cause of a syndrome to successfully describe or diagnose it, to call something a disease does require that we understand it sufficiently well to predict its course and perhaps say something about the pathology that determines it. Disease implies a disordered structure or function sufficient to threaten normal health—some physical or chemical change in a body organ that may become irreversible if it goes too far. But there is a complication here. The homeostatic mechanisms that regulate and defend the chemical balance of the body themselves react to changes that threaten health. Dynamic physiological responses that limit and repair the disease process, seeking to return normal function, automatically come into play. Common examples of such dynamic defense mechanisms are the increase in white blood cells during an infection, or the rise in stress hormones that occurs when we face a threatening or challenging situation. Thus the early signs of mood disorder commonly include emotional disturbance reactive to the developing illness, much as a fever is an early sign of influenza.

The disease process, and the body's defensive response, both alter the usual behavior of bodily systems, including the regulatory systems of the brain. Frequently these processes can be detected by changes in blood chemistry or through other measurements—including tests that measure behavior. These tools extend the physician's fundamental skills of history taking, observation, and physical examination. Hence, in the ideal situation, medical diagnosis is based on the information gathered from all these sources and the physician's basic knowledge, acquired through education and experience, of the disturbed mechanisms that characterize the different diseases. Diagnosis is important because the more precise the diagnosis, the more accurately can the physician choose the most appropriate treatment, and the more accurately can the outcome of that treatment be predicted.

Despite the extraordinary advances in medical science and diagnostic testing, there are still many diseases for which we do not know the specific and ultimate cause: most cancers, asthma, or rheumatoid arthritis, for example, fall into this category. Probably in such cases, and in mood disorders too, there is no single "cause" sufficient to explain the disease but rather there are many factors, including an inherited—that is, a genetic—vulnerability that can place somebody at risk. The relative contribution of some of these factors may be known, but the mechanisms through which the illness begins and expresses itself in the body may not be precisely understood. This is particularly true for most neurological and psychiatric disorders of the brain, in part because the brain is a difficult organ to investigate. Nature clearly intended it to be that way. Not only is the brain carefully protected from physical harm by the bony skull, but it is also buffered from the general chemistry of the body by a special barrier—appropriately named the blood–brain barrier—which ensures that sudden changes in the body's chemistry will not overwhelm the brain's own delicate chemical balance. The brain itself is very tightly regulated in its habits and a small change in its activity can have a penetrating influence upon everything that it controls. Changes in the body and in the environment that profoundly impact brain regulation—generally we call these *stressors*—will dramatically drive behavior toward disorganization, while events that foster regulation and organization will return harmony and balance. There are many complex systems that ensure this precise regulation of the brain is protected—so, for example, the brain is given preference over other body organs in the supply of oxygen, glucose, and the materials essential to its daily activities.

These protections that surround the brain present something of a problem when one is trying to identify the disturbed regulatory mechanisms, and the adaptive brain physiology, that determine the disabling syndromes of mania and melancholia. In organs such as the liver or kidney, harvesting the blood that bathes them—or even directly sampling through biopsy the living tissue itself—can be very informative and an aid to diagnosis, but this is not possible in the brain. While advancing knowledge of the genes that control brain systems, and the technologies that permit the imaging of the living brain, holds great promise, we do not have laboratory tests—comparable to those available to differentiate anemia or to define a category of cancer—that can identify the disturbed brain chemistry of a melancholic or manic episode. Nonetheless, over the last three decades we have acquired a valuable understanding of the brain mechanisms that are disturbed in the mood disorders. These advances have been achieved through a thoughtful integration of the neurosciences and pharmacological research. But, the cornerstone of our knowledge remains clinical observation, methods similar to those Hippocrates taught centuries ago, which over time have led to more accurate diagnoses of the major syndromes of mood disorder. The development of this classification has required the careful study of the varieties of mood disorder and the observation of their natural course, with and without treatment, both within families and over the lifetime of an individual.

Mania and melancholia both cluster in families. Most of us under the right circumstances can fall into a single episode of despair, and repeated depressions may occur in one individual without a strong family history of the illness. However, in most cases where mood disorder becomes a chronic, repetitive illness, others in the family suffer also. In some families, several close relatives may have multiple episodes of depression without mania, suggesting that recurrent unipolar illness may be inherited, but it is where mania and depression occur together, through many generations of a family, that the evidence for genetic transmission is most compelling.

That these illnesses are passed down through the generations was recognized long before we understood the nature of genes and the genetic code. "The disease that the father had when he begot his son appears ... in the manners and condition of the [young man's] mind," noted Robert Burton in his *The Anatomy of Melancholy,* published in 1621. Reprinted many times, *The Anatomy* is still worthy of careful reading, although today in most bookstores you will need to look for it in the English Literature section. Robert Burton, like John Moorehead, was an ordained priest. He entered Christ Church, Oxford, at an early age and spent his life as a scholar and teacher, writing occasionally about the human tendency to eat too much, but mainly about melancholia. Burton considered *The Anatomy* his life's work, written to "help others out of a fellow feeling," for he was himself a melancholic. In keeping with his time, Burton believed "the sin of our first parent Adam [to be] the impulsive cause of these miseries" and acknowledged the power of devils—although probably with tongue in cheek—telling one tale of a nun who had become possessed after failing to cross herself before eating lettuce. Burton also devotes a full third of his scholarship to the melancholy of lost love, emphasizing the critical importance of social and personal attachment in precipitating mood disorder. Many of Burton's observations closely parallel our contemporary experience, but particularly noteworthy is the connection he makes between predisposing temperament and illness. "Those most affected," he wrote, are individuals "solitary by nature, great students, given to much contemplation, [who] lead a life out of action," or "of a high sanguine complexion." Men and women of such temperaments and those "born of melancholy parents are most subject to melancholy."

Temperament, you will recall, is a set of behaviors that emerges early in life and typifies an individual's emotional interaction with the social world. It is best understood as the emotional nature of the person. Temperament is not a disease and it should not be confused with illness, although certain temperaments, as Burton recognized, cluster in families, and especially so in families where mood disorders are common. Thus Burton's depiction of "high sanguine complexion" in our modern idiom describes an individual of energy and optimism, which, making allowance for the span of nearly four centuries since *The Anatomy* was written, is an excellent characterization of John Moorehead's temperament before he became depressed.

The temperaments predisposing to energy and mercurial mood in Angela and John Moorehead's family seemed to have passed down the generations predominantly in the maternal line. Genetic contributions from their father's family of origin were less clear. Albert Moorehead, their father, had been in the British merchant marine when he first met the vivacious Josephine Piane. While selling freshly squeezed orange juice outside the family grocery store on the Battery Wharf in Boston Harbor one hot June day, Josephine had fallen desperately in love with the handsome sailor and decided to marry him. Much to the consternation of her large Catholic family, Josephine had been pregnant with Albert's child when they did marry, and John was born six months later. Albert and Josephine moved into her parents' house, which still stands in the little thatch of streets that have survived the urban renewal of Boston's North End, before they found a place of their own two doors away. John Moorehead's memories of childhood were happy ones; by nature he was outgoing, curious, and into everything. According to his sister he bounced his way through childhood, dearly loved by his extended family of aunts and uncles. He was also a good student, which surprised his parents as he could hardly sit still, and from an early age was fascinated by the rituals of the Catholic Church, enjoying, as he described it, the "endless cascade of saints, festivals, and food."

Angela arrived four years later and in temperament was the antithesis of her brother. Painfully shy, she preferred the safety of her mother's side to the bustle of Piane family life; even her aunts and uncles had trouble making her smile. However, in John's presence she relaxed, and as they grew older he became her guardian and patron saint. But then, as she reached adolescence, and especially after John left for the seminary, Angela's emotional nature began to change. Family and Church seemed less important as a gregarious side of her character developed, causing some family alarm. "She is going to be just like her mother," complained Albert Moorehead, after one of his extended periods ashore, "devastatingly beautiful and charming, and totally moody and unpredictable." The whole family agreed that Josephine, Angela's mother, was moody, disturbingly so at times. A generous woman, usually expansive and loving (and the "best cook in Boston" according to John Moorehead), she could suddenly turn and heap criticism upon those she loved. After such emotional outbursts Josephine would fly to her bed in tears, and sometimes stay there for a week. Reasoning, pleading, plotting, persuasion: nothing worked. There was no explaining these episodes. "It's because she's Italian," offered Albert Moorehead. "The whole Piane family is like that." Her husband was correct; Josephine Moorehead (née Piane) was certainly Italian but, independent of any ethnic stereotyping, it seemed highly probable from the descriptions of her behavior that she was also of cyclothymic temperament (thymic from *thumos,* which in Greek means "spirit").

Cyclothymia, and Galen's concept of constitutional temperaments, have regained prominence in contemporary psychiatry through the writings of Professor Hagop Akiskal at the University of California in San Diego and Doctor Athanasio Koukopoulos of Rome. Both have emphasized that the temperaments, although well characterized for centuries, are now neglected in medical practice. The evidence suggests that certain temperaments are closely associated with the bipolar illnesses of mania and depression and may be found in a large percentage of the population. Whereas manic-depressive disease, in its most serious form, occurs in approximately one in every hundred people, it is estimated that behaviors characteristic of the bipolar spectrum of temperaments occur in four or five times that number—affecting perhaps twelve to fifteen million citizens in America alone.

The predisposing bipolar temperaments, like Kagan's descriptions of shy and extroverted children, and the pessimists and optimists of Seligman, are most easily recognized by reviewing an individual's typical emotional behavior over an extended period of time. Only when followed over months and years, or assessed carefully in retrospect, do the characteristic patterns of temperamental interaction emerge. Analyzed in this way, the bipolar, or thymic, temperaments can be divided into three main patterns. *Dysthymia* (literally translated, *dys* means difficult or impaired, hence dys-thymia is "impaired spirit") has much in common with the shy and pessimistic profile—the introvert for whom life is a constant struggle, and who habitually withdraws from the world as a difficult, frightening place. The *hyperthymic* on the other hand (*hyper* meaning "above the measure," or beyond the commonplace, in the original Greek) is the optimist, the extroverted Tigger and the sanguine temperament of ancient description. John Moorehead was of hyperthymic temperament, a man of action who invariably got up in the morning intending to get things done. *Cyclothymia,* the third bipolar temperament, is a mixture of these two forms (*cyclo* means "circular"), and expresses itself in a continuous oscillation between unbridled optimism and crushing pessimism, as John's mother, Josephine Moorehead, had experienced.

For many, these thymic temperaments inhabit the shadows of everyday existence and their profiles are not readily defined, or necessarily associated with pathological behavior. Those of depressive, dysthymic cast are often conscientious, meticulous workers—the civil servants, the diligent managers of detail, who may seem to have little fun in life but are the backbone of our institutions and social organization. The hyperthymics on the other hand are the individuals who apply themselves successfully to business, the politicians, those who enjoy being with the crowd and who see themselves as people of action and the leaders of their communities.

In the successful Piane family, these temperaments and other stigmata of emotional dysregulation, including serious bipolar illness, had been passed down through many generations. John Moorehead described how his mother, Josephine, would invite the whole neighborhood to celebrate the beatification of some obscure saint, and then, only hours later, would condemn herself as a wretched sinner, swearing that her family was destined to suffer in purgatory for its transgressions against her. This is characteristic of those with cyclothymia, to alternate between moods of being expansive and creatively carefree and periods of contraction and irrational anger. Cyclothymia, which sometimes is extended to include an "irritable" temperament, in muted form mirrors the cyclic pattern of manic depression, with sleep and energy changing in the synchrony with mood. In fact, as the family story would have it, in their native Campobasso, in the mountains north of Naples, the Pianes had been known for their passion and energy—men of leadership and violence and women of fierce attachment and headstrong opinion. At least one grandfather had shot himself after losing his high political office, and an uncle, about whom John Moorehead knew little, had died in an institution.

Hyperthymia is different. To say that one *suffers* the hyperthymic temperament is to distort the truth; most of us would be inclined to view its attributes as gifts. John Moorehead's own stable and classically hyperthymic temperament was in stark contrast to his mother's moods; he was an individual who for most of his life had been blessed—as his mother described it—with remarkable energy and an expansive vision of how the moral education of mankind could be improved. It is true that some individuals through overconfidence become improvident, offensive, and meddlesome, but John Moorehead had managed to bridle these less desirable qualities. When I learned more about his personal history it became clear, as is commonly the case in hyperthymia, that he had known periods of anxiety and dysthymia, especially as a younger man. However, because they usually retreated within a few weeks, he had dismissed them from memory until our discussion. It was actually following one of these periods that he had left the seminary in favor of a more conventional teaching career. He recalled being troubled by the monotony of the cloistered life and the apparent intellectual constraints, although never to the point of wanting to forgo the Jesuit order. It was these early experiences that had triggered his interest in John Cassian's writings on the sin of acedia. With the advantage of hindsight, these episodes seemed to him as minor depressions, and I agreed. In my professional experience, such isolated slowdowns are of little significance alone, but in the context of a severe midlife depression, following a lifetime of short sleep and hyperthymic energy, they give support to a diagnosis of bipolar illness.

Hyperthymic persons such as John Moorehead become depressed more frequently than the population in general. Of those who suffer severe melancholia, about 30 percent, mostly men, have a predisposing hyperthymic temperament, and when they become ill they are very intolerant of the drained energy and sense of numbing which the experience brings. Without good medical care many commit suicide, which was why I took very seriously John Moorehead's suicidal ruminations and his sister's concern. Of others who experience melancholia, approximately one-third are of dysthymic temperament and are more likely to suffer the depression at an earlier age than the general population. The evidence is building that dysthymia emerging in childhood is a very strong predictor of serious depression in adolescence, with approximately a third of these young people later developing manic depression. This suggests that the bipolar temperament may be the mild expression—a *forme fruste*—of the genetic vulnerability that seeds the polar illnesses of mania and melancholia. However, in my experience, these behavioral patterns and the details of the family history are frequently neglected by physicians, and the patient's complaints are commonly stigmatized as "character problems," "neurosis," or "borderline behavior." Hagop Akiskal, in one of his early studies, followed for several years one hundred individuals who had been referred to his clinic with such wastebasket diagnoses, and found that over 40 percent of them developed specific episodes of mood disorder, including severe depressions and mild mania.

These findings—that disturbed temperament cohabits with mood disorder and may predispose to it—are not new, but confirm the work of Emil Kraepelin, from a century ago. Kraepelin, a German psychiatrist, was the principal architect of the contemporary diagnostic classification in psychiatry. Working first in Heidelberg and later in Munich, during the last years of the nineteenth century, Kraepelin described in detail the syndromes of mania and melancholia and the spectrum of behaviors that lie between them. Kraepelin's observations were made possible by a humanitarian movement that had begun in England and France a century earlier. A French psychiatrist, Jean Esquirol, one of the most talented and systematic observers of his time, had been the first to suggest that sad melancholy was a primary disorder of the "passions" rather than impaired reason. Later, other Frenchmen noted the periodic nature of these illnesses—that they tend to come and go over a person's lifetime—and recognized the alternating relationship between mania and melancholia. The French described the illness (with poetry to an English ear), as *"folie à double forme,"* but it was Professor Kraepelin who most vigorously pursued an objective classification. Kraepelin considered the temperaments to be "fundamental states," "peculiarities in the emotional life [that] are not limited to individuals who suffer from attacks of manic-depressive insanity" but "are observed with special frequency as simple personal peculiarities in the families of manic-depressive patients."

Mood disorders are episodic illnesses which ebb and flow with time. It is important to emphasize, as Kraepelin did, that most people who suffer manic-depressive disease, especially if the episodes are infrequent, experience little or no residual signs of illness after they return to normal health. Of one thousand cases seen by Kraepelin in Munich at the turn of the century, he reported that only 37 percent had "permanent peculiarities," and these seemed not to result from the illness itself but rather reflected the predisposing temperament, in keeping with our contemporary observations.

Many of the contributions Kraepelin made to our understanding of mood disorder came from his careful study of the longitudinal course of illness in numerous individual patients. Some years ago, when I visited the institute that he had directed at the University of Munich, I had the privilege of reviewing some of his patient records. There, in a deliberate copperplate handwriting, were preserved his many detailed observations—including long verbatim reports that he transcribed of the patient's own experience—describing the natural history of manic-depressive disease. Many of these descriptions, perhaps some of the most lucid ever written, will be found in his book *Manic-Depressive Insanity and Paranoia,* first published in English in 1921.

During Kraepelin's time there were no effective pharmacological treatments for melancholia or mania. The patients stayed within the safety of the hospital, with good nursing care and spiritual counsel, until their illness improved. How the disease unfolded in its natural course was therefore readily apparent to the trained observer. The less severely ill were probably not seen in Kraepelin's institute, but cared for in sanatoria or by their families. Of those who did enter the hospital in Munich, Kraepelin records that approximately 50 percent suffered from various states of depression without any evidence of mania—what we now call unipolar illness—while one-third had a combination of both manic illness and melancholia. Kraepelin classified what he observed according to the frequency and timing of the episodes and the severity of the symptoms, which for many patients included psychotic distortions of reality. Frequently when depressed they believed themselves diseased and unworthy (as had Claire Dubois) or, during mania, possessed of some special power (which Stephan Szabo experienced during his manic episodes). The illness almost invariably followed a complex periodic course for months, sometimes years, but with intervals of relatively normal mood. Although over time the episodes of illness grew more severe, there was little evidence of mental deterioration between episodes.

On the basis of these observations, Kraepelin distinguished the mood disorders as a category of brain disease, distinct from another group of illnesses with a deteriorating course, where characteristically the capacity for normal reasoning did not return between episodes. He called these latter disorders *dementia praecox*—what we now call schizophrenia. Based on this distinction and his other observations, Kraepelin decided that unipolar and bipolar mood disorder were members of the same family of diseases. However, on the authority of more recent research, we now question this particular aspect of his classification.

As Kraepelin himself noted among the patients entering his hospital, there is a striking difference between the number of people who suffer depression alone (unipolar illness) and the number of those who experience depression in association with mania. Most recent studies put the prevalence of unipolar depression in the general population at around 7 or 8 percent while bipolar illness is approximately 1 percent. Even if we include the predisposing disorders of temperament as evidence of bipolar disease, the prevalence of unipolar illness still exceeds that of manic-depressive disease by a substantial margin. There are also other important differences between the two disorders. Those individuals with bipolar illness have more relatives who have mood disorder, and certainly more who have mania, than do those who suffer unipolar illness. Bipolar illness tends to declare itself much earlier in life, when one is in the late teens or early twenties, whereas unipolar depression on the average emerges almost a decade later.

Unipolar illness, in contrast, is more commonly associated with other medical ailments, and grief, severe life stress, or any experience that threatens to exceed coping capacity and overwhelm our emotional balance tends to trigger it. Both depression and manic depression repeat themselves at intervals, frequently in association with stressful events, although sometimes with apparent autonomy. However, this repetition is *invariably* true when mania is part of the clinical picture, whereas approximately one-third, or more, of unipolar depressions are a once-in-a-lifetime occurrence. Hence, manic depression appears to be a disease of distinct recurrent profile and genetic lineage, whereas unipolar illness is probably a mixture of disorders, with different causes, that have a similar presenting syndrome.

This distinction between bipolar and unipolar disorder is confirmed by two other observations from research. In genetically identical twins, when one twin has manic-depressive disease, the chances of the co-twin also developing the same illness are over 70 percent, whereas the risk of an individual developing unipolar illness, when the identical twin already has suffered such an episode, is only 43 percent.

This suggests that a major vulnerability to the development of manic-depressive disease is genetically transmitted, whereas the occurrence of unipolar illness depends more on contributing factors from the environment.

The second observation, which complements the first, is that clear differences have emerged between unipolar and bipolar disorder in their response to treatment. Lithium is an effective treatment for mania and prevents relapse. However, antidepressants when given alone, without lithium or another mood stabilizer such as an anticonvulsant drug, while effective in treating and preventing unipolar depression, will often precipitate mania in those individuals with bipolar illness and even increase the number of illness episodes. Furthermore, lithium, which is often an effective antidepressant in bipolar depression, fails in most instances to alleviate unipolar depressive episodes. Taken together, this evidence suggests that bipolar and unipolar illness are two distinct classes of disorder which are born of a specific genetic vulnerability, have a distinct natural history, and respond differently to treatment.

Unfortunately, however, because unipolar and bipolar illness share a final common pathway of physiological disorganization in the emotional brain, the early symptoms can be very similar and lead to diagnostic confusion. Commonly a vague and inconclusive depression is the first sign of illness, even in those who later develop obvious bipolar disorder. Indeed, a recent survey of five hundred bipolar members of the National Depressive and Manic-Depressive Association bears testimony to this difficulty. In the average experience more than three physicians were consulted, and almost a decade lost, before individuals who suffered manic depression were accurately diagnosed. In practical terms this means that many individuals suffered unnecessarily, often receiving the wrong treatment or no treatment at all for an illness that, with optimum care, can be more effectively treated than virtually any other complex ailment in contemporary medicine.

John's sister, Angela Moorehead, had been a victim of this diagnostic confusion as a teenager and her experience had contributed, in part, to her brother's suspicion of psychiatrists and psychotherapists. As Angela had described at our first meeting, there was a twisted irony to her brother's midlife depression for until then it had been he who had comforted her through periods of depression. In a subsequent interview, when I was exploring the family history, Angela described to me how she had been very shy in childhood. She had found adolescence particularly painful, especially after a move her family had made from central Boston to the suburbs. From the description she gave of her high school years, it was clear that she had hated her new home from the beginning. To make matters worse, shortly after the move John had left for the seminary, and during the first winter her world fell apart. She was just fifteen and a little heavy for her age. Making new friends was difficult; all the other girls in the Catholic high school seemed to have their own clique, which on particularly bad days Angela saw as an exclusive arrangement designed to keep her at bay. She was a poor athlete, but sports were mandatory, and she came close to tears as she described how, after hockey games, her body ached from head to toe. Most evenings she would take to her bed and sleep, sometimes for ten or twelve hours at a stretch. There seemed nothing to do but eat. Looking back, as she described this period of her adolescent life, it was clear to me that during that winter she had experienced her first major depression.

As the winter and the depression wore on, her mother had blamed herself for the move, and Angela had cried repeatedly. Then, with the coming of the summer, for no particular reason things seemed brighter. She fell in love with the boy that her father had hired to mow the lawn and she got drunk for the first time. This triggered a minor family crisis, with both her mother and John worrying about her welfare, although her father—home on an extended shore leave—good-naturedly dismissed the problems as just part of growing up and being a Piane.

But the gloom and depression returned with the beginning of the fall semester, and in the last years of high school Angela felt alienated and out of place. Her grades were erratic, and only with the help of a family friend did she gain admission to a two-year college in Maine, hoping to become a physical therapist. Once there, however, she missed her family deeply. Plagued by repeated episodes of stomach and muscle pain for which nobody seemed to have an explanation, during the winter semester of her second year, when the weather was particularly cold and gray, she dropped out of school without a diploma. She had gained weight again and felt fat, miserable, and dispirited. Returning home,

Angela entered psychotherapy and made remarkable progress. By summer she had fallen in love with her doctor and was planning to become an airline stewardess. A pattern of cycling mood, similar to that of her mother, seemed to be establishing itself.

Angela remembered that summer very well. "Mother was worried that I was beginning to demonstrate her own erratic behavior," she told me during our interview, "but I remember it as the first time I was really happy. Yet, in truth, for most of those years after I dropped out of college I was chronically miserable—not suicidal or anything, just sad about my life and with a low-grade chronic discontent. I tried psychotherapy and it helped a little. The doctor said I was 'borderline,' although I'm not sure what that meant. I learned a few things about my family and my problems with men, but nothing really changed. Subsequently I saw a lot of doctors and took a lot of tranquilizers, but most of them made me feel worse.

"John was in fact my biggest support; he was my emotional regulator. I called him all the time when I was down. And the Boston winters didn't help. Every fall, with the first chill and those telltale streaks of color in the trees, I would develop the appetite of a horse, guiltily stuff myself with chocolates, and find it impossible to get out of bed. On winter mornings the espresso machine was more important than my toothbrush. I was depressed and angry all at once. By February I was usually a basket case and twenty pounds overweight, unless I took a vacation somewhere warm. Actually that's why I became a travel agent, so that I could find all the bargains in the perfect sunny places. John decided I had some form of cabin fever. A lot of people get that living up here in New England; you read about it all the time. But then, about ten years ago, I managed to get transferred to San Diego and things have been a lot better since. I'm still miserable from time to time, and still have problems getting close to people, but I'm a lot happier."

Almost two-thirds of those who experience depression never receive any professional help; they suffer through repeated episodes as best they can. Sometimes the story ends in tragedy, even suicide, while some people find a peculiar adaptation of their own, as Angela Moorehead did. Frequently the anguish is not labeled as illness but as some ongoing blemish of character—a monkey on the back—that must be carried and endured. Had Angela lived a century ago, in Kraepelin's time, her story would probably have been little different. It is unlikely that her disability would have been severe enough to take her to the Munich clinic, but doubtless she would have visited several spas and perhaps a sanitarium in her search for help. Kraepelin himself would probably have considered Angela of depressive temperament—one who suffered a "permanent gloomy emotional stress in all experiences of life"—or as Angela described it, a low-grade chronic discontent. Today we call this *dysthymia*. Dysthymic individuals are given to brooding and worry but, as I have described, they can often be also conscientious and self-disciplined, as was Angela. Kraepelin observed that they are frequently "charming and cheerful when stimulated by external circumstances, but when left to themselves return again to their own introspective meditations on the wretchedness of life." Such wretchedness usually falls short of meeting the contemporary diagnostic criteria for depression, although, as Angela experienced, a true cloud of melancholia can descend during times of stress and challenge—what some psychiatrists now refer to as "double depression."

But on close inspection there is more to Angela's story than dysthymia. First, while she was chronically miserable for most of the time, she also described periods when her misery lifted and she was "truly happy," when life seemed to flow before her with special promise. She associated these moments particularly with the spring and summer months when "I was always in love with something or somebody. I'm a very emotional person and I get attached easily. It's the breakups when things don't work out that I hate." Angela considered these periods of passion her "normal self," the way she would be consistently if she could just shake her chronic anguish, while to others she appeared "accelerated." During these "truly happy" periods her sleep decreased as her energy and ideas grew. She took on tasks that she would never complete, bought things she didn't need, and made friends whom later she didn't want to keep. These periods of acceleration that Angela found so wonderfully compelling and exciting in fact represent the other side of the bipolar coin—periods of mild hypomania punctuating the dysthymic misery. It is this pattern of long episodes of depression, sometimes deepening to melancholy but relieved by periods of hypomanic energy, that American psychiatrists have named bipolar II to distinguish it from the more dramatic form of manic depression, bipolar I, where both poles of the illness are marked by severe social incapacity and sometimes psychosis.

The second important observation is that as Angela reached maturity her periods of severe depression developed a markedly seasonal pattern, confined largely to the winter months, with improvement in the spring. While she was living in Boston the profile of her bipolar illness was particularly tied to the seasons. After her move to San Diego these cycles of dysthymic depression and springtime hypomania became less seasonally distinct, and much more manageable, suggesting that the seasons did influence the timing of her mood cycles. This synchrony of depression with the darkest season of the year has been noted for centuries. In Galen's time, seasonal melancholy was thought to be influenced by the rise of the planet Saturn in the winter sky. "Of the seasons of the year, the autumn is most melancholy," emphasized Robert Burton in his *Anatomy of Melancholy* in 1621. The majority of those who suffer winter depression—what is now called seasonal affective disorder, with the descriptive acronym SAD—are women. And, as in Angela Moorehead's family, there is frequently a strong history of mood disorder, and alcoholism too in many instances. There are also other clues that clearly identify SAD as a member of the bipolar rather than the unipolar family of illnesses. While it is not an absolute rule, most bipolar individuals sleep excessively and gain weight, just as do individuals with SAD during their periods of winter depression. By contrast, in unipolar depression the appetite is commonly poor, with a loss of weight, shallow sleep, and early awakening in the hours before dawn.

Hence SAD appears to be a special version of bipolar mood disorder, where the periods of depression are distinguished by their sensitivity to seasonal changes of climate. From recent research we know that it is the changing seasonal pattern of light and dark that drives the timing of SAD; it is likely that Angela Moorehead could have achieved a similar relief to that afforded by her move to California by improving the lighting in her Boston home. I shall say more about light therapy for winter depression in a later chapter, when I describe some of the ways in which our brain clocks help regulate the body.

Seasonal affective disorder is an example of how the genetic vulnerability to mood disorder can express itself differently, depending upon environmental circumstance. You will recall that in genetic science a distinction is made between the specific genetic template that is inherited—the genotype—and the form in which this template is expressed in life, which is called the phenotype. The genotype is the genetically coded vulnerability to emotional dysregulation that is passed from generation to generation; in the case of SAD, the phenotype is expressed in synchrony with the seasonal cycles of light and dark. In fact, to some degree all bipolar illness is seasonally responsive, with mania having a seasonal cycle in markedly seasonal climates, being most prevalent in the late summer. This emphasizes again something easily forgotten, even in this age of promised genetic engineering: When it comes to the brain and the central nervous system, neither genes nor the body proteins that they encode are immutable structures.

Thus when I describe the spectrum of illness that lies between the polar peaks of mania and melancholy, the "terrain" of mood disorder, I do not wish to portray some harsh landscape of immutable granite, but rather a picture of sandstone and dunes—some geology more easily shaped by the ebb and flow of environmental change. Diagnostic categories are phenotypic descriptions. As sandy cliffs and dunes have much in common, so each diagnosis has common characteristics, and each individual case is shaped by prevailing circumstance.

Ideally, in diagnostic description the physician strives to be as objective as possible, recognizing that as we learn more about the neurobiology and genetics of these illnesses, our definitions will become more precise and more useful in guiding treatment. There is nothing magical about the diagnostic categories themselves. They are tools forged from scientific agreement that help define what is common in the experience of mood disorder and its treatment. Through this common agreement it is then possible to clearly identify groups of individuals who, in sharing a similar set of abnormal behaviors, share also a common pathway of disturbed communication among the neurobiological systems that regulate mood. It is to the evolution of these systems that I now turn, to their anatomy in the emotional brain, and to how their function is intimately connected with the challenge of everyday life.

Six

The Legacy of the Lizard

The Anatomy of the Emotional Brain

The human brain has evolved and expanded to its great size while retaining the features of three basic evolutionary formations that reflect an ancestral relationship to reptiles, early mammals, and recent mammals. In an evolutionary sense countless generations apart, the three neural assemblies constitute a hierarchy of three-brains-in-one, a triune brain.

PAUL D. MACLEAN
The Triune Brain in Evolution
(1990)

J ust as muscles grow and bodies change with exercise, so do brains. The brain that guides me now is not the same as that which guided me at my birth half a century ago. In youth it seemed an attractive idea to become a ski instructor, and had I done so rather than studying to become a physician—or had I played the violin, as did my father— today the anatomy of my brain would be different. The changes would be subtle, certainly nothing that could be seen with the naked eye if the brain were sliced in half, but profound nonetheless. Some of the billions of nerve cells inside my head would have developed other connections, as is necessary when we are learning new skills. My brain would be different, and I would be different too.

The same mutable biology, which over a single lifetime enables me to continuously learn and remodel my behavior, over millions of years has allowed the evolution of the human brain. Cells communicate among themselves—and did so long before brains evolved—by releasing messenger molecules from their cell walls. These chemical packages, upon arrival at the door of an agreeable neighbor, plug into receptors rather as a key fits into a lock, and once inside they have a powerful influence over the activities of the new household. The cells of the brain, the neurons, are no different in this fundamental principle. Neurons continue to pass information to and fro via messengers— in the brain called neurotransmitters—but in their search for a more efficient way to process that information, they have undergone remarkable adaptation in their structure and function. In fact, the intricate physiology and complex anatomy of the human brain is best understood as the ultimate result of this evolutionary quest for the efficient processing of information.

Each human brain, during its development from a collection of primitive embryonic cells, in many respects retraces the fundamental stages that have occurred in the brain's evolution over millions of years. During the late nineteenth century, when such an idea first became popular, anatomists described this as "ontogeny recapitulating phylogeny" (the development of the individual reflecting the evolution of the species). Such an evolutionary perspective can be valuable in understanding how the human brain is organized, and why we share many of our behaviors with those of other animals while being superior in intelligence and intellectual accomplishment. Thanks to technical innovation and the explosion of knowledge in recent decades, we are now beginning to dissect the molecular machinery of the brain and its systems of communication, and can take pictures of its activity while a person is thinking, seeing, hearing—and even suffering, as during melancholia or mania. In this chapter I shall describe some of these advances and explain how they assist us in understanding the anatomy and function of the emotional brain, both under normal circumstances and during periods of emotional disturbance.

When the brain is actively processing information, we call it *thinking*. The cells that think are the neurons, and their supporting cells—functioning much as does a trellis in a rose garden—are called *glia*. Lacing through this trelliswork is a delicate net of blood vessels that are the supply lines for the brain's energy and vital nutrition. Thinking is a collective enterprise and is dependent upon the health and vigor of all these elements. The neurons themselves, however, are especially adapted to the purpose. They live in close-knit communities and each is equipped with spidery tendrils—called *dendrites*—that intimately entwine with those of the neighbors to create local networks of communication.

Neurons come in different shapes and sizes and exchange information with each other through special junctions on the dendrites, known as *synapses*. The more synapses a neuron has, the more complex is its conversation with the neighborhood and the greater the volume of information it can handle. Furthermore, we have learned from recent research that, although the basic growth and organization of the brain is determined by genetic instruction, the number of synapses and connections developing within the neuronal networks varies relative to an animal's active engagement with its environment—and the amount of "thinking" that is required. The glia and blood vessels, in their supporting roles, grow too. Just as physical exercise helps build stronger muscles, when it comes to building a sophisticated brain, exercise in thinking and an environment that necessitates learning new things can be important stimuli to neuronal growth.

Over countless generations this brain plasticity, which in the face of environmental challenge improves the machinery of neuronal communication, has enabled the more successful creatures to survive. Over millennia, brains have matured in sophistication from the simple aggregation of a few communicating cells to the networks of billions of neurons that collectively support human behavior.

How in the development of the individual human brain each of the billions of neurons migrates to its rightful position and establishes functional networks is an astonishing story, the details of which we do not entirely understand. In the human embryo the brain develops from the ectoderm, the outermost layer of the fertilized egg. Subsequently, guided by a set of instructional genes in the genome library, the architecture of what will become the adult brain unfolds very quickly. Aided by the glia, the embryonic cells differentiate into mature neurons. They proceed through several stages, including migration, when they cluster together in distinct regions, developing the specialized communities that later will become the nuclear centers of the mature brain. Then the dendrites are formed, synaptic connections established, and axons grown.

Axons are the electrical superhighways of the brain. They grow from the neuron's cell body and are long, like a miniature elephant's trunk. Through them information can be accurately and rapidly conducted over long distances by an electrical pulse called the *action potential*. This efficient architectural innovation is essential to the increased speed of signaling that is a key element of survival behavior. The axons are what we speak of as "nerves," the glistening strings that run throughout the body conducting information to the brain from the skin and special senses of the eyes, ears, and so forth, or delivering operational commands to the arms and legs. Within the brain itself, axonal transmission performs a similar function. Axons make up the anatomically distinct "white matter" of the brain. They link the "gray matter" of the neurons, which live in specialized neighborhoods, into a series of functional systems much as different villages in parts of Europe have been connected over many centuries to become the neighborhoods of great cities.

Once the migration of the neurons is complete, these specialized neighborhoods of the brain undergo a process of "pruning," guided in part through genetic programs of "cell death" and in part through environmental interaction. Thus the number of neurons in each community is reduced to lean efficiency for the optimum transfer of information. Many refinements of the dendritic and synaptic connections are made at this stage, which in human beings extends beyond birth into the first few months of postnatal life. In their ultimate development, of course, the brains of different creatures are different, although in general the genetically programmed stages through which they pass have similarity. While the brain of the snail is different from that of the lizard and the lizard brain is distinct from the rabbit's, they all have something in common. The behavior of human beings is more complicated than that of other animals—in planning, abstract reasoning, the manipulation of objects, the use of language, and so forth—but nonetheless we share in common with many creatures such behaviors as sexual courtship, pleasure-seeking, aggression, and the defense of territory. Hence it is safe to conclude that the evolution of human behavior is, in part, reflected in the evolution and hierarchical development of other species.

Recognizing these interesting common denominators, Paul D. MacLean, a distinguished scientist at the National Institute of Mental Health in Washington, D.C., has extended the study of embryology and evolution to include behavior by comparing behavior and brain structure across different species. MacLean proposes in his book *The Triune Brain in Evolution* an archeological classification of the human brain based upon an evolutionary hierarchy. He divides the human brain's evolution into three stages: first, the reptilian brain, which embodies an essential core of survival functions, as we see today in the lizard; second, the ancient mammalian brain, which is the first layer of cortex surrounding the lizard brain and responsible for the social and family behavior of mammals; and finally the new mammalian brain, with its massive cortical development, which has made possible reasoning and abstract intelligence as found in the primates, particularly ourselves. MacLean proposes that over evolutionary time each advance in development has resulted in a new layering upon the pre-existing brain structures, in a hierarchical fashion, with anatomy and function becoming intimately entwined. This evolutionary amalgam reaches its pinnacle in the human brain, which MacLean has called the triune brain, emphasizing that it has a power to process information far greater than the sum of its three contributing parts.

I always have found MacLean's ideas helpful in thinking about the evolution of human behavior, and particularly about the anatomical roots of emotion. But it was unexpected when an opportunity to explore his ideas presented itself firsthand. During the early stages of planning this book I was fortunate to be a Fellow at the Center for Advanced Study in the Behavioral Sciences at Stanford University in California. Known affectionately as the Center, this monastic retreat sits high on a remote hill overlooking the Stanford campus. It is a magical place. The buildings of redwood and stone owe much to their California heritage, but in conception and form, as a place of scholarship devoted to the thoughtful analysis of human behavior, the Center closely resembles a medieval cloister. Around a central courtyard, which doubles each noon as a communal eating hall, are clustered the fifty individual studies of the fellows in residence, the library, kitchen, and meeting rooms. On some days, as I sat working in dappled sunlight, I reflected that the only thing missing was a bell declaring vespers, or the sight of a vineyard following the curve of an adjacent hill.

My own cell of study was on the southern edge of the monastic square, facing unspoiled countryside. I had chosen it carefully, hoping to profit from the solitude of the place. Nature in her abundance, I had decided, would help me think—and she did, but not in the way that I had expected. As I sat down to write I soon discovered that solitude, as with most things, is relative. The fellows and staff were not the only creatures living on Center hill. In fact, among the fauna, the human contingent was in a minority. At first there were noisy jays and squirrels. Then over time I became acquainted with shy companions like the jackrabbits and a family of deer who came each morning to eat the young grass sprouting at the edge of the irrigated shrubbery. Occasionally I caught a glimpse of smaller, more frenetic, animals—voles, field-mice, and the like—hurrying about their business, tails up in nervous alarm. The hill was alive with a rich variety of mammalian species. And there were reptiles, too—the lizards would skitter away from their sunny perches in the stone wall as I passed on my way to lunch at noon. All around me, I began to realize, were gathered representatives of the human brain's hierarchical history. It was an opportunity to compare and contrast the behaviors that MacLean describes.

It is the reptilian part of the triune brain that is concerned with the machinery of the body, the systems for life's maintenance, and primitive survival. This is true for the lizard as well as for ourselves. In the human brain the bulk of these reptilian structures are found in the medulla—meaning "pith"—where the long spinal nerves that travel together to serve the limbs and trunk first pass into the protective vault of the skull and expand to become the stem of the brain. There are found the homeostatic mechanisms—the autopilots—that regulate the activities of breathing, digestion, and the speed at which the heart pumps.

As I learned from my daily observations, lizards exhibit primitive social behaviors. Most of the earliest reptiles on this planet have died out, and MacLean argues that it is the brain of the lizard that bears the closest resemblance to the brain of those species that evolved into mammals. Furthermore, he asserts, it is by studying the activity of the lizard that we can discover the reptilian legacy to human behavior. Lizards live in groups and are fiercely territorial, aggressively threatening and harassing others; they even kill intruding strangers. The lizards at the Center were about six inches long, striped, and greenish brown in color, and seemed to have a well-developed social hierarchy. One of the largest creatures, presumably the dominant male, usually took a high perch, commanding an unrestricted view—somewhat reminiscent, it seemed to me, of human urban behavior. (This emphasized, too, that some of the "social" behaviors we associate with mammals have roots much earlier in evolution.) Despite my best effort at stealth, he would usually sense me coming and rise up on his forelimbs, fingers splayed, head up, and jaw held forward—looking rather like a human baby learning to crawl—before skittering off.

The lizard's daily routine was clearly dictated by the sun. Lizards have the same regulatory clocks synchronizing rest and activity that warm-blooded mammals such as ourselves possess, but because they are cold-blooded they have difficulty precisely regulating body temperature—which explains their love of sunbathing to raise metabolism before the activities of the day. Occasionally I would catch them foraging, or hissing at each other—part of the social displays of greeting, courtship, and aggression, as I learned from my reading of MacLean—but by late afternoon they had usually disappeared, presumably to some favored nook for the night.

The lizard comes into the world genetically equipped to do everything it must do to survive, for lizards possess a comparatively primitive brain with only a rudimentary cortex and little capacity to learn. The word *cortex* comes from the Latin and means "outer shell or husk," as in the bark of a tree or the rind of a melon. It is the growth of the cortex of the brain that is associated with the capacity to learn. The first layers of cortex that developed during the evolution of mammals wrapped themselves around the centers of the primitive reptilian brain, and are called the "great limbic lobe" (from *limbique,* meaning "limb or border").

The first mammals emerged during the Jurassic period some 200 million years ago, when reptiles were dominant. What characterizes mammals, in addition to being warm-blooded, is that they have mammary glands—hence the name—and they nurture their young; the voles, squirrels, deer, and rabbits that live on Center hill are all mammalian species. The parents and infants of these species communicate with each other, usually by vocal expression, fostering closeness and mutual attachment. The lizard, by contrast, knows nothing of parenting activities. While lizards have a well-defined social community, they do not have families, as do mammals, nor attachment toward offspring and each other—a fundamental difference in behavior. Mammals also have the ability to learn, initially through parental guidance and later through experience and from the behavior of others. Thus mammals engage in repetitive learning behaviors (which in the young we call *play*) and through this practice develop new skills of coping and adaptation. Lizards, on the other hand, are very poor students and have only limited ability to adapt to a changing environment.

The prerequisite for the development of all these mammalian behaviors in the evolution of the brain is the great limbic lobe. In union with the reptilian brain centers responsible for primitive social behavior and body homeostasis, the great limbic lobe forms a brain amalgamation that represents the second stage of evolutionary development—what MacLean has called the *ancient mammalian brain*. An alternative designation, more commonly used and also suggested by MacLean, is the *limbic system,* or limbic alliance. Both are scientific names for the collection of brain centers that I have been describing as the emotional brain—the part of the brain concerned with emotion and social and family attachment.

It is a mistake to think of the brain as a single organ. Rather, it is a series of specialized organs—neighborhood clusters of gray neurons which over centuries anatomists have labeled with Latin and Greek names—that perform harmoniously together. In the past few decades, as neurobiology has gathered momentum, it has been possible to identify distinct subsystems of brain function. The "visual system" was an early example. Various types of neurons—from the light-sensitive fibers in the retina to the neurons in the brain cortex that integrate the information generated by what the eyes see—collaborate with each other to create vision, and the concept of a visual *system* is now widely accepted among scientists and the public. Visual function is distributed across several participating brain centers. If any part of the system is damaged or dysfunctional, the experience of "seeing" will also be disturbed. A similar system supports speech, hearing, and so on.

The limbic system, which generates emotional behavior, is also one of the subsystems operating in the brain. However, the idea that emotion is a physical phenomenon has been a more difficult idea for people to accept. One reason for this is that the boundaries of what we recognize as visual experience—or speech, or hearing—have been much easier for scientists to agree upon than the boundaries of emotion. Like the visual system, the limbic system is an alliance of brain centers, but in this instance the alliance is bound together not in the common purpose of generating an accurate visual image of the environment, but in the service of a more global, and less easily defined, responsibility—that of processing information and guiding behavior in the best interest of preserving life and species. The ancient mammalian brain, or limbic alliance, has three major functions—functions that become disturbed in disorders of mood. The first is to ensure the development of the next generation through attachment, nurturance, and learning new skills; the second is to monitor the changing social environment and communicate need—what I have been describing as emotion; and finally to oversee the reptilian brain's smooth running of the body's physical machinery, continuously tuning it to sustain optimum physical vigor and harmony with the planetary environment (the housekeeping functions of the body).

Families, and emotional attachment, are thus an ancient mammalian invention. In human beings, the power of the new cortex, or new mammalian brain, with its capacity for language and the abstract manipulation of ideas, tends to obscure our ancient mammalian heritage. But in our emotional life it is the limbic alliance—the ancient mammalian brain—that primarily dictates our behavior. Its integrity is essential to operating in a complex social environment. It is through the emotional attachments of family life and daily social interaction that we are most aware of the operations of the limbic alliance. Most of the housekeeping functions of the alliance have been automated in the reptilian brain during evolution, and we are not conscious of them until they become disturbed. The response to flight across time zones (jet lag), and the complex physiology that determines seasonal behavior, including winter depressions such as Angela Moorehead experienced, are both examples of disturbance in the housekeeping homeostats. The overall integrity of limbic function, however, is dependent upon the active collaboration of these different brain centers and accurate, unimpaired communication among all of them. Change occurring in any one brain center will influence the activity of the whole limbic alliance. Furthermore, anything that distorts the harmonious interaction can result in the appearance of unstable emotional expression—as in the mood disorders—and, commonly, a loss of the usual patterns of the body's housekeeping homeostats controlling sleep, appetite, energy, and the hormonal rhythms.

Grasping a visual image of the brain and remembering the names of the parts of the limbic system is not easy, but is important in understanding the anatomy of mood disorders. As Stephan Szabo liked to observe when we were studying together in medical school, neuroanatomy is something like learning the detailed cartography of the Lesser Antilles without the benefit of having visited them. Stephan, I discovered early in our friendship, has a visual and creative mind and commonly expresses himself in images and metaphors. During one notable, somewhat maniacal, evening of celebration (following an anatomy exam) that we spent together in the Covent Garden district, then London's vegetable market, the idea occurred to Stephan that many brain structures are strikingly similar in shape to fruits and vegetables. This rather peculiar allusion I have since always found valuable, and I have perpetuated it here—although for most readers it will also be helpful to consult the drawing I provide (see page 298).

The human brain is about the size of a cantaloupe melon, and anatomically it is essentially divided into two halves, each a mirror image of the other. The surface of each of these halves, or hemispheres, is folded upon itself in a complicated, wrinkled pattern that serves to divide each hemisphere into four distinct lobes (like the lobes of a grapefruit but without the regularity), each with a different name and function. The frontal lobes (contained within the prominence of the forehead) and temporal lobes (down around the ears and looking a little like the flaps of an American football helmet) are the most important in emotional life, while the occipital lobe, at the back of the brain, integrates vision. Between the frontal and occipital, and above the temporal cortex, is the parietal lobe. The right parietal lobe is involved in building sensory maps of the body and the left in developing language skills. The frontal, temporal, occipital, and parietal lobes, with each lobe covering the limbic lobe of the old mammalian brain, are part of the larger cortex—just as the lobes of the grapefruit are parts of the whole—which together constitute the powerful hemispheres of the new mammalian brain.

If you now imagine that we have cleaved the cantaloupe into two equal parts, the intricately entwined centers of the limbic alliance, like a cluster of fruits replacing the seeds, will lie cradled in its center, encircled by the great limbic lobe. In such a construction, the limbic lobe is the first layer of cortex, at the fleshy heart of the melon, merging at its borders with the frontal, temporal, and parietal regions of the new mammalian cortex. Prominently placed at the center of the limbic alliance are two large nuclei (a nucleus being a densely packed community of neurons) about the size and shape of small plums, sitting together on each side of the midline (remember, the two halves of the brain are mirrored images). All incoming information from the body and the special senses, plus the brain's outgoing messages, are processed through these structures, to which early anatomists gave the name thalamus. In Greek, *thalamus* means "inner chamber" or "secret room," a rather charming acknowledgment that such a strategically placed nucleus was important even though the early anatomists had no understanding of what that importance might be. We now know that the thalami are the relay stations of the brain—the brain's telephone exchange—accurately guiding all information about events occurring around and within us for appraisal and decision, and similarly steering many outgoing instructions.

Passing backwards over each plumlike thalamus, then curving beneath and forward again, are structures shaped rather like two chili peppers. On the upper surface of the thalami their thin ends lie in close proximity, but the bodies of the peppers diverge as they pass deep into the temporal regions of the great limbic lobe. (The name *temporal* is rather confusing; lying beneath the "temple" of the skull, the lobe is named temporal because here on the scalp with passing age the hair first turns gray.) This intriguing configuration, which early anatomists named the hippocampus, because its shape reminded them of a seahorse, is the emotional brain's memory transducer, consolidating the information passing through the thalamus and adapting that which is important for long-term storage in the frontal region of the limbic lobe.

At the very tip of each hippocampus—as the blunt end of the pepper expands into the region of the temporal-limbic lobe—is found another important group of nerve cells described as the amygdala. These inconspicuous nuclei, which in their clustering are about the size and shape of almonds (which is the meaning of *amygdala* in Greek) are intimately connected to all the other limbic centers. When I was in medical school, the functions of the amygdala were obscure, but today, from recent research in humans, monkeys, and other mammals, their fundamental role in emotion and social communication is beyond doubt. Each amygdala (we have two, one in each hemisphere) has extensive connections with the information centers of the thalamus, the hippocampal memory transducer, and the frontal region of the great limbic lobe, where emotional memories are stored. In the first weeks after birth the amygdala serve as the crucibles of primary emotion, interpreting feeling in the faces of others and communicating emotional need through facial expression. It is a guardianship that persists throughout life, for the amygdala are also the sentinels of the limbic alliance, continuously passing judgment on the emotional significance of all information received.

Each amygdala is also intimately connected to the hypothalamus, a small cluster of nuclei that lie beneath the thalamus (hence *hypo*-thalamus, meaning "under"-thalamus) rather like a tiny bunch of cranberries. One cranberry dangles by its stalk, free from the rest, and is technically outside the limits of the brain. This is the pituitary, the operations manager for a cascade of messenger hormones that are initiated in the hypothalamic nuclei and eventually travel via the bloodstream to stimulate the endocrine glands ("endocrine" meaning that the glands secrete directly into the bloodstream), thereby influencing every organ in the body, including the brain itself. Although the hypothalamus is less than five percent of the brain in size, through the endocrine system that it controls it orchestrates most of the body's housekeeping needs (except for the spontaneous heartbeat and steady breathing). Thus it is in the hypothalamus that the brain clocks and temperature regulators are located, plus the appetites for food, sex, aggression, and pleasure.

The hypothalamus lies at the core of the reptilian brain, and is equipped for independent operation, but in mammals it functions as an integral part of the limbic alliance. Its regulatory responsibilities are profoundly influenced by the behaviors of attachment, social interaction, and—in human beings—by abstract ideas, including personal hopes and fears. Through these hypothalamic connections with the emotional monitoring systems of the limbic alliance and the new mammalian cortex, the hormones released from the pituitary, and the endocrine system they influence, provide an important index of the activity of the emotional brain, including how the alliance performs in stressful situations and in mood disorders.

These, then, are the principal brain centers participating in the limbic alliance: the thalamus, hypothalamus, hippocampus, amygdala, and the great limbic lobe—the ancient mammalian cortex—which wraps itself intimately around them. How these centers of the limbic alliance communicate and work together as a system has emerged slowly from several distinct areas of research.

One of the ways we have discovered what these centers do is through the investigation of emotional expression in normal human beings and animals—the work that was started by Charles Darwin over a century ago. Another is the careful study, in the Hippocratic tradition, of the experience and behavior of those who have suffered epilepsy, strokes, and other brain injuries. A famous example in this category is the story of Phineas Gage. In 1848, Phineas miraculously escaped death in an industrial accident when a three-foot iron bar passed through his skull and removed the left frontal lobe, including part of his limbic brain, radically changing his emotional and social behavior. Before the accident, the 25-year-old construction worker had been a diligent foreman, managing a group of men, while after recovery he was restless, impulsive, and profane, paying little heed to others and exhibiting no ability to plan for his future. His survival was so miraculous, however, that for a while, before succumbing to alcoholism, Phineas traveled around with the P. T. Barnum circus, exhibiting himself as a fairground freak. After his death, his skull ended up in the Harvard Medical School Museum, which is why we know so much about him.

Fortunately, we are now less dependent upon such tragedies to advance our store of knowledge. Both these two areas of research—the study of normal emotion and the study of brain injury—have been complemented in the last decades of the twentieth century by the development of imaging technologies that permit the safe investigation of the living brain. In the remainder of this chapter I shall review some of what we have learned from these lines of research, and how they relate to mood disorders. Then in subsequent chapters I shall explore the important area of stress research (a field of study that received an unwelcome boost from the great wars of the twentieth century) and the hormonal changes that occur in challenging situations. This "hormonal window" into the chemistry of the brain has provided essential information about hypothalamic regulation and how, through physiological homeostasis, the brain balances the body under stressful circumstances, including those that can precipitate mania and depression.

The first breakthrough, integrating a knowledge of brain anatomy with subjective emotional experience in a living patient, came from exploratory surgery for severe epilepsy. Wilder Penfield, the Canadian neurosurgeon whom Claire Dubois had consulted after her automobile accident, was a pioneer in such surgery from the 1940s through the 1960s. He was one of the first to develop a "functional" map of the brain, including some of the limbic centers. In contrast to the scalp, skull, and protective tissues, the substance of the brain itself is insensitive to pain. Penfield developed a technique to explore the area of epileptic focus using mild electrical stimulation in the conscious patient. This meant that individuals undergoing surgery were able to describe their subjective experiences while a specific area of the brain was being directly stimulated. It was through this method that Penfield first charted his famous "homunculus," the distorted image of the body as it is represented in the sensory cortex of the brain. He discovered that important areas sensitive to touch, such as the face, lips, and genitals, have a much larger area of the brain dedicated to them, thus distorting the customary image we have of ourselves when a caricature of the body (the homunculus) is drawn in proportion to the area of brain representation.

Penfield described epilepsy as a disorderly discharge of many neurons that recurs sporadically—"like lightning from a miniature thunder cloud." The symptoms experienced in epilepsy depend upon where in the brain the seizure activity takes place. If it occurs in that part of the cortex that controls the muscles of the arms and legs, then the result, as Hippocrates first described, is a shaking disease. It has been recognized for many decades that a focus of epileptic activity with limbic or temporal lobe involvement frequently is associated with peculiar "psychic" states, including emotionally laden memories, strange smells, or irrational fears. During a period spanning almost thirty years Penfield studied over five hundred individuals with temporal lobe epilepsy and these strange states. Consistently, when the general area of the temporal lobe, including the limbic regions of the hippocampus and the amygdala, were being stimulated, his patients described "moving picture flashbacks" frequently mixed with powerful emotion, including fear and even terror.

Penfield found that these reports of "pictures parading before the patient" occurred mainly when he stimulated the brain area toward the tip of the temporal lobe (which is where the amygdala is located), while stimulation further back provoked memories of hearing music or the sound of somebody talking. He describes in his book *The Mystery of the Mind,* published in 1975, his astonishment when first being presented with such a flashback: "a mother . . . was suddenly aware of being in her kitchen listening to the voice of her little boy outside . . . [and the] . . . neighborhood noises such as passing cars that might mean danger to him." Other experiences commonly reported by Penfield's patients during his exploration of the temporal "region" (covering the prefrontal limbic lobe, the hippocampal and amygdaloid areas of the limbic alliance) included automatic thinking and intrusive repetitive ideas; "observing" one's own activity as alien; or the illusion of "being apart," of looking at what was going on in reality as if from another world. Most important, as Penfield recognized, these "experiential responses came *only* from the region of the temporal lobe, and never from any other part of the brain".

In rereading Penfield's reports of the emotional images and feelings his surgical explorations provoked, I am reminded of how similar many of his patients' descriptions are to the experiences reported during mania, and to a lesser degree in severe depression, and even grief. For example, irrational fear is common in all these states of mind. You will recall the remark of C. S. Lewis, "No one ever told me that grief felt so like fear," and how alienated and apart from her world Claire Dubois felt. Automatic ideas of suicide—what psychiatrists call rumination—plagued her and John Moorehead. A sense of distance and being an observer was a familiar experience to Stephan Szabo, who during mania felt himself a passive pawn in a larger purpose. Also, during his manias, on at least two occasions, Stephan had removed his clothes. "It was not a sexual act," he told me later, "rather the process was automatic and distant from reality. I remember examining and nibbling the leaves of a tree just before the police picked me up." All of these behaviors have great similarity to the sense of fear and the automatic and trancelike states often seen in temporal lobe epilepsy, suggesting that the same areas of the brain are involved.

Because each of us has two brains—or more precisely two mirrored halves—we have two limbic systems. Furthermore, by some necessity of nature, when it comes to managing motor and sensory behaviors, each side of the brain conducts business principally with the opposite side of the body. Thus the left side of the brain controls the right arm and leg and is labeled dominant because the majority of us primarily use the right hand. The two hemispheres of the new mammalian cortex communicate by exchanging information across a large bridge of nerve fibers called the *corpus callosum*. However, we know from studies in individuals where this bridge has been severed—again usually in an effort to seek surgical relief from intractable epilepsy—that the right and left brains specialize to some degree in the behaviors they control. In a right-handed person the left hemisphere is principally responsible for language. It tends to process information sequentially in the order that it is received, while the right brain is more oriented to the three-dimensional world, sorting and analyzing information by visual and spatial patterns.

Given this lateral specialization of hemisphere activity, the question has arisen whether the responsibility for emotional expression or information-processing also differs across the right and left limbic brains, and further whether the balance between them plays any significant part in the illnesses of mania and depression. Despite his extensive reports on patients with temporal lobe epilepsy, because his investigation of the brain was confined to the hemisphere of illness, Wilder Penfield has little to say about laterality and emotional behavior. The clinical case histories of mood changes following brain damage from strokes and head trauma, however, are provocative.

A transient period of depression frequently follows any stroke, but in those circumstances where subsequently an intractable melancholia develops, the brain injury is commonly in the left hemisphere—the verbal hemisphere—and particularly in the temporal and frontal-limbic regions. Instances of "indifferent euphoria"—which is suggestive of mania, or of mania itself—are much less common but when they are reliably reported, the brain damage tends to be on the right, in the spatial hemisphere. These observations are teasing. They suggest that shifts in hemispheric dominance may be important in severe mood disturbance—but because of the general anatomical disruption that usually follows a stroke, most psychiatrists have been cautious in drawing conclusions.

But with the development of brain imaging in the last decades of the twentieth century, this situation has changed rapidly. These case histories have become a bellwether for clinical research. Imaging technologies now make it possible to generate an accurate anatomical image of the living brain and to monitor the activity of specific brain regions. With these technologies, activity across the hemispheres and the nuclear centers of the limbic system can be compared in normal and abnormal mood states.

Key to the development of these imaging technologies has been the creative use of the computer and the mathematical techniques required to analyze extraordinary amounts of information. The revolution began in the 1970s with X-ray computed tomography (CT) where the information from a beam of X rays, passed through a "slice" of brain at different angles, is reconstructed to highlight the different densities of nerve cells and their supporting structures, thus providing a detailed anatomical map of the living brain. With such maps the anatomical integrity of different brain regions can be determined, but nothing can be said about function. However, researchers rapidly realized that the same principles of computerized information gathering could be applied to measuring a radioisotope, such as a radioactively labeled form of glucose. Then, when the glucose was injected into the bloodstream and taken up by the nerve cells of the brain in proportion to their metabolic need, a measure of blood flow and the brain's metabolic activity could be obtained by "photographing" the released radioactive material. With a "camera" capable of precisely recording the trajectory of these radioactive particles as they exited the brain, a functional image of the activity of the brain could be constructed. Here finally were instruments that had the capacity to monitor the activity of specific brain areas, such as the centers of the limbic alliance, while complicated "subjective" states of thinking or feeling were actually in process. For the first time we could begin to describe a genuine anatomy of emotion and mood, even abnormal mood.

The technologies that were the first to emerge have the rather overwhelming names of positron-emission tomography (PET) and single-photon emission computerized tomography (SPECT). These are powerful tools, both for measuring blood flow in the brain (which changes rapidly from region to region depending upon the task the brain is performing), and also for measuring the activity of various receptors that are important in nerve cell communication. These functional imaging technologies have the capacity to illuminate brain systems that generate complex behaviors such as seeing, speaking, or feeling, by analyzing the activity of the participating brain regions while the behavior is being experienced or simulated. Functional brain imaging—so named because it measures function in real time—complements the pioneering explorations of Penfield and goes beyond them. Functional brain imaging is not confined to illness that requires surgical intervention, nor is the investigation distorted by the disturbing presence of an electric probe. However, there are drawbacks. The principal one is that the evidence these techniques gather is not particularly detailed. It's rather akin to what can be learned about a great city from flying overhead at night. The pattern and concentration of light indicates which parts of the city are currently active, and dynamic shifts in city life can be captured by appropriately timed sequential snapshots—especially if you have a working knowledge of the local geography, or in this instance the local brain anatomy.

Each human brain, rather like a face, is slightly different in the details of its anatomy. Selecting the relevant brain centers, precisely defining their anatomical limits, comparing the different regions across the two hemispheres, employing the complex mathematical models required to analyze and interpret the information gathered, plus finding the money for the necessary equipment and financing the studies themselves—all dictate that brain imaging is not a simple science, easily accomplished. Most of the work around the world goes on in major medical research centers such as the McDonnell Center for Studies of Higher Brain Function at Washington University in St. Louis, in the United States. There, since the 1980s, Professor Marcus Raichle, using principally PET, has been carefully exploring the brain images associated with various complex behaviors of mind, particularly language and vision. Under the leadership of the psychiatrist Wayne Drevets, the McDonnell Center has more recently begun to systematically study mood, including depressive illness.

The initial research strategy adopted by Dr. Drevets was a comparative one—scanning the brains of individuals during an episode of melancholia and comparing the images to those of another group who had never been depressed. Subtracting one group of images from the other would then highlight differences in brain activity relevant to depression. To make such comparisons valid, however, it is important that the two groups of subjects have no overlap with each other when it comes to the depressive behaviors under study. Thus Dr. Drevets took great pains to identify a group of individuals with "pure" clinical depression, carefully screening for people who were not only depressed themselves but had an immediate relative—a parent, child, or sibling—

who suffered from the illness. He matched these patients with control subjects who had no personal history of depression, nor any in their extended family.

The first, most obvious difference when the images of the two groups were compared was that in the group with recurrent depression a significantly increased blood flow was present in the frontal cortex, especially in the verbal, and dominant, left hemisphere. This increase was most pronounced in an area of the limbic lobe called the *cingulate gyrus,* the front part of the temporal lobe where the amygdala is located, and a portion of the thalamus. There was also a comparative reduction in the blood flow to the back of the two hemispheres, where vision is processed.

The pattern of activity suggested to the researchers that in depression the brain is generally less interested in things outside the head and more preoccupied with internal emotional affairs, a state reflected in the increased blood flow to the working memory banks of the prefrontal cortex and the emotional sentinel of the amygdala. The prefrontal cortex is the depository of emotional meaning—where memories are stored that have been "coded" to feeling states by the amygdala and hippocampus. Patients with damage to the prefrontal cortex, although normal in mood, have difficulty expressing the appropriate emotion when asked to imagine situations that most of us would consider sad or happy, even when they have little difficulty in solving intellectual problems or planning daily events. Could it be that in melancholia the reverse occurs, with the excessive activity of this "verbal hemisphere" circuitry explaining the persistent rumination which is so frequently described, including recurrent thoughts of suicide similar to those that plagued John Moorehead during the worst of his illness?

Evidence is accumulating that supports this possibility. PET studies show that in a balanced state of mood—what psychiatrists call *euthymia*—the blood flow across the two brain hemispheres in normal individuals is almost equal. But when a sad mood is conjured up, with the help of personal memories or by having someone look at photographs of sad facial expression, overall brain blood flow dramatically increases and comparatively so in the *left* prefrontal cortex, the exact same area found to be overactive in depression. This is a very consistent finding which has been confirmed in several research centers including McDonnell, the National Institute of Mental Health (NIMH), and the Brain and Behavior Laboratory of Professor Ruben Gur and his colleagues at the University of Pennsylvania. By contrast, happiness induced by the same procedure in normal subjects decreases overall blood flow, with a pattern shifting toward the right hemisphere, suggesting that changes in limbic and hemisphere dominance do accompany changes in normal mood.

These studies mapping emotion in the living brain fit well with the clinical case histories of stroke victims where, as you will recall, abnormalities are commonly found in the left frontal region of the brain in those suffering from melancholia after a stroke, and in the right hemisphere when mania is present. Mania is not easily studied by brain-imaging technologies because the procedure demands cooperation and the ability to remain quiet for an extended period of time, qualities not readily available to most individuals in the middle of a manic episode. However, researchers at the University of Iowa studying patients who had developed mania after a stroke or brain injury have reported reduced blood flow—although this was sometimes distant from the part of the brain that had been injured—particularly in the right limbic system structures of the prefrontal cortex, the temporal cortex, and the right thalamus. Subsequently, with SPECT—which is a little less demanding of patients—these same researchers again found a comparative reduction in blood flow to the right frontal and temporal regions during an active mania uncomplicated by brain injury. At the University of Pennsylvania we have found further evidence of this switching of hemisphere with mood in those with rapidly cycling manic depression, a particularly malignant form of the illness where episodes of mania and melancholia follow each other in quick succession over a period of days or weeks.

Taken together, these clinical studies comparing blood flow in the brain during mania, depression, and normal mood confirm shifting patterns of activity, among the major centers of the limbic alliance that are distinct for the three mood states. The studies also provoke questions. For example, does the pattern of activity in mania or depression return to normal when the episode of illness has passed? What of those individuals, such as the people Wayne Drevets studied, who suffer repeated episodes of depression from apparently minimal stress and strain? Do they have some special vulnerability—a neurobiological Achilles' heel—that any environmental challenge quickly discovers and exploits? And if so, does the evidence for such an Achilles' heel persist in the blood flow patterns found after the episode of mood disturbance has passed?

To explore these questions Dr. Drevets extended his investigations at the McDonnell Center by recruiting another group of carefully defined patients with recurrent melancholia. These individuals were similar in every way to the first group, except when the imaging procedures were undertaken they had completely recovered from their depressive illness and were no longer receiving any antidepressant medication; they felt well and on objective measurement their mood was normal. Upon initial analysis the distribution of blood flow among the brain centers of the limbic alliance in these patients seemed to be within standard limits, but then the researchers made a remarkable observation. The amygdala region, particularly of the left limbic complex, still displayed a pattern of increased blood flow compared to the normal control subjects. This suggested to Dr. Drevets that in those who suffer repeated episodes of melancholia the amygdala might remain irritable trigger points, capable of changing the regulatory homeostasis of the whole limbic alliance. To explore this possibility, Dr. Drevets conducted a third study of recovered patients, this time examined while still receiving antidepressant medication. In these individuals the increased amygdala blood flow had disappeared, supporting the theory that these nuclei may be limbic centers of vulnerability in recurrent depression, an encumbrance that is reduced by antidepressant medication.

That the amygdala may be centers of vulnerability in those suffering recurrent depression is an intriguing observation, especially as we know from considerable animal research that these nuclei are important in the processing of emotion. This was first recognized in the late 1930s, at about the same time that Wilder Penfield began his exploration of the brain in epilepsy, when two researchers named Klüver and Bucy described dramatic behavioral changes following surgical removal of the temporal lobe in aggressive monkeys. Following the surgery, normally fierce and wild animals became tame, showing little emotion. Subsequent research in monkeys and other animals indicated that the critical site in the temporal lobe where injury produced this dramatic emotional "blindness" was the amygdala.

Studying the Klüver–Bucy syndrome, as it has become known, has taught us much about the amygdala and their connections to the rest of the limbic alliance, verifying that the amygdala are the limbic centers that assign emotional significance to experience. Animals with the Klüver–Bucy syndrome become indifferent to their environment, losing their usual sexual and dietary preferences. While partners and familiar foods are recognized, they no longer stimulate the same emotional responses. Monkeys, for example, readily eat meat—something they never do normally—and explore their surroundings constantly, showing little fear of sudden noise or human beings.

From other work with monkeys we have also learned that it is neurons within the amygdala complex that recognize and generate the facial expressions of emotion. It is within these clusters of nuclei that the instinctual capacity for the primary emotions is found: the capacity to express fear, joy, anger, sadness, surprise, and disgust—and to selectively identify these in the facial expression of others. You will recall my earlier description of the child hesitating when confronted by the glass-sectioned tabletop, and how he sought reassurance from his mother's smile before crawling forward. With an understanding of the amygdala's role in emotion, suddenly the human infant's fixation with the mother's face makes evolutionary sense. As it does for nonhuman primates, the face provides for each of us a unique badge of personal identification, but it is also the vehicle of expression for the emotional feelings between mother and child that are critical in the development of experiential learning and social behavior. In mammals, attachment and emotional expression between parent and offspring go hand in hand with the development of the emotional self.

The critical role of the amygdala in human emotion is confirmed by an unusual case history reported by Antonio Damasio, Professor of Neurology at the University of Iowa. Dr. Damasio described a young woman who had suffered almost complete destruction of both amygdala because of a rare neurological disease, although without damage to other parts of the brain, including the important memory-processing centers of the hippocampus. When asked to rate the emotional expression conveyed in facial photographs of fear, happiness, and the other primary emotions she was clearly impaired, whereas she had no difficulty recognizing each individual face as representing a unique person. Identifying the expression of fear—an instantaneous response for most of us—was particularly difficult for her, as were the subtle facial cues of emotion that permit a fine-grained response to real-life situations. In fact, Dr. Damasio reported that his patient, despite being of cheerful disposition, had difficulty in social situations and tended to make poor decisions where emotional judgments were required concerning the intentions of other people.

The emotional processing and communication required to manage the complex social situations typical of adult life demands both careful emotional appraisal and a good memory. You will recall that the secondary emotions of shame, pride, and guilt—sometimes referred to as the self-conscious emotions—develop in this way. These emotions differ from those considered primary because they are contingent upon the ability to make intelligent (cognitive) judgments against an acquired set of moral standards. Thus the secondary emotions blend primary emotional judgments with previous social experience in a continuous process of self-conscious evaluation that requires the integrity of both the amygdala (to render emotional judgment) and the memory processing centers of the hippocampus and frontal cortex. Damage to any of these functions will impair the limbic system of emotional assessment and expression, compromising social interaction.

For most of us, the secondary emotions are very powerful, and beyond infancy they play a commanding role in determining our social behavior and the stability of social relationships. In mania and depression this emotional processing is disturbed, and both the processing of memory and the meaning of experience are distorted. Witness, for example, the self-conscious guilt about his family that helped destroy Emmett Jones and the overwhelming shame of John Moorehead as he struggled to explain the symptoms of his emerging disability. This distortion of emotional processing often fluctuates during illness in an interesting way. The distortion is frequently greater in the morning, when other symptoms of depression are also more profound, with an improvement occuring towards the evening hours. Furthermore, we know from the research of David Clark of Oxford University and other cognitive psychologists that memory and mood fluctuate together; the sadder the mood, the more morbid are the memories that the depressed person recalls.

Mania and melancholia are dynamic, fluctuating afflictions which eventually will heal, given sufficient time and a safe haven. Their episodic nature and the changing temporal pattern of symptoms occurring during an episode emphasize that these are disorders of the regulation and control of neuronal behavior. Save for the depressions that may follow brain injury or infection, as far as we know there are no wiring glitches in the emotional brain that trigger and perpetuate melancholia. Rather, the process is some sort of irritable fermentation, possibly a disturbance in the dynamic regulation of the neurotransmitter messenger systems that sustains communication among the neuronal centers of the limbic alliance. The changing blood flow patterns that brain imaging reveals during mood disorder and in recovery presumably reflect this limbic fermentation and disorganization of neurotransmitter activity. The usually precise interchange of messages that sustain communication among the thalamus, hippocampus, amygdala, and the frontal cortex of the limbic alliance is disturbed. New information from the environment is no longer attended to or processed swiftly by the thalamus, and the emotional memory banks of the frontal lobe churn without purpose, as the true emotional significance of ongoing events becomes blurred by the preoccupation with remembrance. The homeostatic systems of the emotional brain have adopted a new steady state, one less responsive to the environment and less adaptive.

What disorganization of neuronal regulation was John Moorehead describing to me, for example, when he characterized depression as "a terrifying loss of self-control, a paralysis, a stroke of the will"? After his recovery, in a meeting which we recorded, John told me how during his illness he couldn't decide about even the simplest of things. "I remember towards the end of the semester," he said, "a colleague was in my office and as he left he asked if he should close the door. I literally couldn't decide. I sat there and finally said in a very low voice that it really didn't matter—it didn't make a difference." John went on to describe how making coffee, brushing his teeth, and other minor decisions became "like career choices, like whether or not to leave the Church." Does such experience reflect the amygdala adopting a different set point—comparable to turning down the thermostat—that slows down the limbic processing of executive decisions?

And what about the internal rumination over morbid past experience? Does this reflect continuous, purposeless communication among the neuronal centers of the limbic alliance in the absence of decision-making? "In depression your normal self shrinks and you have little time for others," John said, in the same interview. "There's a kind of self-preservation going on. The territory of the self gets smaller and smaller until it's totally occupied. So when you talk to anyone you're really not interested in what they have to say because there's no room for them anymore. It's a peculiar and very different feeling from what I usually have."

These subjective experiences of mind reverse during recovery, and the change is apparent to others in the return of social interest and emotional expression. "It's like the nerves regenerate," said John. "Angela seemed to know before I did that things were changing. One morning I got up and shaved—without thinking about it—and she said afterwards that then she knew I was mending. Recovery is like coming out of a dream where you have distorted all sorts of things but in waking up you forget the distortions. All that squirrel-cage thinking [rumination] disappeared. People can look at you and tell the difference. It's in the face—the smile lines return—and you start to talk faster. Actually, as you become reacquainted with yourself, other people confirm that because they start talking to you again, almost as if nothing had happened."

In using the word "regenerate," John captures the imagination. But to my mind, the idea that communication among the neuronal circuits is "reset" offers a more accurate image. To pursue my earlier metaphor, with the thermostat reset the room temperature returns to normal and life proceeds once more at the accustomed pace. We do not know yet how close these observations lie to the anatomical truth of the matter; what we do know is that activity among the limbic centers is poorly regulated in melancholia and states of abnormal mood, and as this dysregulation progresses, disturbances of emotion, thinking, and body housekeeping emerge. Brain imaging for the first time offers the potential to track, as they occur, the shifting patterns of metabolism that must underlie these changes, building a vital bridge between neuroanatomy and subjective experience. As these technologies develop further, one of the exciting challenges for psychiatrists and clinical neuroscientists is the development of a precise functional atlas, integrating our knowledge of disturbed emotion, thinking, and so on, in depression and mania with limbic-brain anatomy and the dynamic chemistry of neuronal communication.

There is also another window, a chemical and hormonal one, through which it is possible to observe the emotional brain at work. The amygdala stand at the gateway between the information and memory systems of the limbic alliance and the hypothalamic organization of the body's housekeeping. These nuclei not only orchestrate emotional expression in the muscles of the face—as is evident in the return of the smile upon recovery from depression—but also, via the nerve centers of the hypothalamus, they coordinate the flow of chemical messengers to the pituitary gland. Through this chain of command the limbic alliance directs the endocrine system of hormones that help coordinate the body's housekeeping and physiological harmony.

These hormonal messengers of the endocrine system can be measured in the circulating blood. Their temporal patterns, because they reflect the brain's oversight, offer an opportunity to track the cellular and molecular systems that regulate emotional balance and keep us in harmony with the spinning planet on which we live. The endocrine system, which includes the stress, energy, sexual, and rhythmic messenger systems of the body—the steroids, thyroid, gonadal, and pineal hormones respectively—is commonly disordered in mood disturbance. Thus, an appreciation of these hormonal systems can help guide our understanding of the links between mood and what we experience, and how mood disorders may be triggered by the circumstances of our environment.

Seven

The Vital Balance

Stress, Homeostasis, and the Seasons of Mood

> The living organism ... has enclosed itself in a kind of hot house. The peripheral changes of external conditions cannot reach it; it is not subject to them, but is free and independent. ... All the vital mechanisms, however varied they may be, have only one object, that of preserving constant the conditions of life in the internal environment.
>
> CLAUDE BERNARD
> *Leçons sur les phénomènes de la vie communs aux animaux et aux végétaux* (1878–1879)

Stress is harmful, or so we read in popular magazines and self-help books. But that is only part of the story. The physical and behavioral "disturbances" of stress are actually signs of the body's autopilots at work, safeguarding us in times of change and challenge. To me, the physiological arousal of the "stress response" is analogous to the whistling and whirring of an airliner's servosystem, efficiently balancing the flaps of wing and tail as the pilot navigates a final approach to landing. In everyday life, the subjective turmoil we associate with stress is not inherently harmful, but a part of normal physiology as the brain tunes the body's internal chemistry, coping expertly with ever-changing circumstance. Stress is vital for healthy adaptation.

The most vulnerable moments in an airplane's flight are during takeoff and landing. This is when accidents happen. To ensure safe passage control, coordination and precise timing are essential, as rapidly changing winds, air pressure, and gravity threaten to pluck the machine from its chosen path. So it is for living creatures: We too are most vulnerable during the stress and strain of rapid change. However, such moments have also made us strong, for it is coping with threat and changing circumstance that has been the backbone of our successful evolution.

Stressful episodes, in moderation, are merely routine exercise—daily calisthenics upon which to grow and thrive. The separation from parents, mumps and measles, bumps and bruises, falling in love, arguments with friends, even finding a parking spot on a busy city street—all are challenges that call forth the stress response in subtle ways. And every time the challenge is met, as we learn from each experience, we become stronger. This we call coping, or adaptation. In fact, Hans Selye, the Hungarian-Canadian researcher and physician whose intensive study of the biology of stress first made it a popular subject in the 1950s, called the changes he observed in his animal subjects the "general adaptation syndrome," emphasizing the central role of stress in maintaining the body's precious balance with the environment. Stress and adaptation, Selye concluded, lie at the physiological core of being alive.

But Selye also discovered during his experiments that prolonged exposure to physical or mental stress can lead to exhaustion and even death. When a task is formidable and overwhelming, especially when it is persistent, seems beyond control, or falls upon an individual already vulnerable—let's say, to depression—illness is frequently the outcome. Under such circumstances of chronic challenge, the stress response persists beyond the point of useful coping, and as adaptation fails, stress itself becomes the damaging agent.

Stressful situations that create personal turmoil commonly precede episodes of depression and mania, and when they occur repeatedly they can sensitize an individual to recurrent disorder. Stress is generally the yoke that binds together these disquieting events—inevitably confronted by each of us—with individual vulnerability, thus triggering illness. In fact, once established, mood disorders can be thought of as special examples of adaptive failure, for they involve the same systems of brain and body that are activated during the healthy stress of adaptation. Hence the study of the hormonal stress response, and the psychology and physiology of adaptation, has become an important research link to understanding how mood disorders may be kindled by the apparently mundane things that happen to us every day, and how the illnesses may persist in vulnerable individuals.

The hormones of the body's endocrine system are particularly important in this linkage, for they offer a magnifying window on the limbic brain and its activity. Therefore I shall explore, in some detail, the biology of these hormones that help keep us safe in the world and how sometimes under chronic strain their action can be maladaptive. I will also return to the stories of Angela and John Moorehead, for their experience helps illustrate how the stress of environmental and social changes, by disrupting homeostasis and disturbing the emotional brain, can precipitate mood disorder in vulnerable individuals. Recent research can help explain, for example, Angela's peculiar susceptibility to winter depression while John, caught in a web of conflict and social change that he could not control, was a victim of melancholia that had been initiated by chronic stress.

The physiological changes that occur during stress are best understood as part of the continuous adaptation essential to staying alive. During our lifetime each of us defies the laws of thermodynamics, maintaining within the body a stable chemical environment (the *milieu organique intérieur* that was first described by Claude Bernard) that is considerably displaced from thermodynamic equilibrium. It is the vital balance of this "internal environment" that your doctor is checking with all those blood tests during the annual physical examination, and it is the reason why we can survive in a wide variety of environmental circumstances. Many animals, including ourselves, are able to adapt and maintain physiological equilibrium over a wide range of temperature and living conditions, from the Arctic to the tropics. This remarkable autonomy and internal stability requires the consumption of energy, and a continuous economic exchange with the surroundings. These are qualities characteristic of an "open" system, in contrast to a system that is "closed"—such as a chemical reaction in a test tube—which proceeds according to thermodynamic law, and eventually exhausts itself. However, regulation is essential to successfully maintain the economy of an open system, and in animals like ourselves it is the brain that is the chief regulator.

Any challenge to a living system that perturbs (causes a deviation from the resting state) will induce a correction (adaptation) designed to maintain the internal balance (homeostasis) and reestablish the system's preferred relationship, or set point, with the environment. These fundamental principles of homeostasis, attained through precise regulation, apply at all levels of brain activity—from the molecular and cellular functions of the neuron to the complex social interaction of individuals and the hierarchical groups that characterize human society—and are essential to the dynamic concept of adaptation. When the brain regulators maintain a stable adaptation, we speak of coping—implying harmony and balance—and when threatening events perturb the regulatory mechanisms to corrective action, the physiological changes that occur are called stress. A practical way of understanding the physiological dynamics of the stress response, therefore, is to view them as tangible evidence of the body's self-correcting homeostats—analogous to the airliner's servosystems—working to maintain equilibrium and find the best fit with the environment. This simple paradigm—of perturbing challenge and adaptive response in the service of homeostatic equilibrium—lies at the core of understanding how stress, in linking environment and individual vulnerability, can precipitate mood disorders. And it is a paradigm that also emphasizes how important homeostatic principles are in the healing of an aberrant mood.

The fundamental behaviors that are disturbed in mood disorder—emotion, sleeping, sex, appetite, energy, motivation—derive from brain systems homeostatically controlled. Some of these behaviors we call drives or instincts, which can be a little confusing as it suggests something linear and of one dimension, not a self-correcting system regulated around a set point. But self-regulating systems they are. Take sleep, for example. When we do not get enough, the "drive" increases and a "sleep debt" is incurred that must be repaid. The greater the deprivation, the more powerful the drive to sleep becomes, and only when we can get to bed and "catch up" is that drive reduced. Unfortunately, for those with a demanding work schedule this self-balancing drive for sleep is all too obvious. The progressive sleep deprivation of the week leads to a game of "catch up" at the weekend, when finally the alarm can be silenced and the debt repaid. Our other drives are regulated in the same way, each with a preferred state of equilibrium—the set point—to which they return after challenge.

The ultimate regulators of these drive systems in human beings, as in most complex animals, are to be found under the custodial care of the hypothalamus, an important member of the limbic alliance that has evolved from the ancient lizard brain. Control centers for body temperature, aggression, the appetite for food, thirst, procreation, the timing of body clocks, are all concentrated in this tiny brain region. These hypothalamic centers maintain the internal environment and successful adaptation through three interdependent systems of communication. First is the directly wired autonomic (meaning self-governing) nervous system that quietly and continuously tunes the body's physiology. Second, via the pituitary gland, the hypothalamus regulates a cascade of messenger hormones, collectively known as the endocrine system, that are circulated in the bloodstream and help orchestrate the chemistry of all the body organs, including the brain itself. (Because of its pervasive role, this is a system of great importance to behavior and therefore the one upon which I will focus most attention in our consideration of mood disorder.) And finally, third, the body's immune system defends against invading organisms and other foreign agents. The immune system is also responsive to hypothalamic command and is intimately entwined with the hormones of the endocrine system, especially the adrenal steroid hormones which play a major role in the chronic stress reaction that occurs in many depressions.

Adaptation to the challenges of daily life—the everyday calisthenics that I mentioned earlier—is orchestrated through hypothalamic oversight of these autonomic, endocrine, and immune systems. Acute adaptation to unforeseen or emergency situations—what hitherto I have called the stress response—also has another name, *reactive homeostasis,* to distinguish it from the brain system of *predictive homeostasis* that adjusts the body's equilibrium to predictable environmental changes. Both predictive and reactive mechanisms are important in mood disorder, for when disturbed, each can significantly increase an individual's vulnerability to illness—as will be apparent in the stories of Angela and John Moorehead. Although they are brother and sister, for our purposes here their experiences are almost entirely distinct; hence I shall take them in turn beginning with Angela's struggle against winter depression and how that syndrome intertwines with a disturbance of predictive homeostasis.

Predictive homeostasis is the brain's clever way of adapting to life on this spinning planet. Living creatures long ago recognized that the earth's daily pattern of sunlight is sufficiently predictable that regulation of the body's homeostasis could be automated and keyed to fit its rhythm. Such careful environmental planning has its origin in the brain's "biological clocks." These are genetically programmed timing mechanisms located in the hypothalamus and which have a cycle approximate to the cycles of light and dark. These clocks efficiently set body rhythms of temperature, energy, sleep, and sexual behavior to coordinate with the daily and seasonal cycles of the earth. Through these brain chronometers, human physiology is carefully adjusted for optimal daytime performance. In health, the core body temperature increases just before sunrise, is highest in the late afternoon, and begins falling again before sunset, to reach its nadir during sleep, with the rhythms of cortisol and thyroid hormones closely coupled. This explains why we are not at our best in the early hours of the morning, a fact underscored by the number of single-vehicle accidents and other tragedies that occur in the middle of the night. The quality of human performance is closely tied to the twenty-four-hour cycle of light and dark, and to the predictable rhythm of the seasons, and it is precisely this comfortable fit with our planetary environment that is lost in winter depression.

Angela Moorehead knew nothing of these details when, together with her brother John, she first consulted me. However, prompted by my questions about a family history of mood disturbance, she described a series of depressions with typical seasonal pattern that she had suffered during the years she lived in Boston. Characteristically, the distress had begun in the fall with a loss of energy and social interest, a preference for sleep and starchy foods, and a mood of gloomy discontent. It usually worsened through the winter months but was broken in the spring by a renewed sense of purpose, sexual energy, and lightness of being. Angela, curious by nature, had come to recognize this seasonal rhythm of mood (without understanding it) and minimized her vulnerability to the depressions by moving south, and to a sunnier climate.

As described in chapter 5, Angela's depressions had begun during her teenage years, but the seasonal pattern hadn't registered in her mind until she was in her late twenties and working as a travel agent. Along with that particular job had come several perks, including the opportunity to travel and, sometimes, to be the "guinea pig" for packaged vacations. During our conversation, Angela told me how toward the end of a particularly cold, but sunny, skiing trip she first questioned whether her depression was "something physical."

"I remember that while I was riding the chairlift one brilliant sunny afternoon it suddenly occurred to me that although my feet were freezing, I felt energetic and happy," Angela recalled. "My depression had lifted during the week of skiing, but shortly after I returned to Boston and my daily routine I found myself totally miserable again." When she confided this suspicion of a "physical cause" for her bad moods to her office colleagues they had laughed at her (everybody feels better during a vacation, after all), but when her spirits improved again, this time after a trip to Florida, she dug out some of her old diaries to see if she could trace a similar pattern in previous years. Mapping out a chronology, she found that it was true, she usually did feel better in sunny vacation spots, especially in the winter, but something else also caught her attention. Even in the winters when she was not particularly depressed, the tone of the diary was distinct, with long and detailed entries much concerned with worldly worries, both her own and those of her friends. By contrast, the summer notes were short lists of everything she was planning or doing. While the summer entries didn't make interesting reading, they were unmistakably the product of a happier frame of mind.

Angela described how she began to pay close attention to her mood, keeping track of its variation through the year, her food preferences, weight, the amount of coffee and alcohol she consumed, and so on. Coffee (in large quantities), she concluded, was definitely a winter habit, but her alcohol intake, as she drank little, had no obvious pattern. She also noticed an irritability just before her menstrual period that seemed to increase during the winter months. Then, in an effort to control her weight, Angela had started jogging in the early morning before work, and maintained the habit diligently throughout one summer and fall. To her surprise, the exercise schedule postponed the onset of her symptoms for several weeks that year. Encouraged, she continued. Winter visits south, and skiing, consistently helped lift her energy and spirits, convincing her that it was not the cold but a lack of sunlight that troubled her during the gray New England winter. Hence, when the opportunity came to transfer to San Diego, it seemed the obvious thing to try and, indeed, although episodes of depression recurred occasionally around times of stress and change, they lost their precise seasonal pattern. "Life seemed less difficult somehow," Angela reflected, "little things, which previously would have upset me, I was able to let go by. I became much less irritable and my friendships are stronger now for that."

Angela is not alone in her experience. Everyone who lives in a markedly seasonal climate undergoes some degree of behavioral change as a normal adaptation to the periodicity of our planet. The limbic brain cleverly predicts the coming of the seasons and sets our behavior accordingly. In a study that I conducted in northern New England—which at a latitude of approximately forty-five degrees is midway between the equator and the North Pole and has a climate similar to that which Angela experienced in Boston—I discovered that during the dark winter months most normal people sleep an hour or two longer, taking naps if necessary to compensate for the demands of the workplace and the alarm clock. Most of us also gain a few pounds, tucking into the sugars and starch like a bear heading for hibernation (our feasts of the winter season are no accident). I found too that the average person is more grumpy in the winter, although this slips away with the coming of spring and a return of the long daylight hours. In fact, summer is the happiest time of year for those who dwell in seasonal climates, with the most energetic period being the late summer—as the weather cools, but before the bite of winter returns.

I have described the winter slowdown that marks this seasonal rhythm of behavior as the "hibernation response," to emphasize how similar it is to the winter retreat of many other mammals. Seasonal affective disorder (SAD), as the winter depressions are now commonly called, is an exaggeration of this drive to hibernation. Women are four times more affected than men and the prevalence varies with latitude, studies having shown that SAD is more common in Alaska than in San Diego. In persons vulnerable to mood disorder, such as those (like Angela) with a strong family history of depression, a markedly seasonal climate disrupts the brain's capacity for predictive adjustment and homeostasis falters, initiating or intensifying affective illness. Especially during the winter months, the victims of SAD experience difficulty in synchronizing their body rhythms, developing a sort of seasonal jet lag that varies in severity depending upon the latitude at which they live. Angela had been fortunate; the opportunity to live in a climate closer to that of perpetual summer had reduced the seasonal adjustments required of her, and minimized her periods of depression.

When I first met Angela I had just returned from a year of research at the National Institute of Mental Health, in Washington, D.C., where I had the good fortune to work with several psychiatrists experimenting with new light therapies for the treatment of winter depression. Clinical experience and a growing body of brain research provided compelling reasons to embark upon such a program. Seasonal disturbances of behavior such as Angela suffered had been described for a long time (you will recall Robert Burton's comment in 1621 that "of the seasons . . . autumn is most melancholy") but never satisfactorily explained. However, by the early 1970s, animal research had firmly established that the seasonal reproductive behavior of many mammals was driven by a sensitivity to the changing patterns of light throughout the year—the first real evidence of what we now call predictive homeostasis. In small mammals, master clocks that control the daily cycle of activity had been found in the suprachiasmatic nuclei (SCN) of the hypothalamus, and it soon became obvious that the same basic system must operate in human beings.

The mammalian brain registers the cycles of daylight as our eyes take in the changing light, and a signal is transmitted via the optic nerves to set the timing of the SCN master clocks. In human beings, these are genetically programmed to run with a daily cycle slightly longer than 24 hours—hence the term *circadian rhythm,* meaning "around one day." Thus the brain's internal clock time is close to sun time but not rigidly so. This "imprecision" in the daily clock is important, for it builds a necessary flexibility into the brain's system of predictive homeostasis that permits accommodation to the seasons. Because the earth is tilted on its axis, over the year that it takes for the planet to travel around the sun the pattern of daily exposure to sunlight at any one point on the earth's surface subtly changes and our brain clocks must be reset each morning if they are to remain synchronous with sun time. It is this continuous correction that permits the SCN to remain in harmony with the seasonal changes in light and dark. Any environmental signal capable of resetting the clock is called a *zeitgeber,* from the German meaning "time giver." Social habits—such as mealtimes, exercise, the drinking of coffee or alcohol—have subtle zeitgeber effects, but sunlight remains by far the most powerful.

These brain chronometers, set each day by the rising sun, in turn set and synchronize—much as a maestro sets time and conducts an orchestra—the daily and seasonal patterns of hormone production. The body's circadian rhythms of temperature, and the release of cortisol, thyroid, and other hormones, are programmed through timed messages from the SCN to the hypothalamic centers that regulate them. Seasonal cycles of energy and reproductive behavior are adjusted by the signals from the brain clocks to the pineal, an endocrine gland about the size of a pea that nestles just outside the brain, between the two large hemispheres of the new cortex. The pineal secretes the messenger hormone melatonin, an important chemical regulator of sleep and seasonal behavior which is carried throughout the bloodstream to all the organs in the body, including across the blood–brain barrier to the brain itself.

Melatonin has been called the Dracula hormone because it is produced mainly during darkness. Each evening, the level of melatonin in the bloodstream rises and then falls again in the early morning, with production being essentially shut off as the eyes signal to the pineal that the sun is rising. Through the long days of winter the nighttime peak of melatonin is extended, shortening dramatically with the onset of spring. This rhythmic production of melatonin provides the body with an excellent barometer of planetary change and guides many seasonal activities, including food intake and the timing of reproduction in most mammals, through inhibiting the sex gland hormones. The pineal's annual rhythm of melatonin secretion explains why "rutting" in the deer herd occurs in the fall and fawns are born in the spring—as are most other mammals with a long gestation—when the sun is warm and the vegetation is plentiful. As others have described, and as I discovered in my studies in New England, human beings also have a spring and summer pattern of birth when living in a seasonal climate, although with the development of domestic microclimates and an abundance of food throughout the year the rhythm has become less distinct. Not so with winter depression, which persists despite our warm and well-lit winter habitats. This raises an important question: If this seasonal syndrome of morbid mood really is triggered by a loss of light's zeitgeber during the dark winter months, why with artificial lighting in the home throughout the year do the Angela Mooreheads of this world continue to suffer?

This was exactly the question that intrigued Alfred Lewy, one of the research psychiatrists that I had met when working at the NIMH. The answer generally accepted at the time was that the human brain must be insensitive to the daily cycle of light and dark and that, compared to other mammals, we have become relatively immune to our planetary environment. To help resolve this question, Dr. Lewy developed a sensitive means of identifying melatonin in human blood, one that could accurately measure the pineal's production of the hormone. Then, together with his colleagues, psychiatrists Norman Rosenthal and Thomas Wehr, he conducted a series of simple experiments, waking normal subjects in the middle of the night—when blood melatonin levels are at their highest—and asking them to read under different lighting conditions.

The researchers discovered a remarkable thing. The lighting in the average home—that favorite lamp by which you read the newspaper each evening—is too weak to influence the daily rhythm of melatonin secretion. Regardless of how modern living conditions have modified the seasonal pattern of births, from the perspective of the brain's chronometers the artificial lighting in most households is equivalent to living in darkness. Indeed, with the coming of the Industrial Revolution and the retreat from an agrarian economy to the indoor workplace, rather than extending our daylight hours we have become a society of high-tech cave dwellers. One study in San Diego concluded that the average office worker spends less than an hour outdoors during the working week. Sunlight, the natural time-giver for the body's hormonal orchestra, has been put aside to be replaced by the gloom of an artificial cave.

For all of us, wherever our home, this increases the urge to hibernate during the winter months. For those who live in seasonal latitudes, and especially for the minority who are vulnerable to bipolar illness, such as Angela Moorehead, the loss of sunlight—the brain's most powerful zeitgeber—fosters SAD and winter depression. Doctor Lewy's research suggests that in most SAD sufferers the timing of the production of melatonin drifts away from its usual cycle during the winter months, rising later in the evening hours and extending into the morning—reflecting a loss of synchrony between the brain's internal clocks and the planet's cycle of light and dark. As Angela experienced, the result is a sleepy brain, a brain suffering a chronic "winter jet lag." There is difficulty waking in the morning, and the body is poorly harmonized for the demands of the day. With the onset of short winter days, as the early morning sunshine weakens or disappears, the brain's predictive homeostasis falters. With the conductor gone, the precise temporal harmony of the orchestra is disturbed. The drift in timing (demonstrated by Dr. Lewy) of the pineal gland's nocturnal production of melatonin, which is an important drummer that helps set the rhythm for the rest of the body, reflects this loss of harmony.

Researchers Lewy, Wehr, and Rosenthal reasoned that for those with winter depression, strengthening light's zeitgeber might improve the hormonal harmony, and the depressive symptoms with it. They constructed a lamp with the brilliance of a spring morning—a light source some ten times more powerful than the average room lighting—and extended the winter day by instructing their subjects to read or work beside this source of artificial springtime for approximately two hours, night and morning. Within a few days, most of the subjects experienced remarkable improvement. Their sleep was shorter but more restful, the craving for carbohydrates diminished, and the depressed mood began to lift. Furthermore, when the bright light was removed, or replaced with one of lower intensity, the symptoms returned.

When Angela came to see me in 1980, it was early in this chain of discovery. The first lamp boxes were large, constructed by the researchers themselves, and similar in size to the ceiling fixtures one finds in offices and grocery stores, containing eight fluorescent bulbs, each forty-eight inches long. Sitting three feet away from such a contraption provided the 2,000 lux necessary to simulate a springtime dawn. But in subsequent years the technology has rapidly improved, for the discovery that light is a powerful zeitgeber for adaptive human harmony has become the foundation of a new treatment—indeed a new industry—successfully managing human hibernation and SAD by adding bright light to the home and office environment during the dark winter months. Relief from winter depression can now be purchased in the form of a powerful desk lamp.

Debates continue regarding the specific biological mechanisms that may explain this action of bright light in reversing winter depression, but it seems clear that the critical time for resetting the brain clock is in the early morning—at the dawning of the day. That is when the zeitgeber is most needed and is probably why Angela, while still in Boston, discovered an improvement in her symptoms when she took up jogging each morning. It was not the exercise alone, but running in the bright morning light, that had helped reset her brain clock at the most critical time, improving the synchrony of her body's diurnal rhythm and with it her sense of well-being.

Some researchers have suggested that the SAD syndrome is mediated by melatonin, which in the mid-1990s became something of a cult drug, selling even better than vitamin C in the health-food stores of America. Claims were made for the anti-cancer effects of the hormone, its ability to prolong life (at least in mice), and to cure AIDS by enhancing immune function. While such magical assertions have yet to be substantiated, the role of melatonin as the hormonal coordinator of circadian and seasonal physiology is beyond doubt. Small doses of melatonin, appropriately timed, appear to be helpful in overcoming jet lag but, as yet, there is little evidence that use of melatonin alone can alleviate winter depression. Rather, it seems likely that the pineal production of melatonin is a chronometric marker of brain activity, reflecting the clock activity of the SCN, just as the hormones released from the other endocrine glands—the steroids, thyroid, and sex hormones—are reflective of the activity of the hypothalamic centers that regulate them. While disturbances in melatonin rhythm may compound and perpetuate the symptoms of seasonal mood disorder, the changes in the timing of melatonin production are the result, not the cause, of the limbic dysregulation occurring in winter depression.

Individuals predisposed to winter depression, and those prone to bipolar mood disorder in general, are especially sensitive to changes in light intensity and to the disorganizing effects of any adaptive demand that requires resetting the brain clocks. This sensitivity expresses itself in various disturbances of mood and behavior. The springtime and the fall—the times of the year when the ratio of sunlight to darkness is changing most rapidly—are when suicide is most prevalent. Mania is most common in the late summer, the increasing energy perhaps driven by the long days of sunshine, and indeed sometimes bright light therapy for winter depression can provoke short sleep and hypomania. For those with manic depression, jet travel will frequently induce a "switch" from one mood to the opposite pole. Chronic sleep deprivation can do the same. On several occasions I have had as patients young students whose first manic episode was brought about by cramming all night for exams or by a sleep-depriving social life, often "enhanced" by the liberal addition of alcohol or stimulants, both of which will disturb the brain clocks. It also happens that a night without sleep, powerfully perturbing the clock and endocrine system it orchestrates, will sometimes "jolt" a person out of depression for a day or two. Stephan Szabo told me how he had consciously used this technique to lift his mood when he was depressed, having first discovered it quite accidentally while on call as a medical intern.

All this suggests that those of bipolar lineage, as were Angela and John Moorehead, suffer a diminished flexibility in the regulatory mechanisms that govern the rhythms of predictive homeostasis, perhaps even abnormalities in the brain clocks themselves. Rapidly advancing research into the genetics of behavior and bipolar disorder may soon specify the disability that such a pedigree bestows. My suspicion is that we will discover not one but a variety of genetic variations, each of which can predispose an individual to bipolar illness, including the milder forms such as seasonal affective disorder or the hyperthymia of great achievement. Probably each variant will be responsible for a slightly unusual genetic code in the library of instructions that builds the machinery of the neuron. The resulting minor deviations—for example, in the metabolism of neurotransmitters, the building of receptors, in the enzymes, or the messenger systems that transport information within the neurons of the SCN—will each disturb the dynamics of regulatory function.

Regardless of the specific abnormality in genetic coding that predisposes to bipolar illness, the result is a diminished ability to adapt smoothly to the changing planetary environment—or to accommodate to the turmoil of chronic stress—and recover homeostatic balance once the challenges have passed. We call this *vulnerability*. Under normal circumstances the brain manages to ignore such vulnerabilities. It is an organ in which billions of neurons are in constant communication, and it has a remarkable resilience and adaptive capacity to overcome adversity. Throughout the life of a human being each individual neuron continues to modify its local networks of interaction to accommodate new challenge. Even disability, such as minor strokes, can be compensated for when necessary. Hence we can be certain that whatever specific functional deficit determines the vulnerability to mood disorder, the neuronal centers of the brain will work to minimize its behavioral effect. This ongoing self-correction—a necessary baseline adaptation—burdens performance, however, and reduces the brain's adaptive capacity under stressful circumstances. What is really referred to when we speak of vulnerability is this diminished capacity to function under stress.

Let me draw an analogy. Imagine for a moment that you suffer with a painful arthritic knee. Despite discomfort, you may make fairly rapid progress when walking over smooth pavement, but faced with a winding cobblestoned path, or the need to avoid a speeding bicycle, the knee fails and suddenly, in acute pain, you lose your balance as the whole leg gives way. In my analogy, vulnerability—the painful arthritic knee—is something that persists over time. Challenge, on the other hand, is the immediate, perturbing, event—equivalent to the bicycle or the cobblestone path—that demands accommodation and threatens one's precarious balance. Similarly, in affective disorder, a slight genetic variation in the code for an enzyme that contributes to temperament, or a stress response easily triggered, or prolonged, because of repeated childhood trauma, are vulnerabilities that increase the risk of illness. Everyday events are the challenge. It is these variables of challenge and individual vulnerability that conspire, in ways unique to the individual, to initiate the dysfunctional state experienced as mood disorder. Tying them together are the dynamic changes that occur in acute stress, or reactive homeostasis.

Reactive homeostasis is another name given to what Selye called the general adaptation syndrome. In minor form, this is the reaction that comes transiently to awareness when, already late for an appointment, you can't find that parking space, or expresses itself as panic when, lost in a strange city at night, you fear for your safety. Episodes of the former are so common that they are soon forgotten, while the moments of panic and alarm we remember as stress.

In sudden alarm, when the ever-vigilant limbic brain perceives impending loss of control, threat, or imminent danger, the custodians of the hypothalamus place the autonomic nervous system on instant alert (that's the tingling sensation running down the back of one's neck upon being startled). Suddenly the heart is beating faster and the blood pressure is rising; the pupils dilate and the adrenal glands squirt the hormone epinephrine (sometimes called adrenaline) into the bloodstream. The face assumes the "pale mask" of fear as tiny vessels in the skin constrict and blood is shunted to the powerful muscles of the arms and legs. Subjectively, we experience a sense of tension and an increased awareness of our surroundings, as we search for opportunities to escape whatever challenge besets us, and become suddenly vigilant about details previously overlooked. Emergency energy supplies are conscripted by the liver, which releases to the blood its store of sugar, preparing the body for rapid escape, or for battle should the situation so demand. This is the mobilization phase of acute stress, taught in every high-school biology class, and probably familiar to you as the "flight or fight response."

But in the crescendo that marks reactive homeostasis, the "flight or fight" phase, is just the beginning—a sudden deviation from equilibrium to accommodate dangerous circumstance. If the threat continues, upon signals from the hypothalamus and the pituitary the endocrine glands release powerful stress hormones to sustain the body's readiness. The production of cortisol, an emergency hormone from the adrenal cortex, is increased (the cortex is the external husk of the adrenal gland; epinephrine, the other emergency hormone, is secreted from the central medulla, or pith, of the gland), preparing the body for the fight ahead and any injury that might occur. Similarly, hormonal messengers surge from the thyroid gland in the neck to stimulate body metabolism. The body's immune system is enlisted and, should injury or infection occur, antibodies and scavenger cells are marshaled rapidly to the wound or site of invasion.

These are the ways in which the brain prepares the body for an extended battle—a "clearing of the decks," much as when, in sailing times, a man-of-war made ready and maneuvered to find the best offensive position. Under these emergency conditions, functions less important are placed aside. Sexual interest, the drive to sleep, the appetite for food, are all inhibited as irrelevant details in the face of threat to life and limb. Physiological arousal may persist for hours, days, or even longer depending upon the persistence of the danger, its magnitude, and whether it can be controlled or dismissed—although, ideally, adaptive efficiency demands that the stress be switched off as quickly as possible. Only then can equilibrium be reestablished, vigilance be relaxed, and the hormonal profile return to reflect the rhythm of the day. Reactive *homeostasis,* then, is more than the acute response to danger, for once that emergency has passed the body must be returned to *equilibrium,* in preparation for any further threat that may emerge.

Through these two intimately entwined mechanisms of predictive and reactive homeostasis, the centers of the hypothalamus regulate the physiology of brain and body. These are the servosystems of life—the infantry of survival—which during times of peace and health go about their business largely ignored by the conscious self. However, when battles continue to rage, excitement persists, and the stress response is repeatedly triggered until arousal becomes a way of life, homeostasis can become impaired. The chronic flood of stress hormones chasing through the body grows damaging and the once-protective physiology of adaptation becomes maladaptive. Under such conditions of chronic strain any one of us can fall victim to illness, but for those caught in a moment of special vulnerability—as was Angela each winter—the collapse comes with lesser provocation. For Angela during the dark days of the year, when her basic physiology had lost its rhythm, the most trivial challenge was magnified; an episode of the flu or even a simple argument or disappointment would trigger sadness and self criticism, increasing stress and initiating a vicious circle of increased autonomic arousal and hormonal release.

Stress also appears to sensitize an individual to repeated episodes of illness once a mood disorder is established. Mania and depression are recurrent in their natural history, often returning in a regular seasonal pattern as I have just described, and certainly the frequency of episodes increases over the life span. Also, in some individuals each recurrence of illness brings greater suffering and a more precipitous onset. To Robert Post, a psychiatrist and the chief of biological psychiatry at the National Institute of Mental Health, this relationship between repeated stress and escalating illness is no accident. In fact, Dr. Post proposes that the chronic arousal induced by recurrent stress increases the irritability of the limbic brain, in a process that he calls kindling. Stress kindles the activity of neighboring neuronal communities much as a fire is kindled.

In the brain, kindling may be understood as a form of aberrant learning—or a vicious feedback loop—where repeated stimulation conditions increase behavioral activity, much as Pavlov conditioned the behavior of his dogs. Kindling was first demonstrated during investigations of epilepsy, when repeated electrical stimulation of the amygdala caused an increased sensitivity to seizures occurring. However, the seizures only occurred when the stimulus was intermittent. Continuous amygdala stimulation, even stimulation every few minutes, failed to produce convulsions, presumably because the limbic system quickly adjusted to the ongoing challenge and reset its regulators accordingly. It was only when the electrical stimulus was applied one or two hours apart and the brain had become sensitized, or "allergic" to the stimulus, that seizures occurred. Later, in similar experiments with animals, it was found that pharmacological agents—notably cocaine—could also produce kindling. In these animal experiments, kindling was capable of initiating long-lasting, possibly permanent changes in neuronal excitement. Indeed, animals sensitized to cocaine in youth retained convulsive susceptibility into maturity.

Dr. Post has suggested that stress, through this kindling mechanism, may play a major role in the onset and the recurrence of affective disorders, particularly in those people who suffer a genetic vulnerability to mania and depression. Repeated stress, kindling limbic arousal much as cocaine kindled seizures in the animal experiments, results in anxiety, insomnia, exhaustion, and disruption of function until a vicious feedback cycle develops. Furthermore, it is repeated stress or chronic stress, where control over the situation has been lost or given up (as in the learned helplessness paradigm of Seligman's experiments with dogs that's described in chapter 4), that seems to be particularly malignant in the kindling of bipolar illness.

As I shall now describe, such was the situation that John Moorehead had faced, and which had eventually overwhelmed him, drawing him down into melancholy.

Eight

Of Human Bondage

Stress, Vulnerability,
and the Feeling of Control

You must be master and win,
or serve and lose,
grieve or triumph,
be the anvil or the hammer.

JOHANN WOLFGANG VON GOETHE
Der Gross-Cophta (1791)

Stress is powerfully influenced by those qualities that determine the self. Our behavioral temperament, the belief system we grew up with, the critical experiences that have molded us—each plays its own part in determining what we find stressful in life. Ultimately it is the personal meaning and significance that we attach to any event, and the feeling of control that we have over the situation, that helps determine physiological arousal and how vulnerable we are to stress.

Hans Selye, a pioneer in stress research, first became intrigued by stress as a link between individual vulnerability and environmental adversity as a medical student, when he had discovered changes in the rat's adrenal gland secondary to a variety of unpleasant stimuli. That had been in the 1930s, but by the time his book, *The Stress of Life,* appeared in 1956, the "guinea pigs" of research were predominantly human. In a burst of enthusiasm to learn more about the interaction of the brain and the environment, individuals in all sorts of extraordinary situations had become fair game for experiment.

Oarsmen in the Oxford and Cambridge boat races, those flying airplanes, parachute jumping, visiting dentists, experiencing a stay in hospital, students taking exams, all fell under scrutiny. In these early studies the daily excretion of steroid hormones in the urine was commonly employed as a barometer of stressful change, and initially it was thought that the physical demand of the stressor was the critical variable. However, rapidly it became obvious that, rather than the strain of the task itself, it was the "feeling of control" that an individual was able to exert over any situation that determined whether or not physiological arousal occurred. Rather than the challenge, it was the meaning of the challenge that appeared to be critical. Thus, in one report, judging by the levels of steroid hormone appearing in the urine, an individual who was expert in cold stress and survival found crossing the polar ice cap considerably less demanding than giving an important lecture at the Edinburgh Medical School.

In this second chapter on the vulnerability to mood disorder I will explore some of what we have learned about stress and this feeling of control in adversity, especially how such feelings are dependent upon the intimacy we share with others. Particularly I shall recount the events which preceded John Moorehead's depression and surrounded his recovery, for his is a story that underscores how powerful are human sentiment and intellectual ideas in determining our emotional behavior.

In *The Stress of Life,* Selye proposed three phases of his general adaptation syndrome—alarm, resistance, and exhaustion. The first two, taken together, equate essentially to what I have described as reactive homeostasis—the autonomic arousal and hormonal release of healthy adaptation. The third phase, exhaustion, corresponds to the kindling of chronic arousal and adaptive failure. Under these chronic conditions, the feelings of foreboding, increased vigilance, and being "wired" that occur during acute stress evolve into anxiety, insomnia, loss of sexual interest, and reduced appetite—all symptoms experienced during melancholia.

Research now suggests that these similarities between the behavioral experience of depression and chronic stress are reflected in a common neurobiology—particularly the activation of the locus ceruleus, the brain's arousal center that lies deep in the lizard brain, and the pituitary–adrenal axis regulated by the hypothalamus. The messengers produced by these systems have important responsibilities within the neuronal centers of the limbic alliance, particularly those in the hypothalamus, hippocampus, and amygdala, all of which are involved in the expression of emotion. It is the neurons of the locus ceruleus that produce most of the brain's norepinephrine, which together with serotonin is thought to be disturbed in anxiety and depression. For example, in a monkey otherwise going about its daily business, electrical stimulation of the locus ceruleus will result in the animal's becoming intensely anxious and vigilant. It will retreat to a safe corner—behavior very similar to that seen in the fear and anxiety of acute stress, and which under chronic conditions can lead to adaptive failure and depression.

It will come as no surprise, given this evidence, that many experimental models attempting to simulate the neurobiology of human depression are built upon the paradigm of chronic stress. The learned helplessness model of Seligman, the tests used in the screening and development of new antidepressant drugs where rats are forced to swim in a bath of water, and infant–mother separation in primates all utilize the paradigm of stress. They have in common the exposure of an animal to adversity that is inescapable and uncontrollable. In each instance, alarm and adaptive resistance is precipitated but because there is no escape—neither flight nor fight is possible—the acute physiological changes of reactive homeostasis extend rapidly to the phase of exhaustion and behavioral withdrawal. The physiology of many of these mammalian stress models is remarkably similar to human depression, and furthermore most are reversed or improved by antidepressant drugs.

Our technical ability to investigate the cellular and molecular systems of the brain, in animals and other creatures, has exploded in the last decades of the twentieth century. These advances have made possible a detailed understanding of the normal regulatory systems of the neuron, how communication occurs, and how it is altered by pharmacology. Building bridges between these experiments and the human experience of depression has been difficult and imperfect. But without such attempts, the application of what we learn about brain anatomy and chemistry from animals, in explaining human disorder, is diminished. The limbic brain, deep in the protective vault of the skull, does not reveal its secrets easily, even in the era of molecular neuroscience and brain imaging. The brain is protected by the blood–brain barrier, and the general chemistry of the body does not directly reflect the dynamics of brain chemistry. Norepinephrine, for example, is manufactured in large quantities outside the brain by the adrenal gland, especially during stress, and is an important messenger in the autonomic nervous system. Hence, merely measuring this emergency hormone in the bloodstream or urine of those who are depressed, or undergoing stress, will tell us little about brain activity.

A partial solution to this dilemma has been to make use of the endocrine hormones as a sort of telescope, or amplifying device, to explore how limbic regulation changes in depression and under adverse circumstances. Although several hormones, including norepinephrine and those hormones produced by the thyroid gland, change during stressful periods a particular focus of research has been the hypothalamic-pituitary-adrenal system and its emergency messenger, cortisol. This was the hormonal system that Selye and others first studied, and it has remained important, especially in the clinical investigation of melancholia. The actions of hydroxycorticosteroids, of which cortisol is a family member, are essential to life. Because they influence many cellular and neuronal processes, including the regulation of fundamental genetic programs within the cell nucleus, they play a crucial role in adaptation, as Selye had initially proposed.

Cortisol manufacture is homeostatically controlled, switched on and off by cortisol in the blood "feeding back" information to the pituitary gland and the centers of the limbic brain, particularly the amygdala and hippocampus. A series of events that begins in the limbic brain is responsible for the production of cortisol. Initiated in the hypothalamus, the brain messenger corticotrophin-releasing hormone (CRH) signals the pituitary gland to release another hormone, corticotrophin (ACTH), which in turn stimulates the secretion of cortisol by the adrenals, small glands that sit on top of each kidney. The pituitary acts as the production manager, regulating the adrenal gland's activity by adjusting the delivery of ACTH according to the concentration of cortisol in the blood. However, the important decisions that determine the sensitivity and set point of the system are made higher up, by the amygdala and hypothalamic centers of the emotional brain. This dual feedback facilitates both reactive and predictive homeostasis. If feedback and regulation occurred only at the pituitary, it would be difficult to accommodate the changing daily and seasonal rhythms, and other long-term considerations, while still remaining sensitive to emergencies. Hence a graph of daily cortisol production has a distinct circadian rhythm with superimposed "spikes" of transient stress that mark the ebb and flow of everyday challenge.

This exquisite orchestration of regulated feedback becomes disorganized and imprecise during periods of melancholia. In many patients, particularly those with agitation and anguish, the circadian profile of cortisol release is flattened so that the usual decline during the early part of the night is lost, while the total production of cortisol over twenty-four hours is considerably increased. There seems little doubt that in depression the adrenal cortex is working overtime; indeed, in some individuals actual enlargement of the gland can be seen on X ray. Dexamethasone, a synthetic type of cortisol that in normal persons fools the pituitary into shutting off ACTH, thus decreasing cortisol production, fails to do so in about 60 percent of those severely depressed, suggesting that regulation of the system is also impaired. Further investigation, using the tools of modern neuroscience, reveals that in some depressed patients production of the hypothalamic hormone CRH (measured in the cerebrospinal fluid that bathes the brain) is increased, suggesting that the limbic centers regulating emotional behavior and the stress response are in a state of chronic arousal, which in turn is driving the excessive production of cortisol hormone.

The metabolism of thyroid hormones is also profoundly changed by stressful circumstances, and during depression. Thyroxin, the principal hormone produced by the thyroid gland, and its more active partner, tri-iodothyronine, are important in sustaining the energy of all the body organs, including the brain. These hormones are regulated through the hypothalamus and, like the steroids, have powerful effects upon behavior. During infancy, should a child be deprived of thyroid hormones, mental retardation results, and in adult life a hypothyroid state is commonly associated with depression. Indeed, considerable evidence suggests that thyroid hormones are essential to normal emotional function, especially under stressful conditions.

However, from a comparison of animal and human research we know there is nothing unique about the fundamental neurobiology of human stress. Once triggered, our physiological reaction to unforeseen threat is little different from that of the young bird when the eagle's shadow falls over the nest. Thus we are faced with a critical question. What perpetuates the state of stress and chronic endocrine arousal that in vulnerable individuals can kindle human mood disorder? Obviously, in those individuals where there is a strong family history of unipolar or bipolar illness, some regulatory impairment of neuronal arousal may contribute to the explanation. However, from studies in primates and other mammals—including ourselves—we know that another critical variable is the level of psychological uncertainty that is experienced—the novelty of any situation combined with the degree of control that we feel we can exercise over it.

While the biology of reactive homeostasis is embedded deep in the reptilian fiber of our being, as the human brain has evolved the variety of situations that can initiate stress and foster its chronicity has increased exponentially. The limbic, or old mammalian brain, you will recall, is the social brain tuned to attachment and social interaction. In ourselves, with the growth of the new mammalian cortex, there has emerged a triune brain vested with an extraordinary power to learn and manipulate abstract information. These are evolutionary changes of enormous significance in determining what human beings perceive as uncertain or beyond control. Challenge is no longer the hovering of eagles, but something triggered by a rich imagination and personal memories. These abstract forms for many individuals cast shadows equally as terrifying as the presence of a primordial beast. What is remarkable in the human experience of stress is not the linking neurobiology, for that is in keeping with our primitive ancestry, but the myriad ways in which learning, experience, and social attachment can initiate the physiological arousal that leads to stress.

Thus, for Claire Dubois, the disappointment of a canceled year in Paris was sufficient to kindle long-standing difficulties with self-esteem and the feeling that she had lost control over her destiny. It was this existential crisis, emerging from the shadows of the past, that had initiated the stress that led to her social withdrawal, drinking, and ultimately to her severe depression. As you will learn from the details of his story, John Moorehead struggled with a different set of demons, within himself and his beloved Jesuit community, than did Claire Dubois. But in the stress that was kindled, the impact of their experience had been similar. Under chronic strain the regulatory mechanisms of reactive homeostasis had been driven beyond balance until John, genetically predisposed to mood disorder, had fallen into a state of melancholic withdrawal.

John Moorehead, the provost at a small Jesuit college in Massachusetts, first had consulted with me, as I have said, upon the urging of Angela, his sister. While visiting him, Angela had become alarmed about his agitation and suicidal preoccupation. During the initial interview, John had avoided my direct questions and was mysterious about the events leading up to his depression. I had not pressed him at the time, for my primary goal had been to gain his confidence, such that he accept some care and professional advice—no easy matter, considering his suspicion of psychiatrists. On the surface, as John explained it, his switch from energy and optimism to anxiety and depression had been triggered simply by too little sleep and a long, exhausting flight from Italy—plus perhaps a bit too much to drink. But something told me, even on the first visit, that there was more. There was something almost Faustian in the way John explained his fall from grace. What twisted hand of fate, I found myself wondering, had transformed John's considerable personal achievement into such a serious depression? Over the next few days, as he found greater trust and confidence in me, the answer became clear. No single event had generated the turmoil, but rather a chain of repeated stressors. Self-doubt, born of conflicted intellectual purpose, and a dramatic shift in his position in the social hierarchy, together with the loss of a trusted friendship, were the stressful life circumstances that had kindled John's vulnerability to melancholia. The visit to Rome had been merely the final straw.

John had traveled to Rome to speak at a Vatican conference, held just before Thanksgiving, some eight months prior to our first meeting. From his description, I gathered it was a prestigious invitation to a symposium exploring a topic which, although obscure to me, had been much on John's mind—the future relationship between the traditional responsibilities of the Jesuit order and the faculty of a modern university. As John explained it, during the 1970s, driven by changing economic times, this was a discussion that had preoccupied Jesuit communities all across America. To remain competitive with nonsectarian institutions for outstanding students, it seemed imperative to many Jesuit scholars, John among them, that the Jesuit colleges must aggressively recruit the best professors available, without regard to their religious persuasion. Many senior members of the established Jesuit order, including those at John's own college, opposed such developments, however, as a threat to their community and the traditional Jesuit education.

Rather dry stuff, I thought to myself, given the troubles of the modern world. But, as John was quick to emphasize, for a highly disciplined organization such as the Jesuits, which had been in existence for almost 500 years and which was devoted to spiritual training and scholarship, these were debates of great emotional and intellectual significance. The issue was particularly urgent at the small undergraduate college where John was a professor. The institution was facing financial difficulty; student applications were down, and some trustees believed that new professors and a broader curriculum would help. John described how initially he had tried to stand aside from the debate, concentrating on his own teaching and writing in medieval Italian history, a subject in which he was a highly respected scholar. "But it was impossible," he said. "Eventually everyone in our small community, like it or not, was caught up in the discussions. We were all very gentle at first. The arguments for and against 'opening' the faculty were neatly abstract and very academic but then, as tempers became heated, opposing camps developed; close friends of many years were suddenly rivals, in fact worse than that; we behaved toward each other as archenemies."

I remember, when John first described this situation to me, how he had clothed it in typical academic regalia—offering considerable detail, little emotion, and an erudite analysis of the intellectual merits of each opposing argument. As I sat listening, somewhat mesmerized by his extraordinary verbal facility, it was obvious to me that John was desperately trying to bury something under the very weight of his words. Finally I interrupted him to inquire about his own participation in the feud and noticed his face flush as he looked away. After a silence he said quietly, staring into space, "I was on the wrong side; I can see that now. I was the leader of the opposition, the champion of those who favored change. It was a mistake, the worst mistake of my life, but that was how I became provost. Accepting that position was a sin of pride."

Intrigued, I pressed home my questioning, ignoring for the moment his obvious emotion. The puzzle was beginning to fall into place. Instinctively I knew that close to the situation he was describing lay much anxiety, perhaps some deep personal sadness or a broken attachment that would explain his sudden lapse into self-depreciation. "If you were the leader of the opposition, who were you opposing?" I asked. "None other than the rector," came the immediate reply. I confessed my ignorance regarding the leadership structure in a Jesuit college and, recovering his composure, John explained.

As rector, Father Frank Herrington had been for many years both the respected leader of the Jesuit community and the administrative head of the college. This was a man long admired as a mentor and trusted friend. "He was almost a father to me during my early years, always encouraging in his rough way," John said. "Before this happened, we were very close." Here then, I thought to myself, was probably the important attachment that lay at the center of his sadness. I nodded in understanding and John went on with his story. "Our differences were extremely painful to me, especially when they became part of the public debate, but at the time I was convinced that I was right. I was certain that the ultimate survival of the college lay in an administrative separation of secular and academic purpose. The creeping reality of the world demanded it. Without such a separation I believed that the intellectual strength of the college and the student body—perhaps even that of our own Jesuit community—would lose vitality and eventually die."

John was soon back to his refuge of intellectual details, speaking in a monotone and staring at his hands, which were tightly clenched together in his lap. He described how public arguments had flared sporadically throughout the spring until, concerned over rising tensions, the trustees took action. The faculty would be opened, they declared. Father Herrington would remain rector and head of the Jesuit community, but in academic matters the school would answer to a provost. These two individuals would report to a president, who would be recruited after a search by the trustees. "Then," said John, "in a move that surprised everybody, the trustees asked me to be the first provost. They gave me twenty-four hours to decide, ostensibly because the announcement had to be made before the students left for the summer. But I think they just wanted to avoid any more trouble. Frank Herrington was away and by the time he returned the deed was done. I tried to see him and explain what had happened, but he refused. We haven't spoken since except in the course of our official duties." John's wrists and knuckles cracked as he stretched his hands, still intertwined, and looked at me. As he spoke, he grimaced. "As the old saying goes, I had won the battle and lost the war."

When I recall John Moorehead's bureaucratic struggles, they always remind me that human beings, despite our unique intelligence, remain intensely social creatures. For each one of us, social interdependence is at the center of life's ebb and flow. Our daily challenges and rewards remain variations upon basic themes of attachment, loss, threat, opportunity, loyalty, and hierarchy—mammalian social behaviors that evolved with the limbic brain. However, because the human brain is infinitely programmable compared to other large brains, through technology and planning we have developed the most complex social civilization of any animal. In consequence, the social situation in which we each live is the most potent variable in the development and resolution of stress, and a powerful emotional modulator. Attachment is a particularly potent modifier. A close bond with a parent, lover, or friend brings a feeling of security and peace, while separation and social isolation increases anxiety and withdrawal. Life as a complex social being is difficult, a truth underscored by the insatiable public appetite for books that promise a formula for social success and personal happiness.

For John Moorehead, in the wake of his appointment to provost, community life rapidly moved beyond difficulty to something unbearable. He was, as he put it, a man "first hoist with his own petard, then tarred and feathered as a social outcast." His impassioned and well-intentioned arguments defending change as necessary for the survival of the Jesuit community, others now held against him as Machiavellian instruments of personal greed, cunningly designed to ensure his own academic advance. In September, at the start of the new academic year, John found himself increasingly isolated from his former colleagues, even those who had once championed him as a leader. As the stress increased, John's usual energy and exuberance disappeared; his ideas came slowly. Some mornings he would sit in his new office in the administrative building just staring at the barren wall. Returning to the cloister each evening—the part of the campus where the Jesuit faculty lived in community—John found himself eating alone. Although an officially designated leader, John Moorehead was left in little doubt from the behavior of those around him that he was a diminished man. Despite his new title of provost he had fallen, not risen, in the social hierarchy and was rapidly becoming a victim of socially generated stress.

We have learned much about this interaction of changing social status and the stress it induces from studying other primates who, like ourselves, live in hierarchical groups held together by strong bonds of mutual attachment. One species successful in a wide range of habitat, including the man-made villages and urban centers of India, is the Macaque family of "Old World" monkeys. These small animals—a rhesus-macaque male is only about two feet tall—have a hairless mobile face and a broad range of primary emotion that human beings have little trouble understanding. They take about four years to reach maturity and with good fortune have a life expectancy of another twenty—plenty of time for complex social development and the nurturing of young animals. Stephen Suomi, a psychologist of the

Laboratory of Comparative Ethology at the National Institute of Child Health and Human Development in the United States, has studied the physiology and behavior of these rhesus monkeys (with whom we share almost 100 percent of our genetic constitution) and has written extensively about their response to separation and the social stress of changing hierarchical position, both under natural conditions and in the laboratory.

In their natural habitat, several score macaques, of both sexes and all ages, usually live together in closely knit interdependence. The long-term social stability of the troop is provided by the females, and a smaller hierarchically organized group of mature males. From his research, Dr. Suomi has found that an animal's place in the social order, as in human societies, is a major determinant of the level of stress and neuroendocrine arousal. The position occupied by an individual monkey in the troop's hierarchy largely determines access to food, territory, and sexual partners, and thus a change in social order generates intense competition. Dominance and status is settled among both male and female contenders by threat, and physical combat if necessary. These conditions generate both acute and chronically stressful situations. Research has shown that during episodes of fighting and confrontation, as in the human stress response, levels of the arousal hormones epinephrine and norepinephrine rise in the blood together with the steroid hormones, particularly cortisol. However, as the outcome of the competition is decided, the neuroendocrine patterns in the blood of the victorious and vanquished animals significantly diverge.

For the victor in the struggle—now the dominant animal—there is a rapid reduction of the stress hormones circulating in the blood as the physiology of the body returns toward homeostasis. However, in the defeated individual, cortisol remains high, and may climb even higher if he or she remains the object of belligerent behavior. The sex hormones, growth hormone, and the emergency messengers of the autonomic nervous system (norepinephrine and epinephrine), however, slowly decline as the beaten animal withdraws from social activity. Such physiological changes, when placed in a broad evolutionary context, have obvious adaptive value for the endurance of the group as a whole. After the initial struggles for dominance, when a clear hierarchy has been established, a period of relative calm and cooperation usually develops within the troop, facilitating general survival. The beaten animal, having declined in rank, withdraws from social interaction, and exploratory behavior virtually ceases. Indeed, in some instances the vanquished monkey, fending alone for itself on the edge of the troop, becomes ill and eventually dies or is killed by predators.

I believe these primate studies serve to inform our own experience and emphasize the stress inherent in changing social fortunes. While I am mindful that far wiser men than I, Darwin among them, have lived to regret anthropomorphic leaps from monkey to man, I find the parallels compelling. Although the human social fabric is immeasurably more complex than that of any other primate society, the warp of the weave is similar. John Moorehead had challenged his leader and, in his own words, "lost the war." Now, vanquished, isolated from his social group and alienated from his mentor, with no clear path to reconciling the uncertainty and helplessness he experienced, John had become increasingly withdrawn. A lifelong sense of belonging had been severely shaken. The invitation to Rome, prestigious though it was, served only to fill his mind with further dread as he imagined senior colleagues, perhaps Herrington himself, questioning him from the audience. Under such strain—and probably already in a state of endocrine turmoil—John believed he had performed poorly at the Vatican conference. At the podium he had become acutely anxious, scanning the auditorium for Father Herrington's familiar face. He was not there. The audience received the presentation in silence, a mark of interest and attention to a healthier mind, but for John in his state of tension and distress, the very stillness had been an expression of disapproval.

After the formalities of the day were complete, as John walked alone across the Piazza San Pietro against a driving winter rain, the thought emerged in his mind that his new role endangered everything that he most loved and admired. Contrary to his earlier hopes and vision, the future now seemed bleak and lonely. Once formed, the idea persisted and on the plane, flying directly to California to celebrate Thanksgiving with his sister, he felt tired, old, and strangely restless. Hoping to calm himself, and perhaps to get some sleep, he drank several cocktails—more than his usual habit. It hadn't worked. By the time he reached San Diego, John was convinced he was developing influenza. "I never did get the flu," he told me, "but looking back I think the discomfort I experienced over Thanksgiving marked the beginning of my depression. I didn't feel sad, just a sense of distance mixed with a rising tension and an aching body, plus I couldn't sleep. At first I put that down to the nine-hour time difference between Rome and California. But the tension and insomnia never really went away. Then, around Christmas, as I began to realize the impossible task I faced at the college, my mood went down, and for the first time in my life I saw myself as a complete failure."

John's experience, where a decline in mood is not the first sign of oncoming depression, is not unique. In fact, some 5 or 10 percent of individuals never experience obvious sadness, but progress through a period of increasing anxiety and disorganization—which they find difficult to pin down as a mood disturbance—directly to anhedonia and the syndrome of melancholia. Sometimes it is sleep that is first impaired, or they find themselves increasingly tense and fatigued, with vague aches and pains similar to the early stages of flu. Philip Gold, a research psychiatrist and chief of the Clinical Neuroendocrinology Branch of the Intramural Research Program at the National Institute of Mental Health, has pointed out that many of these early depressive behaviors mimic the symptoms of chronic stress. Dr. Gold's observations are reinforced by research in animal models showing that, in rats for example, CRH not only initiates the release of the stress hormones of the adrenal cortex but also has direct effects upon the behaviors of the limbic brain.

Painfully distorted thoughts and feelings are so pervasive once melancholia is established that it is often difficult to look beyond the vivid and dramatic language that describes them. However, in the early phase of illness, when primarily the basic housekeeping functions and motivation are disturbed—and verbal descriptions of complex emotion are less pertinent—emerging symptoms are more reasonably cross-referenced to animal models. Decreased motivation to eat and diminished sexual activity, which are characteristic of animals under stressful conditions, are reproduced when CRH is injected in small doses directly into the brain; with larger quantities there is a reduction in sleep and a decrease in the usual exploratory behavior, suggesting increased vigilance and anxiety. Thus, the tension, insomnia, and flulike symptoms John Moorehead had experienced following his trip to Italy

I considered a reflection of increased stress hormone release, and I so explained them to him. For months, as events at the college unfolded, I suggested, he had been under chronic strain. The turmoil surrounding the visit to Rome, plus the demanding flight schedule and the associated jet lag, were together the culminating stressful episode that had tipped the physiological balance and, playing upon an inherited vulnerability to mood disorder, had kindled a melancholic illness.

The details of John's story that I have recounted here were actually gathered over several days, while he was a voluntary patient in a series of clinical research studies. His participation had grown initially out of compromise. John had made it clear to me during his first visit, kept only out of obligation to his sister, that he feared being hospitalized as further evidence that he was losing control of his life. But at the end of the consultation I explained that I too had fears, especially regarding his nihilism and morbid preoccupation. In order to provide thoughtful professional advice, I needed to understand his illness, and the stress that he had been going through, in greater detail. Realizing, despite the possible danger of suicide, that I could not justify hospitalizing John against his will, I explained some of our ongoing research into melancholia, hoping he would participate, and thus accept care. Fortunately, after a lengthy discussion he agreed to evaluation in the hospital's clinical research center. For each of us it was both an opportunity and a compromise; it appealed to John's intellectual curiosity and concern for helping others, while affording me the chance to further evaluate his melancholia.

John was an entirely appropriate subject for the research we were conducting. We were curious whether the hormonal disturbances of steroid and thyroid metabolism that can be measured in the circulating blood during depression had any usefulness in predicting the outcome of antidepressant drug treatment. And I was also interested in whether the addition of thyroid hormones to antidepressant drug treatment could speed recovery for some individuals. The study required both male and female patients, and after the initial evaluation everybody received a standard dose of the antidepressant imipramine for six weeks, plus either thyroid hormone or a placebo. The research assessment involved objective behavioral tests, such as the Hamilton Scale for Depression, and a detailed assessment of family history and any previous illness. After two days, during which the patient acclimatized to the unit and its procedures, we also measured the circadian profile of cortisol and thyroid hormone production, taking tiny blood samples drawn every thirty minutes. This part of the study was designed to provide a general measure of endocrine arousal and was followed by various endocrine challenge tests to determine the sensitivity of the hypothalamic-pituitary-steroid and thyroid systems to feedback regulation.

The behavioral assessment confirmed my clinical examination that John was in the grip of a severe melancholia. Furthermore, his family history and hyperthymic temperament suggested that he was predisposed to bipolar disorder. Later, when our laboratory had analyzed the blood samples we had collected, we found changes reflective of his melancholic state. The levels of thyroid hormone circulating in John's bloodstream were high, a common finding during stress and severe depression. Cortisol also was increased throughout the day with a loss of the characteristic circadian profile, and, furthermore, dexamethasone failed to suppress this increase in cortisol production, confirming that the usually precise hypothalamic regulation of steroid production was impaired. I have seen these abnormal profiles of cortisol and thyroid hormone production in many individuals with melancholia—indeed, they are among the most consistent endocrine changes found in depressive illness, both bipolar and unipolar—but there was something that struck me as odd about the shape of the cortisol curve in John's record. We had collected profiles over five days but the fourth was dramatically different from the others, with several large spikes of hormone production in the late afternoon and evening. At first we were perplexed, for these findings were not explained by any of our procedures or research investigations. But then I remembered that it had been on the fourth day that John's sister, Angela, had visited, and the three of us together had discussed in detail the personal conflicts and social upheaval with which John had been struggling. The explanation for the cortisol spikes was suddenly obvious: Reflected in the profile of stress hormones excreted that day was the emotional turmoil that John had experienced during our meeting.

Work in the late 1960s by the psychiatrist Edward Sachar, a pioneer in the study of the endocrinology of depression, clearly established this intimate relationship between endocrine arousal and the psychological pain experienced during the acute phase of illness. He was one of the first researchers to interpret both disturbances as a reflection of central limbic dysfunction. "The hypersecretion of cortisol in certain depressive illnesses," wrote Dr. Sachar in 1973, "may not be simply a stress response as such responses are usually envisioned . . . but rather another reflection of apparent limbic system dysfunction, along with disturbances in mood, affect, appetite, sleep, aggressive and sexual drives, and autonomic nervous system activity." He had noted that endocrine arousal (measured in his first studies by the daily excretion of corticosteroids in the urine) was particularly evident when a patient in psychotherapy acknowledged the pain and sadness of a broken and irretrievable relationship, an anxious state comparable to the ill-defined fear that accompanies the death of a loved family member or a friend.

For John Moorehead, this feeling of bereavement and loss of control had played an important role in the development of his melancholia. Although John had found it difficult to verbally acknowledge his pain during our discussions, that pain was reflected in the increased levels of the stress hormone cortisol that we had detected in his bloodstream during the research studies. Driving the feelings of loss and sadness was the fractured relationship with Father Herrington, which for John—indeed for both, I learned later—was a primary attachment. The intellectual disagreements that had estranged the two, and later the public alienation of John following the precipitous appointment to provost, had brought profound sorrow to both men, damaging their long-established bond of mutual affection and loyalty. As John freely admitted, Frank Herrington for many years had nurtured his career as a loving father nurtures a son.

Bonds of attachment and loyalty lie at the core of our humanity and social organisation. For each of us, in infancy, the primary social relationship is the attachment we have with our parents, particularly the individual who provides the essential mothering, and this early bond commonly shapes our behavior and attachment to others later in life. We know from careful research that when these primary bonds are damaged or frustrated, normal development is profoundly disturbed. One of the first descriptions of this was by the London psychoanalyst René Spitz, who in 1947 compared two groups of children isolated from their mothers—orphans who were living in a group home and cared for by nurses, and children whose mothers were in prison. Although both groups received adequate nutrition and hygiene, there were profound differences in their emotional care. The children of the prisoners were lavished with affection during the appointed hours that they met with their mothers, while those in the foundling hospital were rarely picked up or held by the nurses, each of whom was caring for some seven children. Over time, these two groups of infants became profoundly different in the way they engaged the world. While in the earliest months their behavior had been similar, by the age of one year those in the foundling home were withdrawn, showed little spontaneous smiling or curiosity, and in general suffered more illness. By the second year the differences were even more striking. The children of the prisoners were talking and actively engaged with those around them, while only two of the twenty-six orphaned children could say more than a few words. This catastrophic retardation of normal development Dr. Spitz called *anaclitic depression*.

Research with young monkeys parallels these findings in humans. In the first days after birth the infant monkey clings to its mother for both warmth and protection, even when not suckling. This intense bond of affiliation is fostered by the hormone oxytocin, produced in the hypothalamus of the mother during delivery and responsible for the letdown of her milk. A similar release of oxytocin occurs in the infant as suckling begins, further reinforcing the attachment. Although they grow rapidly in physical independence, most young monkeys prefer to be in direct contact with their mother or in close proximity, running to her at the first alarm, and any involuntary separation precipitates a high-pitched protest that most human parents easily identify.

In the young rhesus monkey, physical isolation for a period of six months permanently impairs behavior later in life and predisposes to withdrawal and depression. Animals separated from their mothers, but able to touch them through wire netting, are much less disturbed than those who can see the mother but are separated by a plexiglass screen. (Under such conditions, infants have less distress when the mother completely disappears, suggesting that the old aphorism "Out of sight, out of mind" may accurately reflect physiological adaptation.) However, further laboratory studies in the rhesus monkey have shown that if infants are reared by their mother together with peers, the removal of the mother does not generate the same rapid increase in the stress hormones that one sees in an isolated infant.

One conclusion from this is that social bonds with peers can extend, or even sometimes replace, close maternal attachment; in their absence, parental loss definitely causes greater disruption of behavior. This value of a peer group is a simple example of what Seymour Levine, a psychologist and for many years a professor at Stanford University in California, has called "social buffering"—the power of social relationships to modulate an individual's adaptation and buffer physiological arousal. Professor Levine has suggested that uncertainty—essentially a diminished sense of personal safety—is the common denominator of the many psychological and social variables that drive endocrine arousal during stress. A novel situation, because the true nature of the challenge is unknown, generates uncertainty, as will familiar events if they are extremely threatening and beyond control. Strong attachment bonds, especially to dominant members of the community, and social networks among peers reduce this uncertainty by providing social support and potential assistance.

Although neither John Moorehead nor his sister knew the details, what they did know, and which was pertinent to his vulnerability, was that their mother, Josephine, had suffered a period of depression after John was born. Cyclothymic and irritable by temperament, Josephine had retreated to her bedroom for several months and John's care had been provided largely by his grandmother. What scars, if any, this particular episode inflicted is difficult to tell, but Josephine's capricious behavior in subsequent years, together with John's father's absence in the Merchant Marine, may have sensitized him to repeated loss and made peer relationships especially important during his formative years. Fortunately, he was ebullient by temperament, intensely curious and popular in school—indeed a natural leader from an early age. This positive way of engaging the world had stayed with him during adolescence and his student years at the seminary, although interestingly he admitted to periods of "emotional slowdown"—perhaps minor depressions—immediately after leaving home and before he came under the parental tutelage of Father Herrington.

These early attachments and peer relationships during critical periods of emotional development are important because they help to "shape" the dynamics of reactive homeostasis and can subsequently influence the physiological stress response over a lifetime. This has been demonstrated by research across many mammalian species, from rats to primates, including ourselves. For example, rat pups deprived of maternal care during their first weeks have an increased sensitivity to stress as adults and a hyperactive cortisol response. Orphaned children with poor social support—such as those described by Spitz in the foundling home—also have been reported to have increased blood cortisol levels as adults. And the converse also pertains; the handling and careful nurturing of isolated baby rats can reduce the hypersensitive release of cortisol in maturity.

This intriguing line of scientific inquiry supports the argument of Dr. Robert Post that repeated psychosocial disruption may alter an individual's vulnerability to depression and anxiety through "kindling" arousal sensitivity. From studies in animals, we know that repeated experimental application of CRH, the peptide hormone that is released in the brain to initiate the cortisol stress response, can sensitize the amygdala to kindling. This is important, for it suggests that repeated stress—kindling further stress and arousal as does cocaine or repeated electrical stimulation—may have the capacity to modify the long-term genetic expression of behavior.

Stress may thus interact with temperament and increase the vulnerability to mood disorder, especially bipolar illness. Jerome Kagan's research, revealing that human infants differ widely in their response to the same challenge, strongly suggests that variations in individual temperament are encoded in the genome. Stephen Suomi's studies of young rhesus infants have an interesting resonance, identifying a group of animals at three to four weeks of age who are highly reactive to novel stimuli (the Piglet syndrome, as I have previously described it) and experience prolonged and severe elevations in blood cortisol and other stress hormones upon separation from parents and peers. A highly reactive infant appears frightened and clings to his mother when a novel object is introduced, and if alone will withdraw into a corner. At four to five years of age (the period of adolescence for the rhesus monkey), such animals when faced with similar uncertainty become hyperactive and agitated, running frenetically around the cage, rather than withdrawing. The hormonal stress response also remains abnormal compared to other animals of the same age. However, when the young monkey is secure and among peers, or with the parent, there is little detectable difference from the behavior of other animals, the agitation emerging only under stressful conditions. These patterns of behavior run in families—as does human temperament—and appear to "breed true" under stress. When these highly reactive infants have been reared by nonbiological parents, their behavior looks little different from that of the parenting family until they are isolated and frightened—then their response has more in common with the patterns characteristic of their biological lineage than that of the foster family.

Thus stress, genetic predisposition, learning, and social networks weave a dynamic tapestry in determining vulnerability to mood disorder. Usually it is impossible to identify a single factor as the incisive variable, for commonly all are necessary but each alone is insufficient to precipitate illness. This intertwining is readily apparent in the story of John Moorehead, where a man of unusual energy and talent, in apparent deviation from his path, had fallen victim to a set of circumstances uniquely capable of kindling his long-dormant liabilities. However, in truth, John had been deviant all his life—his distinction from the crowd had not begun with his melancholic illness. But his earlier distinction had been of a different sort and one we do not label as illness. John's extraordinary energy and short sleep, his capacity to interpret medieval history in insightful ways and to create images of social reform, had long set him apart as a man of inquisitive vision. These we applaud as a society, for they are the talents of leadership—and commonly the signature of a hyperthymic temperament.

Ever since anybody could remember, John had engaged the world with vigor; "a Tigger from the start" had been his sister's description of him at our first meeting. This, coupled with his curiosity and intelligence, seemingly had made it possible for John to ride above the turbulent moments of his childhood (in contrast to Angela, who suffered every glance from others as an intrusive wound), and from a distance others would have described him as a man who suffered little pain. But within himself John had been sensitized to loss—to his mother's moods and his father's absence—and as he grew older he assiduously avoided situations that might rekindle feelings of helplessness and bereavement. As he described after his recovery, in the recorded interview I have quoted from previously, "I had a distaste, no, almost a phobia, for ambiguity and uncertainty." When his sister's love life had verged on chaos in her younger years, it had worried him deeply, and angered him too, as something beyond his control. The Jesuit order and his life as a disciplined scholar had struck, as John put it, "just the right balance of challenge, routine, and a collegiate group of peers."

Frank Herrington had been the pivot for that life, not only for John but for the majority in the community, and when the intellectual tournaments began around the issue of "opening" the faculty, many had found themselves uncomfortable. "While most considered Frank to be wrongheaded on the faculty issue, nobody wanted to oppose him. It was some sort of oedipal thing," said John. "I was chosen in part because, as a favored and respected son, everybody thought that he would listen to me. Of course he didn't and when things fell apart and I made the mistake of accepting the provost's job I lost everything—including Frank's respect and most of my friends." John had lost both his anchoring attachment and the "social buffering" afforded by trusted peers, in one fell swoop. Within the tightly knit community of the college the general perception of him as a brilliant scholar but now as a self-serving and distant "leader" had garnered him envy and anger, but little emotional support.

It was under this constellation of uniquely personal stressors that John, with his vulnerability to bipolar mood, had swung from hyperthymia to depression—something that I have seen happen in similarly susceptible, and equally successful, men and women in midlife. Stress, no longer contained within the adaptive templates of the self, had disturbed the systems of limbic arousal to initiate the path to illness and melancholia.

The limbic brain—the old mammalian brain—is the organ of adaptation and survival, our defense against an uncertain world. The mechanisms of defense are skillfully layered (or perhaps more accurately, interlaced) one upon the other. Social networks buffer self-esteem through multiple attachments while psychological mechanisms guard the self and the body's internal environment from the disturbing effects of unnecessary stress. Attachment and peer support are essential to each of us as social animals, and these social relationships are shaped by experience and learning. Within the paradigm of homeostasis, attachment and peer support may be understood as one of several defensive social and psychological strategies that protect from the hyperarousal of chronic stress. Learning plays an essential role in shaping the mechanisms of adaptation, and the brain learns at every level of its organization—from the refinement and maintenance of appropriate social attachments to the molecular processes that sustain the industry of the individual neuron. The unwavering objective is a comfortable fit with the environment and control over immediate circumstance—to exploit opportunity and avoid harm. The greater the degree of psychological and social control an individual learns to exercise, the lower the neuroendocrine arousal. Thus in the aggregate, the limbic brain—the locus of the individual self—is a homeostatic system that seeks to shape the immediate environment, just as the environment reciprocally shapes the individual self.

In mood disorder, as the homeostatic mechanisms falter, this adaptive competence is replaced by the vicious circle of turmoil, arousal, and hopelessness. A series of perturbing events—perhaps personal loss, chemical challenge such as alcohol or cocaine, a transatlantic flight, chronic social difficulty—in overwhelming the regulatory mechanisms of defense, kindle stress and, if sustained, also result in exhaustion and adaptive failure. The victim, recognizing the impending loss of control, experiences escalating levels of subjective distress and an increasing disturbance of the general housekeeping functions of the self—concentration, sleep, appetite, and so forth. This pain and disorganization further heightens neuronal arousal and in an interplay with genetically and developmentally vulnerable neural systems in the amygdala and other limbic regions, initiates the pathway to mania or melancholia. As the illness deepens, the usually smooth communication among cells is impaired, with a further breakdown of behavioral coping. The development of misconceptions and inappropriate behaviors that result further distort social interaction, leading to stigma, loss of peer support, even greater turmoil, and rising hopelessness. So is the circle viciously perpetuated, and so it had been for John Moorehead.

John had been in the clinical research unit for approximately one week when Frank Herrington came to see him. The rector had learned of John's depression from Angela and had set out immediately for the hospital to aid in any way he could. I met him only briefly. He was just as John had described—gruff and large. "John's pact with the devil is behind us," he declared as he left my office. "He's too valuable a man to lose." Three days later, in the care of his sister, John left for Massachusetts. The initial evaluation phase of the research protocol was complete and also everybody agreed that John was much improved in mood. His suicidal rumination had disappeared and he considered his sleep more restful although he was still awakening early, as had been his lifelong habit. The plan was to conduct two evaluations by telephone over the next month, before he returned for follow-up and repeated endocrine testing. In the interim he would continue with the daily dose of antidepressant medication—150 milligrams of imipramine and the research capsule that contained either thyroid hormone or placebo.

That was the plan. However, within the month John reappeared, unexpectedly asking that his final assessment be brought forward as he was now completely recovered and wished to spend a little time with Angela in California before the start of the fall semester. He certainly did look well. His lean body, which had seemed so frail during his illness, now suggested an athletic competence; his face was mobile and expressive, and when he spoke it was quickly and to the point. He wanted to describe to me the details of what he had been through and how things had changed in recovery. "I'd be pleased to record it, to help others perhaps," he volunteered. I found myself troubled; was this man becoming hypomanic? Was the imipramine and perhaps the thyroid hormone switching him, as they are known to do sometimes in manic depression, into a hyperactive state? I explained my concerns and described some of the symptoms of mania—the racing thoughts, expansive ideas, and unbridled energy. "I don't think so," he replied after listening carefully. "I feel back to my normal self, but I will keep my eye on it." I asked for more details about what had been going on in his life, about the college and his plans for the coming year. Father Herrington, it seemed, was back in the picture. "Nothing is resolved, but we are talking again and that's what matters," said John. "We have great respect for each other, as you know, and I feel confident we will work it out." He appeared now as the leader his sister had described, again in control of his destiny.

With some lingering concerns I agreed to John's request, bringing the evaluation forward by five days. It was close enough, I told myself, and better than the alternative, which, as John made clear, would be after the beginning of the new semester. However, before he left two days later I gave him my telephone numbers and those of colleagues in San Diego and Boston—just in case.

I was not to see John Moorehead again. When, several months later, the study in which he had participated was complete and we analyzed the results, it was clear that his recovery had been swifter than average—even for the men, who as a group had fared better than the women. (With thyroid hormone added, however, women in the aggregate improved from their depression almost as quickly as the males—an interesting difference which possibly reflects the greater incidence of thyroid dysfunction that women suffer throughout life.) I was curious, and reviewed John's individual file, thinking that he might have received thyroxin, only to learn that in his case the "thyroid" capsule had been a placebo. We could not explain John's rapid recovery on the basis of our pills alone.

At about the same time, the postcard arrived. It was from Florence and carried a simple message. "I'm here writing a piece on Machiavelli," wrote John in a fine copperplate hand. "I resigned from being provost two months ago and with great relief have replaced a life of political intrigue with the study of one. I also remain well, you will be pleased to hear, but I have heeded your warnings and am now hooked up with a psychiatrist in Boston and intend to see her regularly when I get back." The photograph was of an attractive, sunlit corner of the San Marco Cloister, a renaissance convent which, John went on to explain, had housed the first public library.

John had returned to his academic roots. I found myself wondering about Father Herrington and the complex entanglement of culture and attachment that had led to John's melancholic episode—the same labyrinth that was now again his source of strength. The blind alleys and hidden opportunities of such a cultural maze are uniquely human and, as the germ of disorganizing stress, lie at the opposite pole from the raw biological challenge that had so swiftly struck at Melanie Branch, the young woman whose story I am about to tell.

Nine

Pills to Purge Melancholy

Neurons, Chemistry, and the Pharmacology of Mood

It wasn't until I was on medication that I could start dealing with the emotional issues. Psychotherapy has helped me tremendously, but had the therapy preceded the medication, I'd have been on the couch for the rest of my life.

> SUSAN DIME-MEENAN,
> Executive Director, National Depressive
> and Manic-Depressive Association
> *McCall's Magazine,* October 1994

I came forward because I was helped. I owe it to let people out there understand it [depression] can be treated. You can get better and it is not all that difficult if you hang in there.

> MIKE WALLACE,
> American journalist
> *Larry King Live,* Cable News Network
> April 22, 1996

The medical students stamped their feet in enthusiastic applause. Even as I rose to thank Melanie Branch for her special contribution to my lecture I could see eager individuals threading their way, past classmates and book bags, to the podium where she stood. It was a sight familiar to me. The first-year students in the neuroscience class, struggling for a fledgling understanding of mind and brain, invariably found Melanie's frank discussion of her personal fight with bipolar illness provocative and stimulating. It awakened in many of them a genuine intellectual curiosity about these disorders of the emotional self, and in others a deeper concern about the suffering of friends and family, and themselves.

Melanie Branch is one of a number of patients who have assisted me over the past decade with my teaching responsibilities at the University of Pennsylvania. She is a woman of courage and persistence, qualities not easily dismissed by those who cling to the stereotype of the mentally ill as individuals who have failed in life through a weakness of will—not even by medical students, suspicious of individual testimony as information apart from their regular "scientific" diet of body organs and molecular biology. These personal strengths have sustained Melanie through times of exhausting disability, including a manic psychosis after the birth of her only daughter, and they have carried her also beyond a sense of shame about her illness. Melanie is now an articulate advocate for better public education about manic depression.

A cluster of ardent students had gathered around Melanie. "Before you became ill, were you aware there was something wrong with you?" asked one. "How do you feel about taking lithium for the rest of your life?" inquired another. The gaggle of voices grew, each competing to be heard. One woman, somewhat older, wanted to know whether Melanie's moods affected her relationship with her daughter. "May I call you?" pleaded a tall, athletic-looking fellow. "I really admire your courage in talking to us. My mother was just diagnosed with bipolar illness." His pain and confusion were palpable. "We've all known there was something wrong, for years, and it would help me a lot to talk to you—to understand more." Without hesitation, Melanie agreed.

For me, remembering my own introduction to severe mental illness, such moments represent a small miracle. As a young student in England, on the cusp of deciding whether I would apply to medical school, I had worked one summer at a mental hospital, in a country place, close to where I had grown up. It was an experience I shall never forget, and one that stands in stark contrast to my pleasure, almost four decades later, in watching Melanie's spirited interchange with the students after our lecture together.

My memories of those I met during that summer as an orderly in North Wing, one of the hospital's several units for chronically ill men, remain unusually vivid: the elderly Jewish man standing in communion with his rituals, his aid to prayer—an electric junction box—strapped to his forehead; and "Don Quixote," extraordinary for his emaciation—a parchment white body confined to bed while he wandered in his demented mind, raging and windmilling his arms in gesticulation to an imaginary crowd. I can admit now to being frightened by what I saw—frightened and fascinated—especially by Wilfred, the deep-chested farmer of middle age, his full beard tinged with gray. Wilfred, in moments of Biblical rage, could lift a chair and break it on a table as matchwood, then at other times he would be quiet, withdrawn, and politely conversational.

Wilfred's repeated torment was my introduction to the ravages of manic-depressive disease. There was a regularity to his manias, a short cycle of perhaps two or three weeks, similar to what Melanie had experienced during the worst of her illness. In the days immediately before the nurses predicted an episode would begin, Wilfred was given opiates and extra doses of paraldehyde, then the standard sedative for manic excitement. But the confusion and rage still broke forth, together with all their tempestuous violence. Our efforts at treatment in those days were palliative at best. And worse than that, the necessary custodial containment of the disease had confined the man too. When I met him, Wilfred had spent more than a decade in the asylum with little hope of reprieve. Melanie, by contrast, has become her own custodian. She too has known manic psychosis and the doomsday voices of self-destruction; she has been in the hospital—for weeks on one occasion—but thanks to good medical care, improved pharmacology, her own tenacity, and the detailed knowledge of manic depression that she has acquired, she has retained control over her personal destiny.

Melanie, today, manages her mood disorder as adroitly as most individuals who suffer diabetes have learned to manage that illness.

Certainly it is true that stigmata of her disability remain, reminders of her struggle and evident to the educated eye. Her body is heavier than before, stimulated in its weight gain by lithium and the antidepressant medications that she takes to balance her mood, and the associated hand tremor, although mild, is irritating to her—especially in social situations and under stress. But, with diligent self-care and regular medical advice, she now lives a life of relative freedom from her chronic illness. Her daily existence is remarkable in being ordinary. Outgoing and warm as a person, she holds a responsible job in a children's day-care center, is a loving and beloved mother of a five-year-old daughter, and is a public advocate for those who suffer as she has done. In stark contrast to the tragedy that had enslaved Wilfred those many years ago, Melanie has broken the chains of her disability and is her own mistress.

This is the small miracle to which I bear personal witness, and of which Melanie's story is a valuable example: a revolution in the healing of mood disorder. In this chapter I shall explore the developments in psychiatry, and particularly the contributions from pharmacology, that have made it possible for Melanie and many who have known these afflictions to regain health and proceed with a life of meaning and social purpose.

The pharmacological treatment of mania and melancholia is a story intertwined with a revolution in understanding brain biology, and it is still unfolding. The serendipitous discoveries of the 1950s, and the subsequent scientific research, together have radically improved the treatment of these severe illnesses, until the rate of therapeutic success in mood disorder is comparable to that of any other specialty of medicine. With each fresh insight we have gathered new knowledge—about the inner workings of the neuron and about the mechanisms that regulate its behavior—and in turn this has stimulated the development of new drugs, more precise in their action, that have further enhanced the treatments available. But perhaps most important of all, these advances have brought into being a powerful alliance of those who suffer and those who care about that suffering. Today, after repeated onslaught, the medieval bastion of public ignorance that has surrounded these afflictions of the emotional self may finally be tumbling down. At least that is what I like to imagine when the students stamp their feet for Melanie Branch.

Melanie was in her early twenties when she became a patient of mine. Her first episode of illness had struck at lightning speed within a few days of the birth of Althea, her daughter. Althea was a child much welcomed and much loved from the beginning, which made Melanie's rapid disappearance into psychosis particularly difficult for the family to understand. A previous pregnancy had miscarried at thirteen weeks and this pregnancy, too, had been a delicate process with increasing blood pressure confining Melanie to bed during the final months. Then, amidst great family joy, Althea finally arrived, a healthy, snuggling, delightful brown-eyed bundle of seven pounds, and Melanie was very happy.

Perhaps too happy, as Melanie's mother, Gloria, wistfully recalled when we first discussed the events leading up to her daughter's illness. In the first days after Althea's birth Melanie seemed to be enjoying something almost beyond happiness. She had been radiant—indeed transcendent—in her pleasure, bubbling about the delights of motherhood and often too excited to sleep. Even when Althea was quiet Melanie was beside her, rocking the child and humming to herself. "At the moment of her greatest joy," Gloria told me, "she seemed intent on exhaustion." Gloria, who for many years had been a research technician at the medical school (although we had never met prior to her daughter's illness), knew these details because at the time of Althea's birth Melanie and her husband, Alvin Branch, were sharing her home, located close to the University of Pennsylvania campus. The couple had moved in after the earlier miscarriage, largely because Melanie felt better living with her mother after such a trauma. It was also a practical arrangement, providing mutual support for the whole family. Gloria, a widow, disliked living alone and Alvin, who drove for a long-distance moving and storage company, was frequently absent from the family for days on end. Indeed, as fate would have it, he was in Seattle on the night when Melanie's madness first declared itself.

When Melanie first started teaching with me, I had met with her, and with her mother, to review the historical details of Melanie's illness. During our interview Gloria described how in the early hours of the morning, exactly ten days after Melanie and Althea had arrived home from the hospital, she had been jerked from sleep by the high-pitched whine of an electrical motor—the distinct sound of a vacuum cleaner hard at work. Her first thought was that she had been dreaming, but the noise was persistent. She wondered briefly whether Alvin had unexpectedly returned—the green numerals of her bedside alarm clock registered 3:15 A.M.—but dismissed the idea, noting to herself that he would have enough sense to be sleeping at such an hour.

Then, against the background of the mechanical wail, Gloria became aware of the familiar tones of Melanie's voice, in loud lament. Rhyming snatches of hymnal prose were floating through the darkness. Gloria had raised her head from the pillow to listen more carefully. "Onward Christian soldiers . . . I'm marching as to war . . . He will go before me . . . before I go from here. Swing high . . . sing high . . . and then swing low . . . the soldiers too will go . . . so will I . . . all will die . . ." Realizing that these rhymes made no sense, Gloria was gripped by a sudden feeling of alarm. For her daughter, a person not given to religious excess, this was strange behavior. She climbed rapidly out of bed and stumbled downstairs. There, in the brilliance of a dining room lit by many lamps and candles, Gloria found herself blinking before a spectacle that she will never forget.

Melanie stood at the table, completely naked. Singing, apparently in joy but with tears running down her face, she was systematically wrapping her new-born child in strips of cloth that she had torn from her discarded nightgown. On the floor lay the abandoned vacuum cleaner, roaring to itself, a bizarre continuo to Melanie's strident song. Miraculously, despite the noise and swaddling, tiny Althea was sleeping quietly. "It must have been a mother's instinct," said Gloria. "Without thinking, I had silenced the vacuum cleaner and was hugging Melanie as hard as I could. It seemed the natural thing to do. But I was crying, too," she went on, "for in that moment I knew she had lost her reason."

Such a dramatic "loss of reason" following childbirth is given the special name of postpartum or puerperal psychosis. This mixture of intellectual confusion, manic excitement, and melancholy thought commonly signals oncoming bipolar illness—as it did for Melanie— triggered primarily by the acute hormonal stress of the body's changing endocrine balance upon the termination of pregnancy.

Profound hormonal changes follow normal childbirth. During pregnancy, the developmental union between a mother and her growing child is supported by the placenta, a highly vascular organ that develops from the fertilized egg and is clamped to the wall of the uterus. It is via the placenta, through the umbilical cord, that the fetus receives the nutrients essential to life and growth. By the end of nine months the massive quantities of the hormones necessary to sustain advanced pregnancy, and virtually all the sex steroids circulating in the mother's blood—including the estrogen family of female hormones and progesterone—are being manufactured by the placenta. A twofold increase above the usual, nonpregnant levels of sex-hormone production is commonplace. Then, suddenly, with the birth of the infant and the expulsion of the placenta, now appropriately called the afterbirth, the manufacturing source of these sex steroids is lost. Progesterone plummets first, a few days prior to delivery, followed immediately postpartum by a precipitous fall in the amount of estrogen circulating through the mother's body—and through the vital regulatory centers of the limbic brain.

Despite these stressful hormonal changes, in the general population severe puerperal psychosis is a rare disorder, occurring in no more than one or two mothers of every thousand children born. By contrast, what has been labeled the postpartum "blues"—emotional irritability, tearfulness, fatigue, and difficulty sleeping—is common, being experienced by almost 50 percent of women. This is especially true for those giving birth to their first child, where uncertainty about the role of motherhood, breastfeeding, and career and lifestyle changes can induce considerable emotional distress. Hence a woman who has just given birth to her first child is more likely to experience psychiatric symptoms than at any other time during her entire life. But most vulnerable to puerperal mania are those women who have suffered bipolar disorder, or who have a significant history of it in their family. These individuals face a risk some twenty to thirty times that of the general population.

Melanie Branch had such a family history of illness. As I learned from her mother, Melanie's father had been killed in the Vietnam War when a young man, before any propensity he may have had toward bipolar disorder was apparent. However, his elder brother had suffered repeated depressions requiring hospitalization and had been considered "wild" in his youth, having been addicted to cocaine—a frequent complication of mood disorder. Also raising my suspicion of a predisposition to manic depression in Melanie's family were the many stories told about the colorful exploits of her paternal grandfather, apparently an eccentric and successful funeral home operator, who had committed suicide in midlife.

Between 30 and 50 percent of pregnant women predisposed to bipolar illness develop postpartum mood disorder including, in some, the derangement of psychosis. The psychosis usually occurs early in the puerperium, within four to fourteen days of the baby's delivery, compared to the "blues," which appear later, with a predominance at six to twelve weeks postpartum. Also, in those with a bipolar disposition, episodes of psychosis commonly recur with subsequent pregnancies, again emphasizing the special vulnerability of this group of women. A fragmented circadian rhythm, with loss of sleep fostered by the anxiety of caring for an infant with no reliable sleep pattern of its own, helps dysregulate emotional homeostasis. As in Melanie's case, the insomnia, exhaustion, and general stress of the postpartum period destabilize nerve cell communication, compounding the precipitous drop in progesterone and estrogen levels, and triggering a rapid onset of manic excitement, intellectual confusion, and distorted perception. Emil Kraepelin, who in the 1890s first described this combination of confusion and manic symptoms, called it "delirious mania."

When delirious mania occurs in the immediate puerperium, in addition to confused thinking there is commonly an irritable mood with a mixture of excitement and morbid delusions, agitation, sexual preoccupation, and greatly increased physical activity. Women who suffer these mixed states frequently have little memory of the incoherence and strange behavior that overwhelm them, especially during the acute phase of the illness. Hence, today, Melanie has only an hazy recollection of the events that are indelibly imprinted in Gloria's memory. Melanie has no remembrance of how her mother guided her safe return, together with Althea, to the Hospital of the University of Pennsylvania and to the protection of the psychiatric intensive-care unit. Nor does she remember the four electroconvulsive therapies that were required to break the grip of her delirium in the days that followed.

The experience of acute psychosis is terrifying both for those who suffer and for those who witness that suffering. "Keep away! Don't touch me. You're the third one to tempt me." The resident psychiatrist's confidential notes, written at the time of Melanie's hospital admission, the record of her behavior and verbal outbursts, provide only a glimpse of the chaotic thoughts and images that must have been cartwheeling through Melanie's mind during the initial hours of her confinement. "You want me . . . you're after me . . . I can tell. Take me . . . you want my sex. I'm tingling . . . I feel the itching on my skin. Don't look at me! Alvin's away . . . he's in heaven. I'm going there soon. Get away from me. This bed's a scaffold for sex and death. You're evil . . . you want to kill me . . . sell my baby for dissection. I've embalmed her . . . you won't have her. God will take her . . . I heard it. Where is she? Where's Althea? Leave me alone. Let me out . . . "

Reading the resident's account of Melanie's jumbled speech, I can hear again her ringing cries and see the fear in her eyes. Voices had told Melanie to kill her child, before she herself was killed. She had confided in her mother about this knowledge as they clung to each other, there in the brilliance of the dining room. Such was the trust that Melanie had in Gloria that she had permitted herself to be dressed and, with the help of a neighbor, brought to the hospital, but amidst the strangeness and the bustle of the emergency room, her terror had multiplied. Now a tormented soul, traveling through a private world of distorted faces, she had begun screaming and biting, and again tearing off her clothes until for her own safety and that of the attendants trying to restrain her, she had been sedated. Later, when I examined Melanie in the psychiatric intensive care unit, she was hallucinating, a prisoner of vivid but imaginary sensations, and she was still frightened—very frightened—her eyes flicking wildly from my own countenance to that of her mother, in a desperate search for meaning.

What frightens those in the throes of psychosis is what frightens any of us in times of perceived danger—sudden movement, loud noise, and a mistaken understanding of the intention of others. However, in psychosis, perception of the world is distorted and capricious; vigilance is increased, as if the amygdala had abandoned their objective sentinel duties to find danger in every shadow. Under such circumstances even the most well-meaning overtures of care and assistance can bring a combative response, just as threat will precipitate aggression in a cornered animal.

But even in madness a vestige of the normal self remains, observing amidst the turmoil, and it is to this fugitive of objective reason that one must address oneself. "Hello, Mrs. Branch," I said, holding out my hand in greeting. "My name is Dr. Whybrow." For somebody as disturbed in thought as Melanie was that early morning these simple words are not frivolous details, as one may think, but important acts of caring. Even in madness we are each entitled to civility and an honest introduction from those who offer their healing prescriptions. Nodding in recognition of Gloria, who stood beside me, I explained to Melanie that I had already spoken to her mother and understood what had befallen her. It was clear, I told Melanie, that she was suffering an acute illness, driven by the changing chemistry of her body after the birth of her baby, which in turn had disturbed the balance of her mind and distorted her thinking. But she was safe now, I added, and in the hospital. Althea was safe, too, and nobody would harm her. Her mother would stay with her, and her husband, Alvin, who had been notified, would arrive very soon. In the meantime she needed sleep and I had prescribed some medicine to make that easier—a medicine that I hoped would also help to return some order to her thinking.

While I had been speaking, Gloria had moved to sit at Melanie's side, on the bed, holding her hand. Although obviously still distracted by her demonic hallucinations, Melanie seemed less distressed and I remember thinking how fortunate it was for all concerned that mother and daughter were secure in their mutual affection. Small miracles, even with the powerful medications now available, continue to draw upon old-fashioned trust, and the first step necessary in any healing relationship is a genuine alliance between doctor and patient—built by proxy, if necessary, upon some secure attachment that already exists. Hence, with Gloria's coaxing, in small, suspicious sips, Melanie swallowed the medicine I had prescribed for her and drifted slowly into the temporary respite of sleep.

The restorative elixir I had given Melanie is a powerful tranquilizing agent called haloperidol lactate. Haloperidol belies its innocent description in the pharmacology textbooks as a colorless, odorless, and tasteless liquid. As a treatment for agitation and acute psychosis it is a vast improvement upon the pungent ineffectiveness of the paraldehyde that Wilfred had been forced to swallow during his manic episodes those many years before. Haloperidol blocks the activity at the synapse of the neurotransmitter called dopamine, the brain messenger that in severe mania appears to dominate the pathways of limbic communication and to ferment the psychosis. Dopamine is also thought to be operating to excess in acute schizophrenia, helping to explain the overlap of symptoms—particularly the emotional agitation, distorted thinking, and fragmentation of speech—that commonly exists between the two illnesses.

Dopamine is one of the three principal neurotransmitters that balance the chemistry of mood. Norepinephrine, the second of the trio, and dopamine are both members of the catecholamine family because they are produced from tyrosine, an amino acid present in the normal diet. The third, serotonin, an indoleamine, is a close cousin and manufactured from another dietary amino acid called tryptophan. Dopamine, norepinephrine, and serotonin are collectively known as monoamines and, together with another monoamine—acetylcholine, which is more involved in muscle movement and thinking than in mood—they are the key brain messengers maintaining the flow of information across the synaptic junctions of the limbic alliance.

When, for whatever reason, the balance of these neurotransmitters is disturbed, emotional regulation becomes unstable and, in those individuals genetically vulnerable to bipolar disorder, the syndromes of melancholia and mania may develop. The antidepressants and many mood-altering drugs—including many that produce addiction, such as amphetamines or cocaine—similarly achieve their behavioral effect by blocking or mimicking neurotransmitter activity and altering the messenger balance at the synapse. Thus the pharmacology of the different drugs used in the treatment of mood disorder is intimately entwined with what happens within the chemical factory of the neuron and the homeostatic mechanisms that regulate the synapse. To understand the pharmacology of mood, therefore, and why psychiatrists prescribe the drugs they do, it is essential to understand the workings of the neuron, and vice versa.

The synapses, you will recall, are the connecting links between individual neurons, the sites of communication on the surface of the dendrites and the axons of the neuron. These are the brain's decision points, where neurons—the thinking cells of the brain—choose to say yes or no, depending upon their collective wisdom. It is the synapse, and what happens there, that converts the brain from a predictable, dull, and deterministic organ into a flexible, interesting, and adaptable system where choices are made. Intelligence and abstract reasoning—behaviors at the root of our humanity—are synapse dependent, and so too is the complexity of human emotion and the moods we experience.

In the late 1950s, when Stephan Szabo and I were in medical school, the science of mind-changing drugs (with the daunting name of neuropsychopharmacology) and the discipline of neurobiology were both in their infancy. Indeed, although it had been known since the early part of the century that communication points existed between neurons, only in the previous decade had researchers finally accepted that synaptic transmission was neurochemical rather than electrical. Little was known about the molecular complexity of the neuron as a chemical factory, its internal economy, the behavioral purpose of the neurotransmitters, their manufacture, or the precision of the self-regulatory and recycling systems that preserve homeostasis and make possible precise communication at the synapse. Virtually all of our knowledge about these things has emerged during the short span of my own professional lifetime.

What was understood when Stephan and I were in medical school, and what we learned in some detail, is that neurons are cells modified to maximize communication through the transmission of electrical impulses over long distances, with great speed and efficiency. The neuron achieves this by functioning as a small battery and maintaining an electrical charge (the action potential) across its outer wall. Upon stimulation and discharge of this action potential, information passes as a wave of electricity down the axon—the superhighway of the neuron—to the synapse, and after the impulse has passed, the neuron resets the ionic charge in preparation for transmitting the next signal. From the study of the nerves controlling the muscles of the body it was recognized that the electrical impulse triggered the release of chemical "neurotransmitters" to carry information across the synapse, but little was known about the function of these messengers or their distribution in the brain.

However, during those years we were at the threshold of dramatic change. Powerful new technologies made possible the investigation of brain chemistry in laboratory animals, and for the first time we could explore directly the biological activity of drugs that change behavior. Discoveries were also being made that would revolutionize the pharmacological treatment of serious psychiatric disorder. Two drugs still in use today, chlorpromazine, found to be effective in schizophrenia, and imipramine, for the treatment of depression, were discovered in the 1950s through a combination of luck and careful observation. Two others were already in development and struggling to achieve general acceptance. In 1949, lithium carbonate, a common salt, had been reported as an effective treatment for manic excitement. Then, a few years later, drugs inhibiting the monoamine-oxidase enzymes that destroy messenger amines in the body, and which were already in wide use for the control of tuberculosis, were found to have antidepressant properties.

These discoveries stimulated a renewed scientific interest in the "mind-expanding" psychedelic drugs. Since the beginning of recorded history various substances, largely derived from plants, have been known to alter emotion and behavior. Some even mimic psychosis for short periods of time, presumably by transiently disturbing brain metabolism. In the early 1960s I was among the many young scientists intrigued by these ancient herbal potions. Indeed, the first professional article that I wrote during my medical school career was about peyote, the spineless cactus that grows in Mexico and the southwestern United States. Made famous by Aldous Huxley in his book *The Doors of Perception,* peyote was once used in religious ritual by a number of Indian tribes. Mescaline, the active ingredient of peyote, can change an individual's perception of the world, stimulating grand illusions and exalted feelings of power not dissimilar from those occurring in early manic excitement.

This was an experience that the Indians had known and explored for centuries. In my own sheltered English consciousness, what gave their manic-like ritual special significance was a scientific report that mescaline had a similar chemical structure to norepinephrine, one of the three principal neurotransmitters that balance mood. Another sacred plant of the Central American Indians, long favored for its hallucinogenic power, was the little mushroom *Psilocybe mexicana*. Here too, after careful chemical analysis, the active ingredient, appropriately named psilocybin, was discovered to be another monoamine, this time of the indoleamine family, and thus similar in structure to the neuro-transmitter serotonin.

Looking back, I should not have been surprised by these connections between the vegetable world, the chemical messengers of the brain, and the experience of mind. After all, nature is inherently conservative. Some hormonal and brain messengers in the animal kingdom are modifications of molecules that are commonly found in plants. Indeed, the medicines employed by physicians in the nineteenth century were derived largely from an accumulated knowledge of the restorative properties upon the mind and body of various flora and the substances extracted from them. Hence it is not entirely by accident that the human race has discovered the psychic power of the cactus and the mushroom, and the emotional effects—often with pain and illness as a consequence—of other natural intoxicants, such as alcohol, tobacco, opium, and cocaine.

As laboratory techniques improved, the pharmacology of many of these mind-altering substances was revisited and the chemistry of mood became an important area for medical research, spurred on by a quest to understand how the synthetic antidepressants, like imipramine, achieved their healing. Psychiatrists also began to pay closer attention to emotional changes induced by commonly prescribed drugs. Reserpine, for example, which had been extracted in 1952 from a scrubby plant popularly known as the Indian snakeroot, was used briefly in mania as a calming agent, but more extensively in the treatment of high blood pressure. But it was soon discovered that 15 percent, or more, of those receiving reserpine developed major depressive illness and, furthermore, the individuals so afflicted frequently had a history of depression in their family. This raised the possibility that the drug was triggering some established biological vulnerability for mood disorder. Then from laboratory experiments came the news that reserpine *reduced* brain concentrations of norepinephrine and serotonin, both important messengers at the synapse. Monoamine oxidase inhibitors, on the other hand, the drugs effective in the treatment of tuberculosis that had been reported to elevate mood, were thought to have exactly the opposite biochemical effect. These drugs *increased* norepinephrine and serotonin at the synapse by inhibiting an enzyme responsible for the destruction of the monoamine messengers in the cell body of the neuron.

Reports of these early studies eventually reached the corner of the library at University College Hospital in London where I sat pondering the behavioral similarities between mescaline-induced psychosis and the syndrome of mania. The reports provided a fascinating link between my growing clinical experience of mania and depression and an awakening curiosity about brain chemistry. Some established scientists, I learned, saw in the accumulating evidence a consistent pattern where the mood disturbances induced by plant extracts, and the therapeutic benefits of antidepressant drugs, could be explained by neurotransmitter changes at the synapse. They suggested a monoamine theory of affective disorder: that during depression there is too little monoamine messenger available in the brain, while mania is driven by messenger excess. There was a beguiling simplicity to the proposal, and although its value as a scientific hypothesis has faded with time, the idea was an important first step in defining a biochemistry of mood. Science progresses by testing ideas, and it was the monoamine theory of affective disorder that helped stimulate the molecular investigations through which we now have a working knowledge of the synapse and its regulatory mechanisms. The monoamine theory gave hope for a rational drug treatment of depression and mania.

It is remarkable, when one thinks about it, that a simple chemical substance purified from the juice of a plant can profoundly change mood. Such knowledge must remove any lingering doubt that the operation of the brain, at its most fundamental level, is that of a chemical machine. But how does the chemistry of a drug disturb the brain's information processing? In recent years molecular biology—a revolutionary set of technologies that permits scientists to isolate and identify the specific protein building blocks of the brain—has taught us that one clue to this mystery lies in how closely the physical configuration of any molecule—be it derived from plant juice or some synthetic process—imitates the structure of the neurotransmitters operating at the synapse. Some drugs, it would appear, literally hold the keys to the castle of our imagination.

The distinct molecular structure of each neurotransmitter (and of the hormonal messengers such as cortisol, the sex steroids, and thyroxin) ensures that it will fit, keylike, into a specific protein lock, a dedicated receptor embedded in the synaptic membrane of the receiving neuron. This simple physical act is the first step in a cascade of events that guarantees the accurate transfer of information from one neuron to the next. Once the neurotransmitters engage these receptors, they are actively transported through the membrane to the interior of the cell, and any intact messenger molecules left behind are either destroyed or recycled—essentially vacuumed up and stored again in the cell that released them—thus terminating the communication. This recycling process, which in bringing a speedy end to synaptic transmission is essential for precise signaling, is also controlled by an important receptor-like structure called a *transporter*.

Once inside the neuron, the messenger molecules trigger a chain of events (called logically the *second* messenger system) that stimulates the cell's energy turnover and also carries the information into the nucleus and genetic library, where the messenger may selectively influence which genes are active. Drugs that alter mood commonly interfere with the receptor-locking mechanisms of neurotransmission, and some antidepressants, including those of the imipramine family, block the transporter proteins of the recycling process. When the shape of the plant or antidepressant molecule approximates that of the neurotransmitter "key" (pharmacologists call such molecules *agonists*), it will fit the receptor lock and stimulate the receiving cell. An antagonist merely jams the lock without activating it, thus blocking information transfer because the neurotransmitter cannot reach the receptor.

Receptors, which have various regulatory responsibilities in the transfer of information, come in many shapes and sizes but are specific in the neurotransmitter that they will engage. This is because there are numerous neurotransmitters and other hormones operating in the brain, each with its own unique physical structure and metabolic life cycle. Only when the structural characteristics of the agonist messenger is sufficiently precise to "turn the key" in the receptor lock will the receiving neuron activate the second messenger system, perpetuating the flow of information to the nucleus and the genetic library. Getting access to the library is important because there are held the blueprints required to manufacture the protein machinery—the receptors, enzymes, and transport systems—necessary to ensure a healthy neuron and a flexible response to incoming information. The whole chain of events within the neurons is carefully synchronized to support this flexibility and to sustain communication with other neurons. Thus, when an increased supply of serotonin neurotransmitter is required to strengthen signaling to neighboring cells, the genes for the enzymes involved in serotonin manufacture are "turned on," and the rate of manufacture increases to meet demand.

The neurons linking the major limbic centers—of the amygdala, the hippocampus, and the emotional memory banks of the frontal cortex—form distinct communication systems which rely upon specific neurotransmitters to carry their information. It is the relative activity of these systems that helps determine our moods of excitement or withdrawal. Within the neurons of these neurotransmitter networks, the dopamine, norepinephrine, or serotonin messengers are assembled from the raw amino acids of tyrosine and tryptophan, by genetically specified enzymes, and then transported to the membrane wall of the synapse, where they are stored in little packages ready for use. Upon the signal of an action potential traveling down the axon to the synapse, packages are launched into the synaptic cleft and the messengers engage the appropriate receptors on the dendrites of the receiving neuron. As this process repeats itself, over and over again, waves of information ripple through the neuronal networks that connect the limbic centers and sustain communication with the rest of the brain.

Each of the metabolic steps within the neuron is precisely regulated, and thus even a small disturbance of the control systems can have a profound effect upon the behavior of the larger network. Also, although neurons cannot reproduce themselves, through the mechanisms I have outlined, they quickly adapt their local networks to fit environmental demand. Scientists have given the name "plasticity" to this dynamic molecular flux. Thus neurons, by activating the genes required to manufacture the appropriate proteins, can modify the sensitivity and number of their receptors and their signaling to other cells. Thus neurons are continuously tuning their communication, much as the sensitivity of a two-way radio can be tuned to accommodate the strength of a signal it is receiving or to vary the power of its transmission.

Multiply this by the billions of neurons in the human brain, and the billions more synapses, and one begins to appreciate the organ's unique power to process information.

Molecular plasticity makes clear how the brain maintains stability (homeostasis) in the face of challenge and also how one's emotional set point can change—another way of describing adaptation—under the prolonged use of addictive drugs, or the influence of stress hormones. And if we revisit the rapid onset of Melanie's manic psychosis immediately following the birth of her daughter, we will see that the concepts of adaptation and molecular plasticity can help explain the underlying biology of her psychosis. We can also find some common ground between it and the biology of other manic syndromes.

Neurons are very sensitive to rapid changes in their hormonal environment. Hormones, particularly the stress hormones of cortisol and thyroxin, and the sex steroids, help determine the limbic brain's homeostatic set point. Any rapid change in these hormone levels, therefore, demands immediate accommodation, and while adaptation is proceeding, mood is commonly unstable. For example, when high doses of cortisol, given to treat severe asthma or arthritis, are quickly withdrawn, depression frequently results, even in people without previous mood disturbance. Similarly, the discomfort and emotional irritability experienced by many women just prior to their menstrual period is driven by changes in the monthly ovarian cycle of estrogen and progesterone, changes that are multiplied many times following the delivery of a child.

During pregnancy, Melanie had adapted to the increased levels of estrogen and progesterone produced by the placenta, and upon Althea's birth these steroids dramatically decreased, perturbing the limbic brain centers of the amygdala, hippocampus, and hypothalamus. For the individual neurons in Melanie's brain, the physiological challenge that confronted them was similar to that facing an addict who, having adapted to nine months of escalating use of some stimulant, is forced to kick the habit "cold turkey." The analogy is well suited, for the neurotransmitter system that links these changing levels of brain sex steroids and the development of puerperal mania is dopamine, the courier of addiction and the messenger of the brain's reward systems.

Many paths of pleasure in the brain employ dopamine messengers in their reinforcement. When we find ourselves partial to some experience, it is the dopamine system that reinforces the feeling and determines that we will seek it again. Many addictive drugs, likewise, alter dopamine in the brain and stimulate this reward system, sometimes producing manic-like behavior. The "buzz" described by those taking amphetamines (better known as speed), or the "rush" of snorting cocaine, is the subjective "high" of the sudden increase of brain dopamine activity that these drugs stimulate. Cocaine is one of the most powerful dopamine-enhancing drugs. It achieves its effect at the synapse by blocking the transporter of the neuron's recycling system for dopamine, thus dramatically increasing the amount of dopamine messenger available to stimulate neurotransmission. Through this mechanism, cocaine precipitates a transient state of hypomania—with euphoria, sleeplessness, and increased energy—in virtually all who take it. It explains why victims of depression so frequently abuse it—as had Melanie's uncle. Furthermore, in individuals of bipolar temperament, cocaine commonly will precipitate a sustained manic episode, suggesting that a special molecular sensitivity has been triggered.

In the mania of manic depression it is the dopamine system, aided and abetted by norepinephrine, that becomes dominant. Mania developing after childbirth is driven by a similar disturbance of dopamine chemistry, provoked by the drop in estrogen and progesterone which further escalates the dominance of dopamine neurotransmission. Animal experiments indicate that the sudden withdrawal of high doses of estrogen causes an increased neuronal sensitivity to dopamine, and the physiological responses of normal mothers after the birth of a baby reflect a similar susceptibility. In women such as Melanie, genetically predisposed to bipolar illness, this sudden increased sensitivity to dopamine is a sufficient molecular stress to precipitate a fulminating mania. The initial symptoms of a postpartum manic psychosis are commonly more frightening because of the speed of its onset. This gives little time for victims such as Melanie, or family members, to accommodate to the bizarre thinking, or to rationalize the distorted perception.

The norepinephrine system also plays its part during the development of mania. A close cousin of dopamine, the norepinephrine messenger system is the vigilant watchdog of the brain, with special duties under conditions of stress, including the responsibility of tuning the brain's sensitivity to incoming stimuli. Frederick K. Goodwin, an American psychiatric researcher and leading authority on manic depression, has suggested that the norepinephrine system is responsible for the intrusive curiosity, grandiosity, and euphoria of early mania—the enjoyable period of hypomania that Stephan Szabo described, when his world seemed to go so well, and the first euphoric days that Melanie enjoyed after the birth of her daughter.

However, as the mania escalates (or develops precipitously, as it did for Melanie), dopamine neurotransmission becomes dominant, with intense agitation, disorganized activity, confusion, and psychotic perception. Increased dopamine sensitivity also may explain why those in the throes of mania, like people high on cocaine, have increased energy and unusual strength. In addition to the reward system, dopamine helps drive the motor systems of the body and determines the ease with which we move our arms and legs. Melanie, for example, who weighs no more than 115 pounds and stands 5 feet 3 inches tall, needed three nurses to restrain her when she became frightened in the emergency room. Thus, in the treatment of mania the immediate goal must be to slow a brain locked in overdrive, reducing the activity of the dopamine system and resetting the limbic homeostats.

Mania is a medical emergency. I remember my teachers in England—men of long clinical experience, stretching back into the decades before modern antipsychotic drugs—describing patients with fulminating illness like Melanie's who had dropped dead from manic exhaustion. Even with the medications now available, time is of the essence, and I had felt that with particular urgency during Melanie's psychotic episode.

Unfortunately, the respite we had achieved for Melanie following her admission was short-lived. After a few hours of restless sleep she was raving and combative again, ripping at her hospital gown and scolding the air with confused, staccato speech. Increasing doses of haloperidol did little to slow her advancing disorganization as the delirious mania reinforced itself and her mind wandered further into a delusional world. Now wild and frightened, Melanie still recognized her mother but insisted that I was the Dalai Lama in masquerade, and that I had come to remove her brain. Here a seed of memory had been embellished by Melanie's madness, for indeed the Dalai Lama had visited the Penn medical center several weeks before her illness. His purpose, however, was not to remove brains, but, as the student newspaper described it, "to explore the power of brain imaging in the investigation of consciousness." It was clear that despite haloperidol and nurses continuously in attendance, Melanie's medical condition was rapidly worsening.

Antipsychotic drugs such as haloperidol and chlorpromazine are used in the emergency treatment of mania because they are pharmacological antagonists of dopamine. They block the messenger from reaching its receptor site at the synapse, eventually slowing the flight of manic thought and stiffening movement. The initial response of the brain to the presence of haloperidol, however, is to make even more dopamine in an effort to adapt to the sudden blockade of transmission. Thus, time is required for the dopamine neurons to become quiescent and for the increased sensitivity to dopamine to decline. Lithium and anticonvulsants can be added, and working through different molecular mechanisms they both are effective in acute mania. These were possibilities I considered for Melanie, but I knew also that it could still take ten to fourteen days to slow the powerful engine that drove her illness.

An alternative possibility was electroconvulsive therapy. Electrical therapy is a valuable medical procedure used by both the cardiologist and the psychiatrist, and it can be an important intervention prior to starting stabilizing drug treatment. Cardiologists call their procedures cardioversion and defibrillation; psychiatrists call theirs electroconvulsive therapy, abbreviated as ECT. The public image of defibrillation is, correctly, that of a life-saving intervention. Thanks to *One Flew Over the Cuckoo's Nest* and other distorted images of psychiatry, ECT is considered, incorrectly, to be a barbaric intrusion. But the two emergency procedures are similar in principle and practice.

When the heart muscle beats so fast that it is merely shivering, the heart is said to be fibrillating—a state of increased excitability where the blood ceases to circulate in the body. A pulse of electricity, passed through the heart muscle, briefly discharging its electrical potential, stabilizes and resets the conducting system of the heart until the muscle begins once more its rhythmic beat. Mania, too, is a state of increased excitability, but of the brain—a fibrillation of the mind—where information fails to flow in logical patterns. A minimal electrical pulse, passing through the skull and across the frontal-temporal region of the brain, sufficient to discharge the action potentials of the neurons in its path, has an analogous salutary effect to defibrillation of the heart. Just as the passage of electricity through the heart can return a rhythmic pulse, electrical therapy can stabilize the beat of the brain.

Convulsive therapy began in the 1930s as a treatment for severe schizophrenia. The idea sprang from the observation that patients with schizophrenia, who for one reason or another suffered spontaneous seizures, temporarily improved. As it is with the weather, after a storm the days seemed brighter and clearer. Initially the convulsions were induced by medication but, with the refinement of substituting small doses of electricity, it was recognized that those who benefited most were individuals with severe mood disorder. Thus, ECT is now most commonly used in the treatment of melancholia. Probably 80 percent of the patients in America who receive ECT have this diagnosis and nine out of ten will show marked improvement, a significantly higher figure than those treated with antidepressant medication. ECT is particularly valuable when psychosis is profound, and I have seen many individuals with severe depression retrieved from a state of mute animalism by this intervention, to become their normal selves again within a matter of days.

Electroconvulsive therapy is also effective in breaking the flight of mania. This is something the Englishmen of long experience had taught me, early in my career. In fulminating mania, such as Melanie was enduring, ECT is a particularly humane and sometimes life-saving intervention. Although Melanie's life was unlikely to be endangered while she was within the safety of the hospital, I was concerned that the earliest moments in Althea's life, including the precious attachment with her mother, were being severely compromised by the psychosis. Melanie would undoubtedly recover over a period of weeks with dopamine-blocking drugs and then the addition of lithium, or perhaps an anticonvulsant, but it would be a protracted intervention made difficult by her confused and uncooperative state. Ideally, I felt, Althea and she should be spending those weeks together. More rapid results had obvious advantages. In my clinical judgment, ECT gave the best chance of breaking the mania and reducing, from weeks to days, the time required to clear Melanie's mind. I explained this to Gloria—who knew something of physiology from her work in the medical school—drawing the parallel with cardioversion and defibrillation. After consultation with Alvin, racing through Minnesota on his way to the hospital, a short course of ECT was agreed upon.

I had explained to the family that ECT is one of the safest treatments in psychiatry. Its draconian reputation is a relic from before the use of anesthesia, when the procedure was a dramatic spectacle. Without drugs to relax the muscles, a dangerous seizure of the body occurs in synchrony to that of the brain, with the jerking movements of the arms and legs creating a ghoulish display. Modern anesthesia has changed that drama. The therapeutic pulse of electricity is now monitored in its passage through the brain by an electroencephalograph, a machine which harmlessly measures brain-wave activity from electrodes placed on the scalp. In fact, those witnessing the administration of ECT for the first time marvel at how smooth and simple it is—and how totally different from its reputation.

The results of electrical therapy in Melanie's case were gratifying. After the first treatment, even though the haloperidol had been much reduced, Melanie slept. Indeed, she slept for twelve hours—longer than she had done in any of the previous ten days, by a factor of three. The second treatment, just a day later, proceeded without complication, and after the third she sat down for her first meal with other patients on the unit. Her husband visited her that evening, and upon seeing him Melanie burst into tears. For the first time, she was full of questions about Althea and her progress. The psychotic, delusional state was receding, although Melanie continued to be confused at times, with a fragmented sleep pattern and the occasional bizarre question. However, the progress was sufficient that I felt comfortable beginning lithium carbonate, with a view to the long-term stabilization of Melanie's mood. After one further ECT, making four in all, Melanie was decidedly better. Although she had little memory of the events that brought her to the hospital and only a rudimentary understanding of what had happened to her, she had begun to speak positively about the future and returning home. She recognized me from visit to visit, knew she was in the hospital and not on her way to heaven, and was appropriately eager to learn from her mother and the nursing staff about Althea's care. As Gloria put it to me, her eyes shining with tears, "Melanie is back."

A rapid recovery from manic psychosis, as Melanie experienced, is typical for those who receive appropriate emergency care, for the acute treatment of mania is effective in the majority of sufferers, especially when it is the first episode of illness. Once the flight of mania has been successfully broken, however, comes the more difficult task of long-term stabilization of mood. Although in approximately two-thirds of those who suffer bipolar disorder the illness may retreat for months—sometimes years—even without treatment, invariably it will return, and the evidence is that the more frequently it does so, the worse the episodes become and the longer they linger. This is why prevention of recurrence is so important, through education about the illness and the appropriate use of mood-stabilizing drugs.

For many who suffer, and for their families, the true nature of bipolar illness is hard to grasp when mania first appears. It takes a different mind-set from thinking about unipolar depression, or illness in general. It seems a matter of simple logic, for example, that if illness is something that takes away health, when illness is relieved health should return. This is not so in manic depression, which is always an illness of opposites. When mania is removed, depression commonly stands in its place and, in a minority of individuals who suffer a brittle form of the disorder called rapid cycling, this succession repeats itself continuously, with no respite in between. Melanie, in her presentations to the medical students, uses what for me is an interesting metaphor. The management of complicated bipolar illness, Melanie suggests, is similar to the experience of an acrobat navigating a high wire while swinging, pendulumlike, in a cross-wind; careful steps are recommended under such circumstances, but the life-saving necessity is to keep one's balance.

Biological balance is facilitated by neuronal plasticity—the now familiar dynamic flux that underpins the activity of the healthy brain and sustains a flexible response to situations of infinite variety. Although it may seem counterintuitive, research has shown that madness is more rigid and inflexible in its organization than is healthy behavior. This is true of manic depression and also of other illnesses where dynamic regulation is involved, such as heart disease, or epilepsy. Illness is rigidity—the loss of biological flexibility—and, as each of us who has learned to ski or ride a bicycle is aware, rigidity destroys balance. The heart rate of a healthy person is extremely variable, but just before a heart attack the beat develops an inflexibility in its rhythm, losing its mercurial response to thought and exercise. Similarly, in epilepsy the fluctuating brain waves of the electroencephalograph become most regular in their shape immediately before, and during, an epileptic seizure.

We find this same rigidity in bipolar illness. In mania, and in melancholia, a healthy responsiveness to ongoing events becomes replaced by a terrifying determinism. At the height of Melanie's mania, her sleep, thought, and action proceeded as dictated by her demons and the underlying molecular disturbance that had given them life. The progression of bipolar illness has an irregular, but lockstep, independence to it. Moods stand apart from the requirements of the day, and the accustomed balance of life's rhythm disappears. In the wake of mania, depression follows, and, after a variable interval of freedom, mania returns—it is as if some capricious internal pendulum now defines the emotional self. Metaphorically speaking, electrical therapy "heals" in mania and melancholia because, as in heart disease, it perturbs the pendulum locked in aberrant oscillation. But the danger of precipitating a swing to the opposite extreme is always present.

Research has not yet revealed what variation, or variations, in the genome determine the episodic shift in chemistry that drives manic-depressive illness. Possibilities include genetic abnormalities in the dynamic assembly of receptor structures, in metabolic enzymes, in the recycling transporter systems of the synapse, or even in the second messenger systems within the neurons themselves—all of which could impair the dynamics of neuronal regulation. A delay in the usual time necessary to manufacture an enzyme in the dopamine pathway, for example, might result in overcompensation of receptors and other elements so that the system behaves in a surging, oscillatory manner when placed under strain and stress. These deviations in protein machinery might be confined to the neurons of the amygdala or hippocampus, or to one neurotransmitter system such as dopamine or serotonin. While we may eventually discover that these regulatory glitches in genome biology represent a molecular equivalent of the different temperaments, under normal conditions their presence probably has little effect upon behavior. It is when a major challenge to homeostasis occurs that the genetic deviation becomes significant. Then the cascade of regulatory impairment that is provoked by the stress eventually disturbs the dynamic balance among the dopamine, norepinephrine, and serotonin operating systems, and the emotional pendulum begins its wayward swing. Without dampening restraint, an aberrant, locked cycle of behavior can rapidly result.

Mood-stabilizing drugs dampen the pendulum's errant oscillation and help prevent the recurrence of manic-depressive cycling. Through molecular actions that we do not entirely understand, they help return homeostatic flexibility to emotional behavior, and in so doing they restore a balanced mood. Mood-stabilizing drugs are thus essential tools in the long-term management of bipolar illness. Lithium carbonate was the first drug to be discovered with these properties. It was the one I chose to aid Melanie's progress across the high-wire, for, in addition to its effectiveness in the treatment of acute mania, lithium is proven in its ability to defend against returning episodes of illness in approximately 70 percent of patients.

Unfortunately, lithium is not fail-safe in all who suffer bipolar illness. This realization led in the 1980s to the experimental use of a group of drugs already successful in the treatment of epilepsy—another illness where neuronal sensitivity is increased—working on the theory that if mania is an irritable excitement of the brain, then drugs known to diminish brain excitability would probably be helpful in bipolar disorder. The first to be used was carbamazepine, but also subsequently found to be of value has been divalproex sodium. Both had been used in epilepsy for over two decades. Each interferes, through different mechanisms, with the movement of the sodium and calcium ions across the nerve membrane. These ionic shifts are responsible for maintaining the electrical charge—the action potential—of the neurons, and dampening their movement reduces the speed at which the neurons can recover after transmitting pulses of information. Thus, the excitability of a whole network can be reduced and, as a result, the progression of epileptic seizures or the fermentation of manic excitement is contained. The anticonvulsants, as these drugs are appropriately named, have become a valuable addition to the treatment of bipolar illness. Divalproex sodium, better known in America by its trade name of Depakote, is as effective in treating acute mania as lithium is, and it is more useful in those complicated, and rapidly changing, mood states where the symptoms of depression and mania are mixed together.

The usefulness of lithium in mania, in contrast to the logical analysis that has determined the value of anticonvulsants, was stumbled upon in the late 1940s almost entirely by accident. John Cade, an Australian psychiatrist, while seeking a toxic agent in the urine of psychotic patients, was investigating uric acid and urea, both by-products of protein metabolism, as possible candidates. Lithium, when combined with uric acid, made the latter nicely soluble in water, and during extensive experiments injecting this compound into guinea pigs, Cade noticed that the animals became quiet without falling asleep. By an intuitive leap—the sort of which only the human brain is capable—Cade decided to give lithium salts to several agitated and manic patients who were under his care. One of the first patients Cade treated (whom I have always thought from the descriptions had much in common with Wilfred, the tormented farmer I met in England during the 1950s) had been in the hospital, chronically manic, for five years. And yet within three weeks he was "enjoying the unaccustomed and quite unexpected amenities of a convalescent ward," and after three months he was so improved that he left the hospital to return to work and to his family. Doctor Cade had made a miraculous discovery that would revolutionize the pharmacological treatment of manic-depressive illness.

The medical world was at first suspicious of this miracle. How could a simple salt so dramatically change chronically abnormal behavior? The toxicity of lithium in high doses to the heart and kidney, which had become apparent when lithium had been used previously as a substitute for common table salt, was also a concern. But slowly, after two decades of careful experimentation, particularly by Mögens Schou, a Scandinavian psychiatrist, lithium became accepted throughout Europe and later in the United States, where it was introduced in 1970. Lithium is now used widely across the world and has the distinction of not only being a reasonably effective *treatment* of mania, as John Cade had discovered, but, when taken each day, of being capable of stabilizing mood and *preventing* the return of both manic and depressive episodes.

We do not really understand how lithium provides this stability, although we have learned that its actions are many and varied. In common with the anticonvulsants, lithium reduces the excitability of the neuron, probably by changing the dynamics of the ions passing back and forth through the membrane wall. However, it also modifies the signaling of the second messenger systems within the neuron itself, disturbing the energy cycles that sustain the information transfer from the receptor to the nucleus. In addition, lithium alters the balance among the neurotransmitter operating systems of the limbic alliance, strengthening the serotonin messenger system which is important in preventing depression. However, despite this serotonergic action, lithium alone is not a good antidepressant. While it effectively removes the manic highs in most individuals, in some a smoldering melancholy remains. So it was for Melanie.

Melanie's psychosis had continued to clear rapidly following the ECT and the initiation of lithium therapy. Much to the relief of her family, she soon was caring for her daughter under the watchful eyes of her mother and the nursing staff, and shortly thereafter was able to return home. Improvement continued over the next month but she was not herself. While she was diligent and loving in her care of Althea, she had an excessive desire for sleep and little of her natural energy.

As her mother described it to me, during the early weeks after her hospitalization Melanie seemed to have "lost her direction." She complained of no longer knowing who she was, and was ashamed of what had happened to her. Melanie's joys in caring for Althea were dulled by concerns about her illness and its long-term implications for her future. How was it, she had asked, that somebody could become so ill—and so quickly—just at the very moment when she had achieved her dreams? Melanie had been enrolled in the first year of nursing school when she became pregnant, but when complications arose with the pregnancy, fearful of another miscarriage, she had taken a medical leave of absence from her studies. Thus, few classmates were aware of her postpartum mania. Nonetheless, in the wake of her illness Melanie could not bring herself to go back to school, even with Althea, for a visit. Would anybody really believe a story of abnormal chemistry and bad genes? Psychosis was a thing one read about in textbooks—something that happened to other people.

Melanie's emotional pendulum had swung away from mania and, despite the lithium she was taking, had reset itself toward the pole of depression. This is a common occurrence. Lithium with its life-saving stability removes a dynamic edge, previously unrecognized, of mild manic highs, but does not eliminate depressed moods, thus diminishing the accustomed sense of subjective "vitality." Very often, especially if there are also physical side-effects from the lithium—such as stomach upsets or hand tremors—many individuals will want to stop the medication at this early stage, before the true benefits of stabilization are realized. When the lithium *is* stopped, the depression often worsens—or mania precipitously returns—reinforcing that manic depression is an illness that cannot be ignored.

Fortunately, Melanie's side-effects from the lithium were few, confined largely to a fine hand tremor, and, with Gloria and Alvin's support, she was willing to continue. These were painful months for Melanie, and they were the time when the first seeds of her advocacy for those similarly afflicted were sown. We began discussing the questions she had raised, and as frequently as possible Alvin joined us. I took seriously Melanie's existential concerns, and told her so, for to reconstruct life's meaning in the wake of a manic psychosis is solemn business and not something achieved by drugs alone. The lithium, I suggested, should be seen as an insurance policy, a medicine—like insulin's action in stabilizing diabetes—that could help prevent a recurrence of her mania. This would provide Melanie with the stability necessary to get her education back on track and to begin some thoughtful reconstruction of her life after what had happened to her. And in the meantime, given Melanie's residual symptoms of depression, I firmly recommended the addition of an antidepressant, as further aid to her reconstructive effort. Melanie was skeptical; she did not want to take additional medication. Pills reinforced the fact that she was ill and she preferred to forget that. It would be much later, only after the successful treatment of her depression, which included joining a recovery group and meeting others who were successfully managing their lives, that Melanie's confident self would return.

The choice of an appropriate antidepressant requires of the physician a familiarity with psychopharmacology, a certain empiricism, and an alliance, ideally of mutual trust, with the patient. The pharmacology of mood, and especially the treatment of depression, has become a complicated subject in the years since I wrote my review of peyote for the medical school magazine. There are now many antidepressant drugs from which the physician may choose. However, what is still painful and frustrating to those who are melancholic, and to their family members, is that none of the medications, new or old, improves depressed mood immediately. And the first medicine chosen is not always the one that successfully reverses the depressed mood. Thus, in the alleviation of depression by antidepressant drugs, confidence is required between doctor and patient. While success is ultimately achieved in over three-quarters of those treated, it is not something instantly attained. Most antidepressants take three to six weeks to achieve a maximum benefit. This is because the drugs do not directly replace some lost chemical in the brain, but achieve their effect by disturbing the melancholic set point of limbic behavior and changing the activity of neuronal networks in a two-step process that takes time to occur.

Imipramine, for example, the venerable mother of antidepressant drugs, first *perturbs* the set point of melancholic homeostasis by blocking the neuron's transporter for both norepinephrine and serotonin, thus disturbing the recycling system for these messengers and increasing the supply of the neurotransmitters in the synaptic cleft. The second phase is one of *adaptation* as the neurons collectively work to reestablish equilibrium with the imipramine molecules in their midst and in doing so find a new set point, usually with an improvement in the melancholic symptoms. Thus, the ultimate behavioral effects of imipramine in changing depressed mood are not simply the result of increasing the amount of brain neurotransmitter, as the monoamine theory of mood first proposed, but rather a combination of the drug's *action* at the synapse and the brain's *reaction* to its presence.

We have progressed in the design of antidepressant drugs since the discoveries of the 1960s. Although the molecular principles of antidepressant action of the new drugs remain familiar, the focus of their activity has become more specific. Central to this evolution has been a growing confidence that serotonin is an important modulator of depressed mood, and the development by pharmaceutical companies of antidepressant drugs that specifically target the serotonin system of neuronal communication.

Professor Arthur Prange of the University of North Carolina—a colleague with whom I have conducted research over many years—had proposed in the 1970s that a reduction in serotonin activity gives "permission" for mood disorder to develop. When Dr. Prange and I were working in Surrey, England, together with Dr. Alec Coppen (who then directed a Medical Research Council research unit studying mood disorders), we had found that drugs which block the activity of serotonin neurons in the brain made mania worse—much worse. However, supplementing the diet with tryptophan, the amino acid from which the neurons manufacture serotonin, improved the treatment response in both mania and depression. Subsequently, other scientists discovered that reducing tryptophan in the diet can exaggerate depression, including seasonal depression, and that those suffering mania and depression have less of the breakdown products of serotonin in the cerebrospinal fluid of the brain—a finding that persists even into recovery and may be an indicator of genetic vulnerability.

It is clear from this and other evidence that serotonin neurons serve as the moderators in the limbic alliance, balancing the dopamine and norepinephrine systems in their actions. Studies in the chemistry laboratory show that the vigor of the serotonin system is necessary for the other messenger systems, and some of the regulatory centers in the hypothalamus, to operate efficiently. Whereas dopamine drives the reward systems of the brain, and norepinephrine is the watchful guard dog, serotonin is the friendly neighborhood constable, a moderating voice that reduces fear and anxiety and lifts depression. In the absence of serotonin's calming effect, bad things happen in the limbic alliance—the sort of things that Melanie had experienced. When the activity of the serotonin system is reduced, sleep becomes fragmented, aggression increases, and any propensity toward mania is easily triggered. In melancholia, ruminative melancholic thoughts increase, including the obsessional preoccupation with suicide. Indeed, several postmortem studies of the brains of those who have succeeded in killing themselves suggest a disturbance of the serotonin system immediately prior to death.

The central role of serotonin in the modulation of depressed mood became clearer during the late 1980s when antidepressant drugs were introduced that selectively inhibit the recycling of serotonin at the synapse. The most famous member of this family (known as the selective serotonin re-uptake inhibitors—SSRIs for short), trades under the name of Prozac™ (fluoxetine is its chemical name) and has been hailed by some as an elixir from the gods. Prozac in the public mind has rapidly become the aspirin of emotion, the preferred anti-inflammatory agent for almost any behavioral pain. By 1996 the drug had been prescribed for an estimated fifteen million people in the United States alone. In fact, research shows that fluoxetine is no more effective than imipramine in alleviating depression, and is probably less effective than the tricyclic drugs in the treatment of melancholia. However, because a standard dose is required, it is simple to prescribe. Also, Prozac's initial side-effects are less troublesome (although not in the longer term, when significant sexual dysfunction is reported) and thus people are more willing to take it. And, indeed, Prozac and the other SSRIs have been found effective not only in depression, but also in other discomforts, such as panic, and the obsessional behaviors where serotonin plays an important moderating role.

The SSRIs—a growing family of drugs, including sertraline and paroxetine—are significant because they represent a new generation of "designer" antidepressants, created from a knowledge of how the brain works and of the molecular biology of the neuron. Many older antidepressant drugs, particularly the tricyclic family of which imipramine is representative, are diffuse in the regulatory mechanisms they disrupt, disturbing neurotransmission in the dopamine, norepinephrine, serotonin, and even acetylcholine systems, which explains why some of them make the mouth so dry. The tricyclic antidepressants also vary in the combination of neurotransmitters they perturb. Some, such as desimipramine—a derivative of imipramine—challenge particularly the norepinephrine neurons while others, of which amitriptyline is an example, tweak predominantly serotonin. It is because of this broad action, and the resulting unwanted "side-effects," that while the tricyclics remain effective and economically attractive agents in the treatment of many depressions, they are considered less "friendly" than the SSRIs.

Selective serotonin re-uptake inhibitor agents were the first drugs developed in a rapidly expanding effort by the pharmaceutical industry to produce medications that specifically target individual neurotransmitters. A growing knowledge of the molecular biology and physical configuration of the messengers themselves, and the protein components of the receptor and signaling systems of the neuron, will accelerate this effort in the future. New designer drugs are now appearing that precisely engage both the serotonin and norepinephrine systems, but without the troublesome side-effects of the older tricyclic antidepressants. Venlafaxine (with a trade name of Effexor™) is the forerunner of this novel class of medications, and the early evidence suggests that it is highly effective in treating depression.

For those who suffer bipolar illness, however, there is an important caveat to the use of all the antidepressant medications. In patients with manic depression, perturbing multiple messenger systems in the brain, as is required in the treatment of depression, appears to increase the risk of switching the patient directly into mania. Twenty-five to thirty percent of individuals with bipolar illness, when given tricyclic medications, for example, develop manic symptoms, and sometimes mood may switch from one state to another, in a continuous cycle. This had been a consideration of particular concern to me in the care of Melanie Branch, who suffers a particularly brittle form of bipolar illness.

Just as the details of people's faces distinguish us and display our moods, so are the details of our brain chemistry unique. Unfortunately, however, there is as yet no simple photograph that can capture the chemistry of mood as photography captures it in the face. Thus it was impossible to determine the optimum antidepressant for Melanie, a priori. Given her episode of mania, the odds were that a drug perturbing only the serotonin system was an appropriate place to start, carefully combining the antidepressant with lithium to minimize the risk of precipitating a manic recurrence. In fine-tuning the chemistry of mood, it is a general principle that the dopamine and norepinephrine systems work together essentially in opposition to serotonin. It is the balance among the three that creates the harmony of mood, but the optimum balance is different among individuals. Knowing specifically how a drug exerts its action is helpful, but because of the idiosyncrasy of brain chemistry and our ignorance of the specific regulatory defect that drives bipolar illness, for each individual the pharmacological treatment of depressed mood remains a miniseries of empirical scientific experiments through which the doctor and patient, working closely together, discover the best fit between pill and person.

The SSRI which I prescribed for Melanie did not completely eradicate her depression. It has been Melanie's burden that her illness is particularly unstable—a malignant variant of manic depression that affects fewer than 15 percent who suffer the disorder. Over the several years we have worked together there have been periods of stability when, in addition to her priority of being Althea's mother, she was able to finish her nursing degree. During those times the management of her illness has been comparatively easy. At other times blips of manic excitement have returned, distorting her judgment and driving her to inappropriate behavior. It is at these moments that the support of her family has been vital, together with the restraining voices of her friends in the local chapter of the Depressive and Manic-Depressive Association. Melanie has worked closely with these colleagues, in mutual support over several years, and credits much of her success in managing her illness to what she has learned from the collective experience and education that the group offers.

Melanie's story is not typical. There are many, such as Stephan Szabo, who, once stabilized on lithium and coming to terms with the fragile nature of their moods, live comparatively free of recurrence. However, the good news is that Melanie, with the medical knowledge currently available and the skills she has taught herself from experience and from the experience of others, has been able to manage bipolar illness in its most malignant form. And she has done so with tenacity and skill. It is this accomplishment which makes her advocacy so important to others—and her message so dramatically compelling to the medical students.

"There is no magic bullet for these illnesses," Melanie emphasized during her presentation to my neuroscience class. "The hardest and most important lesson to be learned is that the answer is not to be found in a pill bottle. This is not to say that medications are not helpful," she added. "In my case they have provided me with the necessary stability and relief to move ahead and take care of myself. But they are a tool that I use to make a better life.

"The important thing to remember is that no individual chooses to be mentally ill. It is something I would have thought of as obvious," said Melanie, "but apparently not. It was not a choice I made for myself, nor was it for thousands of others like me. Yet this diagnosis of manic depression, I have discovered, carries with it in our society punitive consequences, ranging from no health insurance to job discrimination. After my daughter was born and I became ill, I was both victim and perpetrator of such stigma. I was so afraid of how my life would be damaged if anyone were to discover my secret that I didn't tell anybody, even my friends, for almost two years. I felt from what others said, and the jokes they made, that if my secret *were* discovered my opportunities would be limited, or perhaps people would be afraid of me. Either way, despite hard work, perseverance—all the personal values that remain part of me despite the illness—I felt that if I confessed my mania I would be ostracized. Therefore I hid it. But as I struggled to recover, and met others doing the same, I became increasingly angered by the stigma and society's pathetic misconception about people who suffer from serious mental illness. I decided I could never change the attitudes of those around me, but on the other hand I was not going to perpetuate their ideas by living in shame and fear of what society might consider unacceptable. I had to give up my secret, regardless of the consequences, to reconstruct my life and that of my family."

I agree with the medical students in my neuroscience class. Melanie Branch is a remarkable woman.

Ten

Thoughtful Reconstruction

Adaptation and the Care of the Self

Learn the ABC of science before you try to ascend to its summit.

IVAN PAVLOV (1849–1936)

Come, come, it's only the passions that make you think.

MME. DU DEFFAND TO
HORACE WALPOLE
(1697–1780)

The management of troubled mood requires dedication, education, and personal insight. Beyond the contributions of the pill bottle and the sophistication of the brain sciences, self-knowledge is the key to regaining one's health. Melanie Branch had attained this wisdom of self-understanding and recently, as Stephan Szabo and I sat playing chess, I was reminded of its importance.

"You know," said Stephan, having pinned my bishop and rook with one swift move of his knight, "compared to physics and chemistry, biology is inherently messy. Things are much more influenced by the laws of chance. That used to worry me, which is why I left medicine and went into physics, but now biology intrigues me as more challenging, more of a puzzle. It is living with this illness that has changed my mind."

Stephan was visiting me, on his way to stay with his sister and brother-in-law at their home in Baltimore, and after supper we had turned to chess and reminiscence. Whether Stephan's conversation was a clever trick to distract my attention from our chess game wasn't clear, but, if so, the ruse was working. His move had come as a surprise.

"I used to think there was a pecking order in the levels of science," Stephan continued, "that some are more worthy of investigation than others. After all, if physics and chemistry determine the laws of biology, then ultimately through them we should be able to explain behavior. I felt that the only way of discovering the truth about life was to study the fundamental disciplines, those that are predictable when reduced to their elemental parts. That's what I was taught as a boy in Hungary, and that was my father's viewpoint. But I had not taken into account that living creatures learn and stake out an independence from what happens around them. Physics alone does not help us understand how, over the lifetime of one organism, behavior becomes more predictable and autonomous. When I was younger, in my head I was at sea most of the time. But now, through trial and error, I have learned to establish some order and control in my life."

And certainly control over the principles of chess, I thought to myself, struggling with an intelligent response to the threat of the knight. But then Stephan's talents at chess had never been in question. Early in our friendship, however, I had always found his strident reductionism about behavior irritating. I had put it down to a sensitivity about his illness. But that evening it was clear that Stephan had shifted his ground. His opinions seemed much closer to my own.

When it comes to understanding emotional behavior and abnormal mood states, no single scientific discipline can serve. Many levels of interdependent discourse—many steps upon the scientific staircase—must be considered if progress is to be made. This is especially so when we are advancing our scientific knowledge, but it is also true when considering the appropriate treatment for each individual patient. Information must be gathered at every level of objective inquiry and artfully assembled if the staircase is to be climbed. After all, the disciplines of molecular genetics, anatomy, neurochemistry, physiology, psychology, pharmacology, and so forth are arbitrary divisions of our own manufacture. They are a conceptual expedience, based on history and technology. And yet for many, physicians and lay persons alike, there is something especially attractive about knowledge gained from the biological pieces of the intellectual jigsaw we have created. Fragments of information obtained from the technical investigation of biology are presumed to say something fundamental about the development of the illness and its management.

I voiced my thoughts to Stephan.

"Yes," Stephan replied. "And you will be surprised to learn that I agree with you. I have no really satisfactory explanation for our blind faith in technology. Perhaps knowledge derived through a complex machine, which is increasingly difficult for the average person to understand, has greater credibility than that gained through careful personal observation. I also think that physicians, and especially some psychiatrists, are prone to confuse information with cause when it comes to the biological investigation of the brain, rather than recognizing it as another part of the puzzle. The biological perspective is a welcome shift from a few years ago, when many psychiatrists considered Freud's dynamic theories of intrapsychic life to be a complete explanation, but I wonder whether the swing has gone too far."

Stephan paused, studying the chess pieces before him.

"From the patient's standpoint, biology is only part of the story," he continued. "We know, for example, that I have a special vulnerability to manic depression. And I also know from my reading that if the distribution of the blood flow in my brain had been measured by a PET scan during one of my depressions probably it would have differed from what is going on in my skull at this moment. Now I find such studies genuinely exciting, but unusual patterns of blood flow do not explain my madness, nor do they guide me in what I can do about it. Blood-flow studies are not a sufficient explanation of cause. The manias and depression I suffered were usually precipitated by grief. Certainly, that's what triggered the first episode—trying to cope with the death of my father, and especially that he had committed suicide.

"So if we're looking for an explanation of the *origins* of my illness, a combination of that dreadful social situation plus my family's genes is to me the most reasonable, rather than a change in blood flow in my brain. Measuring blood flow represents an important level of biological inquiry and may tell researchers something about where to look for abnormal brain activity once the depression has been triggered, but it tells us little about what pulled the trigger in the first place. And that's what the patient wants to know. Then there's a hope of gaining greater control, and perhaps a chance for prevention."

I nodded in agreement.

"But, while I'm here on my soapbox, there's something else," added Stephan, raising his hand as if anticipating interruption. "Assuming biology to be the only significant part of the puzzle, as I did for several years and especially after I agreed to try lithium, implies certain things about treatment—it's like the promise of a biological silver bullet. It makes it too easy for patients to take themselves off the hook of personal responsibility and to behave like victims, which is not healthy in the long run. I see it as similar to taking insulin for diabetes and then ignoring the need for an appropriate diet.

"And the same holds true in reverse, at least for an illness like mine. Psychotherapy by itself is not the answer. You can't think your way out of mania. I know; I've tried. The secret to successful self-care is a combination of pharmacology and self-knowledge. Medicine remains an empirical science. It's too early to build monuments to simple solutions to chronic illness, whether the disorder is diabetes or manic depression. This illness has taught me to respect the pluralism of science, in medicine at least, but it has also taught me that there is no substitute for personal responsibility and intelligent pragmatism when it comes to caring for oneself."

I moved the rook out of range of the black knight, positioning it carefully to cover my bishop. It seemed unlikely that Stephan would exchange the knight, one of his most flexible pieces, for a mere bishop so early in the game. But, given Stephan's accomplishments in chess, I knew that nothing should be ruled out. What *was* unexpected, and a pleasure to me, was that Stephan now seemed intent on extending the commanding intellect with which he had so obviously mastered the game of chess, to the successful management of his illness. I sat back in my chair, reflecting on his comments.

It had been almost a decade since Stephan had begun lithium therapy, and there had been little contact between us over the intervening years. In fact, it had been my research for this book, and our meeting in London, that had rekindled the friendship. Clearly he had gone through a period of thoughtful reconstruction since the early years of his illness when, at least in public discussion, he had essentially denied its existence. After finishing his physics degree Stephan had taught for a while but then entered the electronics industry. Although he seemed happy with that choice, it was obvious from the thoughts he had expressed about the optimum care of mood disorder that he had maintained an active interest in medicine, including psychiatry.

For several years I, too, had been thinking about optimum care for these illnesses, and how it is frequently compromised by patients and families being unaware of the breadth of treatments available, and the positive results that can be achieved. Unfortunately, the mind–body split remains evident in psychiatry and in the behavioral sciences, particularly when it comes to treatment. As Stephan had implied, abnormal behavior is a vector of molecular, intrapsychic, and social forces. Ideally, in recognition of that intertwining, any therapeutic intervention should be directed, without theoretical bias, to best serve the needs of the patient. A prejudice to give drugs alone and exclude psychotherapy or, vice versa, to presume in the face of contrary evidence that analyzing the behaviors of bipolar illness or recurrent depression will eradicate the vulnerability, smacks of arrogance, not wisdom, on the part of the therapist. Given the present tools available to help those with mood disorder, the evidence is overwhelming that judicious pharmacological intervention plus a program of self-education—which is the core of any psychotherapy—is the best approach to treatment and prevention.

In my own mind, and in describing to my patients the care available to them, I no longer divide treatments into psychological and biological categories. Rather, building upon the concepts of emotional homeostasis, I ask myself how the intervention that I am proposing will enhance emotional adaptation and ease the burden of mood disorder for this unique individual. If I can satisfy myself with the answer to that question, I present the advantages as I see them to the patient and explain how the intervention may help him or her gain greater control over the illness.

As I have emphasized, the limbic brain is a self-correcting system. The homeostatic principles that govern such systems are of fundamental importance throughout biological and behavioral science, from the molecular regulation of the genome through to social adaptation. When emotion, thinking, and the body's housekeeping functions are disturbed in mood disorders, it is the regulatory process—the pendulum swing—of adaptation and limbic homeostasis that is ultimately impaired. Thus, within these dynamic parameters I have established three categories of intervention which I call *perturbation, facilitation,* and *modification* of the adaptive process.

I explained to Stephan how my thinking had evolved. One way of understanding ECT (electroconvulsive therapy), I suggested (as I have described in chapter 9), was to see the intervention as a short-term strategy *perturbing* the aberrant swing of the pendulum, and stirring the melancholic brain once more into self-adjustment. As the new set point becomes established, a process that occurs over days or weeks, health is commonly regained. Sleep deprivation, where staying awake throughout the night will sometimes jolt individuals out of a depression, is conceptually similar. And so too are some treatments (many of which actually worked, albeit briefly) that were used in the early nineteenth century, such as the "ducking-stool," or the "tranquilizer" chair of Benjamin Rush. Both of these latter treatments cause reactive homeostasis, and the immediate release of the stress hormones of the body. The ducking-stool involved plunging an unsuspecting individual into cold water, while Rush's tranquilizer chair is the most complete restraint of a patient's movement ever devised—and in that regard is no tranquilizer, for we know from animal experiments that restraint is profoundly stressful.

Antidepressant drugs initially perturb in the same way, but with greater precision, and then over an extended period of time *facilitate* adaptation through their blockading presence, promoting regulatory adjustment in specific enzymes and transporter systems. Such interventions, which enhance the normal adaptive processes of resilience and restitution, I proposed to Stephan as the second group of treatments. Supportive counseling in grief and depression, hospitalization to protect an individual from self-destructive impulses or an acutely toxic environment, self-help groups such as those organized by the Depressive and Manic-Depressive Association, and other forms of sustaining friendship and social assistance, are all examples of this category of care.

The natural hormones of bodily adaptation—the stress hormones, sex steroids, and the thyroid hormones—are also vital in facilitating recovery from mood disorder. The thyroid hormones, which are powerful modulators of mood, can speed response to antidepressant drug treatment especially in women (who suffer a greater incidence of thyroid disease throughout life). Thyroxin, when used as a therapeutic drug rather than a replacement hormone (much as steroids are used in the treatment of asthma, arthritis, or severe allergy), can be helpful in stabilizing the rapid-cycling form of bipolar illness. The addition of estrogens, valuable in some women with post-menopausal depression, falls into the same category of facilitating the body's natural adaptive resilience.

The long-term *modification* of adaptive capacity is my third, and final, category of intervention. Psychotherapy and self-education fall here, as does lithium, and perhaps the anticonvulsants. Lithium, when taken continuously, modifies the excitability of neuronal networks and stabilizes the pendulum's swing, preventing the kindling and escalation of manic episodes. The psychotherapies harness the ability of an individual to learn new coping strategies, promoting self-education and social adaptation.

Stephan had been listening attentively. "I think I see where you're going, and I like the idea," he told me. "You're using time—your categories divide acute, intermediate, and long-term interventions—and the simple physiological principles of adaptation to break down the watertight barrier that exists in the mind of many people that biology and psychology are separate. Essentially I was trying to say that earlier. It's something I hear all the time in the support group I belong to in London—the idea that only drugs can help because this is something in the genes, or everything but talking therapy is a crutch.

"I mean no offense," he continued, with a puckish grin that suddenly reminded me of the Stephan of earlier years, "but the names of your clusters are a little academic. What you need is a jingle—you must recall those things I made up in medical school, to remember the anatomy of the muscles in the forearm. While you were talking some doggerel started running through my head, something like: *Here are the keys to adaptation; prodding, assistance, and modification.* What do you think? Use it, if you like it. Not as good as the old days, I'm afraid. I think rhymes came easier to me before lithium, which remains one of my little regrets."

Stephan looked down at the chessboard. "The prodding idea—perturbation—I find particularly appealing. Like taking your bishop, for example," he said playfully, picking up the piece and replacing it with his knight. "I guess we should put that in the category of prodding; a move that demands your adaptation, in an otherwise sluggish situation.

"More seriously," Stephan added, "what about the impact of personal experience in the prodding category? It's not just in long-term psychotherapy that we learn important things. Episodes of illness are in themselves perturbing, but sometimes they are also therapeutic. That has been true for me. In coming to terms with my illness, the turning point was the disastrous visit I made to America several years ago—when I came to see my sister Margaret in the middle of a manic episode. That was so painful for everyone involved that I finally acknowledged I couldn't think my way out of this scourge alone—that I needed medical help and guidance to take control of my life."

Stephan had made an interesting point, for indeed traumatic events in the course of mania or depression will sometimes trigger new purpose in those who have previously tried to dismiss their disabling mood swings. I had known about Stephan's infamous manic visit to America from Margaret, who had described it to me in a letter, written at the time of the crisis. Apparently, one hot August day Stephan had arrived in Baltimore, unannounced and directly from London, with nothing more than three small carpets and a tennis racket. The carpets were

Stephan's present for Julia, Margaret's daughter, but otherwise his proclaimed purpose was to play tennis.

Three dramatic and exhausting days had followed. After staying up all night and making several hundred dollars worth of telephone calls, buying two bicycles and another tennis racket, Stephan had become, to use Margaret's words, "completely unraveled." His initial energy and good humor had dissolved into anger, with criticism of his father, of Margaret and her life in America, and hostility toward the world in general. Michael, Margaret's husband and a physician at Johns Hopkins, had managed to get Stephan an appointment with a private psychiatrist. Stephan had agreed to go, but then had walked out of the consultation in a fury when it was suggested that he needed medication. "Stephan was convinced I had betrayed him," wrote Margaret. "On the way home, as I was stopping for a red light, he began shouting that if he didn't love me, he would kill me for what I had done. I was terrified, and angry with myself for driving alone with him. Then, before I knew it, the door was open and Stephan was running through the heavy traffic, screaming and waving his arms. He could have been killed."

Stephan had been returned that night by the police, who had been looking for him at Margaret's request. His flight had come to an end at the waterfront, where he had created something of a public sensation by removing his clothes in preparation for a midnight swim in the harbor. "Something happened to Stephan that night," Margaret wrote. "I think he was roughed up a bit in the struggle when he was caught, and the police surgeon sedated him. The dose must have been pretty high. He just crashed. He was lying on Julia's little bed, this huge man, like a helpless child huddled up into himself. When he woke up he wouldn't talk to me, and then I got angry. This is all such a waste of a brilliant mind. But both Michael and I know there's nothing we can do until he decides to do something for himself. For the next few days we fed him Italian food and Thorazine, until he was together enough for Michael to fly with him back to England. Quite honestly, I never want to see him again."

Margaret's angry reaction, born out of frustration and love, is a common experience for the family members of those who suffer manic depression and melancholia. Even when the difficult behaviors and accusations are recognized as illness, the pain and emotional chaos that they breed is real, destroying intimacy and caring. The subtle distortion of communication and the irritable blame, which so often occur in chronic smoldering depression, are frequently more difficult for family members to bear than acute, bizarre episodes such as the one Margaret describes. The tragedy of these disrupted attachments is reflected in the statistics about those who suffer these illnesses, with evidence of frequent divorce, poor job performance, social isolation, and even criminal acts being commonplace.

What is often most difficult from the standpoint of family members is the fright, apprehension, and sheer frustration of not knowing how to help. Tolerance cannot always be sustained and inevitably tempers flare and battles rage. There eventually comes a moment when they literally do not know what to do next, and thus do nothing, or actively seek to exclude the suffering individual from the family. Such desperate moves, however, usually serve merely to multiply the guilt already present. Margaret, for example, had always been sensitive to the truth that when Stephan first became ill, after their father's suicide, she was already living in America and was not in England to support him. Under such circumstances, faced with a conflicted sense of broken responsibility and personal guilt further compounded by the refusal (or inability) of the sufferer to participate in any constructive solution, the seeds of anger flourish.

For Stephan, the Baltimore experience had been a watershed. He was deeply attached to Margaret and once alone again in his London flat, reflecting upon the debacle of his visit, he recognized that he had crossed some line in his relationship with his sister. Margaret's tolerance had been exhausted. Stephan was soon in a deep depression. For years, despite accumulating evidence to the contrary, he had insisted that his moods and periods of disability should be accepted as the turbulence of a creative mind. Now he was scared. Finally, looking for counsel, he had turned to me, both as a professional and as an old friend. We had many intensive telephone conversations over several days until, at my urging, Stephan sought the help of Hans Brüner, who had been one of our teachers in medical school.

Hans, during my early training, had taught me much about being a psychiatrist, and about psychotherapy. An accomplished endocrinologist, after a personal psychoanalysis he had turned in mid-career to psychiatry, and to the study of psychosomatic medicine. A man of unusual compassion, and with an intuitive understanding of people, he blended medicine, pharmacology, and psychotherapy with a common-sense hand. While still a boy, Hans had fled from Vienna with his parents, at the beginning of the Nazi occupation. Thus, like Stephan, he was an immigrant who understood the complexity of a life in translation. Stephan immediately found affinity with the older man and developed a deep bond of affection. For the first time, Stephan began to seriously rethink what it was that ailed him.

"Hans really was my second father," Stephan told me. We had put aside the chess game; with my defense collapsing and Stephan's pieces dominating the center of the board, the inevitable outcome was clear. "Except," he continued, "there was a difference in those early months. Hans was much more tolerant of my arrogance, and idiocy, than my father had ever been. It took me six months to agree to take lithium. Six months of grinding depression, which as Hans repeatedly observed, were totally unnecessary. Sometimes I would visit my mother at the weekends and go into what had been my father's dressing room, where the sun shone and it was warm. I would lie down on the rug hoping I would die, afraid that I wouldn't and scared to death that I would. From the experience of those terrible moments, plus Margaret's stern letters warning of future tragedies, I knew I was in trouble. But I was still convinced that taking medication and being labeled as a manic-depressive would be the end of my career.

"But after the second or third visit to Hans, I felt as if someone had thrown me a life preserver. For the first time I thought that help might be available. That's what kept me alive during those six months of depression. It was the attachment, the lifeline to Hans that first assisted me—what you call facilitation in your scheme. It was my attachment, and my confidence in him, that eventually convinced me to take lithium, and then as my mood swings diminished, to really begin thinking about my life in some reconstructive fashion. Obviously, my relationship with Hans could never, alone, have broken the back of the illness, but it was the human thread that stitched things together for me in the beginning." Stephan paused, and then added whimsically, "Strange, isn't it. I never thought I would hear a physicist talk like this."

Not so strange really, for human attachment lies at the core of any psychotherapy. Whether the goal is short-term assistance or the long-term modification of behavior, all psychotherapies exploit the two fundamental mammalian behaviors of social attachment and an ability to learn from experience. There are literally dozens of contemporary psychotherapies, each bearing its own name in a struggle for recognition and survival. And although in their promotion the proponents emphasize unique aspects of each, in fact all draw upon these two core behaviors. The "dynamic" therapies, originating with Freud, consider emotional attachment—specifically the relationship of the patient with the therapist—as the fulcrum for change. The behavior therapies, drawing upon the concepts of behaviorism that were developed by Pavlov and Skinner, consider the paradigm of learning as paramount in modifying behavior, and believe attachment to be of little consequence. The cognitive therapies of Beck and Seligman, although commonly called the cognitive-behavioral therapies, in practice lie between these two polarities. Cognitive therapists, in my experience, are more willing to concede that the relationship between the patient and the therapist plays some role in the patient's successfully learning new adaptive skills. Thus, in truth, regardless of theoretical construct, human attachment and learning (in part through the vehicle of that attachment) invariably emerge as the important engines of behavioral change in the practice of any psychotherapy.

The *dynamic* psychotherapies derive their name from Sigmund Freud's model of human mental function. Freud considered abnormal behavior to be a product of conflictual forces within the psychological self (*dynamic* being another word for *force*). According to psychoanalytic theory, conflicts exist at varying levels of conscious awareness, with some being entirely unconscious, and this disharmony is played out in the therapeutic relationship—technically called the "transference"—which the patient develops with the therapist. Attachments, both positive and negative, to parents and other important people who have molded the personality of the patient, are recapitulated in the transference relationship. It is through this recapitulation, with the guidance of the therapist, that the "core conflictual relationships" are brought into conscious awareness and resolved, promoting healthier patterns of behavior.

In classical psychoanalysis the therapist strives to be a neutral figure, fostering an accurate reflection of the patient's conflicts in the transference attachment. Through a technique called "free association," where the patient is instructed to voice his or her thoughts in a continuous stream, and through information gathered from dreams, the unconscious conflicts are revealed and "interpreted" by the analyst. No advice or suggestions are offered by the therapist in classical psychoanalysis; the goal is for the patient to reshape his or her life using the insights achieved during the treatment, which may last for years. Analysis is thus of marginal value in the treatment those who are critically ill, or of limited financial resources. However, the dynamic principles that Freud set forth have had a profound influence on the general practice of psychotherapy, especially in the United States, and have influenced public thinking about what motivates behavior in everyday life. That early experience plays a critical role in shaping adult behavior, and that unconscious forces can influence what we think and do, are concepts which have left an indelible mark on Western culture and literature.

Hans Brüner was a skilled practitioner of dynamic psychotherapy. On occasion, he would joke about it with the medical students as his last remaining loyalty to Vienna. "Hans knew his business," said Stephan. "But he also knew that psychoanalytic psychotherapy is not the answer to manic depression. In the long-term modification of my behavior, lithium has been fundamental. In fact, Hans insisted that our "work" together, as he called my desperate visits to his office, would not become productive until I was willing to dampen my mood swings with lithium. And he was right.

"Eventually I did agree to start taking lithium, but I hated the way it made me feel—as if I were in a cage, or stuck in an elevator which was between floors and wouldn't move up or down. That was initially a terrible experience for somebody who had grown accustomed to the air under his wings. It was the reason why I stopped the drug, but then returned to it after another high developed. Now I say to myself, when the pressure is on at work, or I'm planning a transatlantic visit like this one to see Margaret and Michael, 'What a Godsend to know that I won't flip!' That's why I stay on the lithium. To me it is comparable to the medicine my mother takes for her asthma. When she's on inhalers twice a day, she doesn't get pneumonia and no one has to drive her to the emergency room in the middle of the night. But she hates taking medication, and after she's been well for a few months she thinks the inhaler is some sort of unnecessary madness—until she stops it and, bang, gets into trouble again. I've learned my lesson. Besides, I'm one of those lucky individuals with the profile of illness, acute mania followed by depression, that responds well to lithium. For me lithium *is* a Godsend and so I take it religiously, every day."

Stephan paused in his reminiscence and seemed pensive. Hans Brüner had died suddenly of a heart attack some two or three years previously and I asked Stephan if he was thinking about Hans's death.

"Yes," Stephan confessed, "I was thinking about Hans. The period after his death was a tough time for me and I was depressed for a while, although nothing like I had experienced before—no insane ideas about suicide or anything like that. A genuine period of mourning. It was encouraging actually, to go through a 'normal' period of grief. It was the first time I had weathered the loss of a close personal attachment without falling apart. It convinced me that it was possible to manage this illness, after all. As you may remember, I had stopped seeing Hans as a patient several years before he died, but he had been instrumental in helping me turn things around and I was deeply saddened."

Stephan drew a deep breath. "Looking back, I realize that I had been pretty sick when I first went to Hans. During those early months I couldn't see anything objectively. I was jaded and pessimistic one day, and unreasonable and demanding the next. My relationships with my friends and family were in tatters. Hans helped me, particularly with my feelings about my father. I was carrying a lot of baggage in that department. Growing up I had wanted to be closer to my father, and more like him. He was always the center of attention and I felt on the edge of things, excluded from his friends, and from the family in general. Everyone seemed to be smarter than I was. Margaret was my father's favorite, and being older she seemed to fit in, while I was shy as a boy, and desperate for attention. I would go into a room and say a timid 'Hello' and, when nobody answered, I would automatically think that I was disliked. Hans helped me to see those early experiences as one reason why I needed to be the best at everything, and why I was always boasting. I remember him telling me that Winston Churchill was similar as a boy—timid and sickly, but always full of bravado. That briefly lifted my spirits—sort of being best by morbid association."

I laughed. Stephan hadn't lost his sense of humor, but I was also curious. Attention seeking and expansive behavior are common in mania, especially in its early stages, and I wondered whether there was a connection between Stephan's striving for attention and the kindling of manic excitement. I asked him what he thought about such a possibility.

"Undoubtedly there was a connection," Stephan replied. "In fact, I don't think it was possible to untangle them—at least not after late adolescence, when my mood swings started. Before then I would with-draw if I felt excluded, but by the time I was in medical school I think one problem fed the other. For years, before I went on lithium, I was driving my vulnerability to mania by staying up late, overworking, drinking, and generally abusing myself. I continuously needed to excel to feel accepted. During those years I suspect my behavior drove the ill-ness, and vice versa." Stephan smiled. "But not all such striving is bad, you know. That was how I became good at chess. Playing chess was one of the only things my father and I did together. He liked to play, and I became good enough to even beat him occasionally. So when you and I were in medical school and I was teased about being the knight of The Seventh Seal, I was secretly pleased. As you may recall in Bergman's film, the knight, to stay alive, challenges Death to a game of chess and gives him a pretty good run. Now that's really being at the center of attention.

"It was when my father and I played chess that I felt closest to him," Stephan continued. "I still think about him when I play, even now. I sometimes wonder what it must have been like for him after the Hun-garian uprising, when he left his homeland and lost his social context. His suicide came as a great shock to me, but now I realize he must have been profoundly depressed. My mother tells me that he had been de-pressed occasionally as a young man, but had covered it up, which makes me wonder whether he was bipolar. Probably so, I have con-cluded, considering my own illness.

"I think what I learned from Hans is that we *define* ourselves through relationships. No individual is an island, no matter how invincible they may appear, not even my father. When he was forced to leave Hungary he lost his 'connectedness,' so to speak, and his world collapsed. And, after he killed himself, mine did too. The cycle of attachment and loss is at the core of our emotional being."

Stephan's experience is also mine, I reflected. Social attachment is the most powerful modulator of stress and emotion—a primary mammalian drive. As John Bowlby, the English psychoanalyst (and a colleague of Hans Brüner), has emphasized in his three-volume work *Attachment and Loss,* repeated grief can sensitize a child to depression, and loss later in life can reawaken that sensitivity. Bowlby was one of the first to draw parallels between research in monkeys, and the withdrawal that follows separation in these animals, and the depressive withdrawal occurring in human infants when a nurturing attachment is disturbed. He did much to move Freudian theory beyond the narrow concepts of transference, instinctual fixation, and repressed memory to the importance of disturbed relationships in initiating depression.

This shift in emphasis to interpersonal relationships and the consideration of cognitive patterns, as Stephan was describing in his work with Hans Brüner, has helped foster the development of new psychotherapies designed specifically for depression. It is an evolution in psychotherapeutic practice similar to the evolution that has occurred in antidepressant drug treatment. The shift was, in part, a reaction to the lengthiness of psychoanalytic treatment and to a frequent lack of success in achieving rapid symptom relief. Several time-limited therapies, including interpersonal psychotherapy and supportive-expressive psychotherapy, have been developed. Each focuses upon problems of self-distortion and how these distortions, through conflicted relationships, can initiate a vicious circle of depression and disturbed behavior. The best known of the new therapies, and the most widely employed across the globe, is cognitive-behavioral therapy, which as I mentioned in chapter 4 was conceived and developed by Aaron Beck, a professor of psychiatry at the University of Pennsylvania.

Cognitive therapy emphasizes positive thinking. The basic postulate is that, given accurate information, the brain can "think" its way back to health. By harnessing the powerful intelligence of the new mammalian cortex, the brain can learn to be objective about itself and replace old destructive schema with new, and constructive, thinking patterns. The cognitive-behavioral therapies differ from the psychodynamic therapies by placing greater emphasis upon the active participation of the patient—including mental exercises and homework—in adopting new thinking strategies to tackle specific problems and symptoms. The ultimate goal in developing these new cognitive models is to regain personal control of the social environment through a program of self-education, which Martin Seligman has described as "learned optimism."

Cognitive-behavioral therapists believe that conscious thought, rather than unconscious motivation, determines behavior. Thus, with the proper guidance, individuals can *learn* to change their attitudes and behavior toward other people. Cognitive therapy is more actively structured, time limited, and goal directed than dynamic therapy, with the therapist seeking an active intellectual collaboration with the patient. The basic proposition is that specific ideas—the now-familiar schema— have been adopted as we learn about ourselves and the way the world is organized. Some of these ideas are inaccurate, distorted by the circumstances under which they were acquired. Nonetheless, they become automatic representations of what we believe the world to be, a kind of self-fulfilling prophecy, continuing to distort our emotional vision, regardless of the truth of the matter.

These automatic thoughts are in many instances the prelude to depressed mood, fostering negative thinking about ourselves, others, and the world in general—what Beck has called the *cognitive triad* of depression. Thus the vicious circle is established, disturbing our emotional view of the world and creating stress, which in turn generates further disturbances of emotional expression and interpretation. The goal of cognitive therapy is to break this cycle, and to provide patients with problem-solving skills. The time spent with the therapist is extended by personal "homework" when, through cognitive reeducation, patients learn how to monitor their own thoughts and behavior, instruct themselves in novel circumstances, and objectively evaluate situations when things go wrong, or cause anxiety. The goal is not only to change behavior but also to revise the perceptual schema (the automatic thoughts, or attributional style) that drive the maladaptive behavior, until ultimately the therapist is no longer required, and the patient becomes his or her own mentor. Some of these therapies have been compared, head-to-head, with antidepressant drugs in the treatment of acute episodes of depression, and found to be effective, especially in milder illness.

This is especially good news for those who suffer a single episode of depression, and where the illness falls short of the profound disturbances in emotion, thinking, and body housekeeping that are characteristic of melancholia. It also confirms, and follows logically from our understanding of the limbic apparatus as a homeostatic system, that ultimately "psychosocial" and "biological" interventions both facilitate a common pathway of regulation. In treating minor episodes of depression it is unnecessary to employ the perturbing power of the antidepressant drugs, if reeducation can achieve the same result. This is especially so if antidepressants, as some psychiatrists have argued, can sensitize and destabilize limbic pathways, even in patients without bipolar illness.

In *recurrent* depression, without mania, the story is different. Professors Ellen Frank and David Kupfer and their colleagues at the University of Pittsburgh, in a carefully conducted study where patients with recurrent unipolar depression were followed for three years, found that a maintenance dose of tricyclic antidepressant provided a significant advantage in preventing reoccurrence. Those individuals who, after their initial recovery from an episode of depression, did not receive antidepressant medication, lapsed rapidly into illness. These findings have been confirmed by other researchers and resonate with the brain-imaging studies of psychiatrist Wayne Drevets, which I described in chapter 6. Doctor Drevets found that in patients who had recovered from recurrent episodes of depression, and who were no longer receiving treatment, there were residual disturbances of blood flow in the region of the amygdala. In a subsequent study Dr. Drevets determined that it was only in those patients receiving maintenance antidepressants that the limbic blood flow returned to a normal distribution.

However, of particular interest in the Pittsburgh study was the observation that those patients who received interpersonal psychotherapy *in addition* to medication had the best outcome of all patient groups. Eighty-four percent of those individuals who received a combination of psychotherapy and the antidepressant drug imipramine remained well for the duration of the three-year study. In these results we see the multiplying power of combining what are generally described as "biological" and "psychological" interventions. The superior outcome of their combination emphasizes that, while operating at a different level of the scientific staircase, each treatment contributes in a significant way to emotional homeostasis.

Manic depression is the most predictably recurrent mood disorder. Approximately 95 percent of those who suffer mania will experience recurrent manic or melancholic episodes throughout their lives, and before the discovery of lithium many became irretrievably ill. Thus, it is safe to conclude that, for the majority of sufferers, lithium, or some other mood stabilizer such as an anticonvulsant, will be required to modify neuronal excitability and moderate the illness. Antidepressants, too, may be necessary should melancholy become profound. But is that enough? Can bipolar illness be stabilized by medication alone?

I asked Stephan his opinion.

He thought for a moment. "Let me put it this way," he said. "There are some fortunate individuals whose illness responds dramatically to lithium—I, myself, have been lucky in this regard—but my sense is that everyone who struggles with manic depression can improve his or her life by understanding the illness, and managing it aggressively. The comparison to managing a life with diabetes, in my opinion, is a valid one. Both are chronic illnesses, and in both instances the disordered homeostasis—of sugar metabolism, or of mood—can be modified by medications that intervene at the cellular level. But that is only the beginning. If, with diabetes, one eats cream cakes every afternoon, takes little exercise, and insists on drinking a bottle of wine with dinner, the illness will soon be out of control. Insulin alone does not ensure health. Some thoughtful reconstruction of one's daily habits is also required.

"So it is with manic depression. Lithium in most instances will modify the profile of the illness, dampening—even eradicating—the manic highs and softening the plunging depressions, but swings of mood do not disappear. To complete the task, most of us also require a diet for the soul. Fortunately, in practice this differs little from sensible nourishment for the rest of the body. I have learned to control my caffeine and alcohol intake, like the diabetic, and I avoid other substances

I know will disorganize my mood, such as cold medications. I try to maintain a social rhythm, exercising a little each day and letting my energy run off in the evenings by reading or listening to music. And sleep is very important. For me it is the bellwether of changing mood. When my sleep shortens I know I'm getting high, and if I'm fatigued it's a sure sign that a depression is approaching. Hence I try hard to be in bed at my regular hour each night, and to stay there for seven hours. Then, when my sleep changes, I have a benchmark from which to understand the alteration that I am experiencing."

Stephan paused. "By now you must be thinking that my life is all routine and deadly boring. It is not true. I have a stable circle of friends, an interesting job, and like everyone else I still have my passions. Life *is* emotion, after all. The limbic brain is who we are. Some time ago, in one of my more maudlin moments, I decided that God had provided our gigantic cortex simply to teach or torture us. Having endured several years of torture I've now opted for teaching my cortex how to cope with my illness. So, perhaps life *is* boring by comparison to my past experience, but I prefer my daily existence to the roller-coaster I once rode.

"At the practical level," Stephan continued, "when it comes to managing my life, I'm now in the cognitive camp. I don't see manic depression as a 'thinking' disorder; I'm with you that it's a limbic *system* problem, and that thinking is only part of the picture. But from the patient's standpoint, in coping with this illness I find that 'thinking about how one thinks' is an essential tool. It provides a sense of personal control and complements the wise use of medication. I know from experience that in the early phases of mania, or at the start of a depression, my pattern of thinking, and my emotional coloring of events, is distorted—in subtle ways at first, but distorted nonetheless. Learning the skills (which cognitive therapists teach well) to identify these subtle changes really helps in managing mood, and in keeping ahead of the natural swings of the illness. And the same skills help to identify those situations which are stressful and need some special attention. So I monitor my thinking patterns as an index of my emotional balance—rather like checking the blood sugar level in diabetes."

Listening to how pragmatic Stephan had become, I found myself recalling what he had said at the beginning of our conversation, that he used to think that physics would eventually explain everything, and how living with the manic depression had changed his mind. I asked him to elaborate.

"I wasn't being mysterious," Stephan replied. "I simply meant that no physical or chemical explanation will ever be found that will negate my experience of this illness. The subjective reality of what I have been through is part of me, and that will never be changed. I learned to accept that reality, and even to appreciate it, in my psychotherapy with Hans. He taught me the true meaning of accepting my 'self'—to work with what I have and to build for the future. Through the insights that I gained about myself, and others, I came to understand that none of us is perfect, and that basically, despite this illness, I am a decent human being. The rough and tumble of my childhood, even my relationship with my father, was not that out of the ordinary. If it weren't for the peculiar vulnerability of this illness I might have been considered rather dull—just one of the crowd. But either way I'm still unique. That's what I learned from Hans, and from this illness. I didn't learn it from the science of physics.

"Psychotherapy is open to objective study but it is not a science, as Hans would remind me on occasion. But, within the context of a warm relationship with another human being whom I admired, and through psychotherapy, I was able to grow into myself and learn to manage my illness. I think you psychiatrists call that a therapeutic alliance. I believe it's an alliance that is essential to getting well—a sense of mutual respect and trust. Indeed I suspect it is part of any healing relationship that works. I have it now with the psychiatrist who manages my lithium, and who has taught me the cognitive tricks I was describing, and it's there too, although in a different form, among the members of the Manic Depression Fellowship group that I attend. There I can walk into the room and I don't feel that immediately I must explain myself. There's a sense of mutual support and caring that is unique among those who have heard the angels of madness.

"Manic depression brings you close to the marrow of existence. At times it is compelling in its absurdity, promising life at a more passionate level. But it is an underworld, a world where it is difficult to learn the truth of things, and where it is dangerous to live. Only in standing back from that realm have I come to recognize that life's true beauty is in its balance and its harmony. I am content now, in the place where I have found myself.

"But I'm getting carried away," Stephan said. "This has been like old times, except that I am exhausted and must go to bed. It's almost three in the morning in London. I've learned to respect the power of jet lag in stirring the chemicals of the brain, so I will see you in the morning."

I sat for a while, looking out at the lights of the city. Stephan hadn't really changed, and the realization gave me a good feeling. Despite illness and struggle, Stephan remained his essential self. I was happy.

Epilogue

The Human Carnival

If the carnival is a metaphor for existence, then it is emotion that brings color to the pageant. Emotion is an artist's palette for each of us, supplying tints for mood and memory. The moods we paint are shaped by the thoughts and conditions that justify them, for mood is our social connection. We share our moods, we enjoy them, we suffer them. We trust our moods as an intimate part of the self.

But moods go wrong. And when they do, they significantly alter our familiar behavior, changing the way we relate to the world, even changing our perception of who we are. In their most aberrant form, these are the disordered states of mind we call "mania" and "melancholia," the ancient masks of the self that for centuries have been persecuted and misunderstood. Even now, as we reach toward genuine knowledge, these illnesses continue to thrive, often striking those who seem most able in our society.

The symptoms of mood disorder are widespread, and some researchers believe that depression has grown more prevalent in modern times. Many people have a few symptoms; a few people have many. Repeated surveys in a Swedish community suggest that the incidence of mild and moderate illness is growing, especially in young adult males. A current argument is that we are entering an age of melancholy.

To embrace the evidence that mania and melancholia are syndromes found throughout human society—transcending culture and language—it is not necessary to join such debate. There is no disagreement that mood disorders are prevalent and persistent across the world. The question is *why*?

For answers, Randolph Nesse, a psychiatrist at the University of Michigan, John Price, an English psychiatrist, and a number of other scientists believe that we must look beyond the suffering of the individual to the evolutionary significance of emotion and the adaptive value of the behaviors that are magnified in mania and depression.

Is there a greater purpose to pleasure and pain?

The personal search, Americans are told, is for life and liberty. We are in pursuit of happiness. We would rather avoid sadness, for it makes us uncomfortable. During a recent interview with a magazine writer, I was asked what hope I could give to those who suffer the "blues." "Not serious depression," my interviewer explained, "but the blues of everyday life. In the future, will antidepressants eliminate sadness, just as fluoride has eradicated the cavities in our teeth?" The short answer is no, for antidepressants are not mood elevators in those without depression, but the question is provocative and significant for its cultural framing. In many Western countries, the pursuit of pleasure has become the socially accepted norm.

Behavioral evolutionists would argue that our increasing intolerance of negative emotions and dependence on mood altering drugs perverts the function of emotion and is maladaptive for the species. If we have learned anything in the past few decades about the brain and about emotional behavior, it is that transient episodes of anxiety, sadness, or elation are part of normal experience and actually serve a dynamic purpose in the development and maintenance of human social order. The emotions are not aberrations of mind but barometers of experience that have been essential to our successful evolution. Seeking ways to blot out variation in mood is equivalent to the ancient mariner throwing away his sextant, or the airline pilot ignoring his navigational devices. Emotion is the homeostat of life, an instrument of social self-correction—and when we are happy or sad, it has meaning.

If we consider the cross-cultural persistence of mood disorders from this evolutionary perspective, some interesting ideas emerge. Perhaps mania and melancholia endure because they coexist with behaviors that serve a greater human purpose, attributes that have had survival value for the individual and thus, indirectly, are useful to society. Did the pain of depression, for example, once offer the profit of conserving limited resources during times of danger and deprivation, and does the excitement of mania spring from the unusual energy necessary to rapidly exploit opportunity? In short, are the thymic temperaments and the vulnerability to mania and melancholia a legacy of human success—aberrant combinations of a pool of genes that has conferred evolutionary advantage?

There is some evidence to support this. Mood disorder is not random in its distribution. The thymic temperaments and the syndromes of mania and melancholia are more prevalent in the general population than can be explained by chance genetic misprints. Furthermore, considerable evidence suggests that bipolar illness clusters more commonly in prosperous families, where members of the professions, political leaders, and successful creative artists are significantly overrepresented.

Winston Churchill's family line provides an example. Mr. Churchill, who was dogged by depression throughout his life, summoned the vision and strength at a time of crisis to lead a nation, as had his great ancestor the Duke of Marlborough, who generations before him had also suffered from depression.

Over the next decade or so, with sophisticated tools for molecular dissection increasingly available, it will be possible to explore some of these questions directly at the genetic level. I believe we will discover that the final common path of molecular disorder, which disrupts the regulatory systems controlling mood and limbic communication, will be determined by a number of genes. Rather than one molecular misprint, it seems likely that the vulnerability to bipolar disease results from a combination of molecular deviations, a situation comparable to that which is emerging as the probable explanation for the genetic vulnerability to Alzheimer's disease.

But whereas Alzheimer's is a degenerative disorder occurring late in life, bipolar illness emerges early and is a disorder of communication between neurons, a chemical imbalance of information processing that is exquisitely responsive to the social environment. Furthermore, I suspect that different permutations in the number, or combination, of genes within the same genetic constellation will code for illness in some individuals but in others for the energy and stamina that we associate with leadership. Thus we may find that carrying one susceptibility gene for mania results in energy and optimism, while carrying two such genes predisposes a person to psychosis and a loss of the selective advantage. Such genetic multiplication would help to explain the occasionally observed "noble" intertwining of manic-depressive illness and creativity in families of high achievement.

The advantages of the generative energy of hypomania are obvious to all those who have experienced or witnessed it. The characteristic acceleration and clarity of thought are not commonly viewed by those who suffer as illness but as gifts to be preserved. To the evolutionist, the potential fecundity of increased sexual desire makes hypomanic behavior—and the energy and drive expressed in the hyperthymic temperament—an important adaptive strategy for species survival. Hypomania, it can be argued, is good for the individual *and* for the social group, which may explain why the genes that seed it have endured—and perhaps even increased in prevalence over thousands of years.

Several suggestions have been made to explain the survival value of behavioral depression. Withdrawal and low energy is the flip side of mania, the built-in braking system required to return the behavioral pendulum to its set-point after a period of acceleration. In the infant, of course, the protest that follows a broken attachment and which preludes withdrawal has survival value in attracting the attention of the infant's parents. Such caring extends inconsistently into adult life and to the social support that is commonly provided during illness or misfortune. The behavioral evolutionists have also suggested that depression has adaptive value in helping to maintain a stable social hierarchy. After the fight for dominance is over, the vanquished withdraws, no longer challenging the authority of the established leadership. And, for the defeated individual, such withdrawal provides a respite for recovery and an opportunity to consider strategic alternatives to further bruising battles.

Thus emotional flexibility is the key to survival and to safety. The vision and energy to exploit opportunity must be balanced by the wit to withdraw. The swings that mark the manic and melancholic experience are musical variations upon a winning theme, variations that play easily but with a tendency to become progressively off-key. For a vulnerable few the adaptive behaviors of flexible social engagement and withdrawal unravel under challenge and stress into the polar syndromes of mania and melancholic depression. These disordered states are dissonant and maladaptive for the individuals who suffer them, but their roots draw predominantly upon the same genetic reservoir that has enabled our evolution as successful social animals.

These musings will become increasingly important as we learn more about the genetic library that shapes complex human behavior. All human societies have ways of acknowledging and managing behavioral deviation—the punishment of crime, the care of suffering, the celebration of success—but, historically, deviations of mind have fallen betwixt and between established categories. Manic-depression is considered an illness in the United States, but those who suffer it rarely receive medical benefits comparable to those afforded other grave illnesses. American society has consistently discriminated against those with mood disorder.

A growing library of genetic information will soon force us to re-visit this traditional prejudice. The deliberation will test our humanity. If the thinking of our evolutionist scholars is on target, then risk-taking, out-in-front Wall Street traders, temperamental opera stars, highly competitive athletes, and the Stephan Szabo's of this world all have much in common. In a few years, when DNA screening by a "genotyping–computer chip" has replaced fingerprinting, we may be surprised to find that these behavioral stereotypes all reflect deviation from the "emotional genetic norm." When that day comes will we still applaud those who fall into the first three categories, while depriving and stigmatizing those in the last?

We are not there yet, but science is moving rapidly. Several research groups across the world are now combing the genome of extended families carefully selected for their members' vulnerability to manic depression or to recurrent depressive illness. The early fruits of the effort have been informative. There is no single gene, operating under simple Mendelian rules, that dictates bipolar instability. The long arm of chromosome 18 may be important, although the region identified contains many genes and no specific candidates have emerged. Even when they do, it will take time to identify the precise regulatory functions for which the candidate genes carry responsibility.

There will be no miraculous medical breakthroughs, but for those who suffer recurrent mood disorder there will be many benefits from the new genetic knowledge. An understanding of the regulatory misprints that contribute to the instability of emotional homeostasis will aid the design of new drugs—improved mood stabilizers and antidepressants that can modify the long-term expression of the disorder. Genetic screening ultimately will permit early intervention for those at risk. Eventually, prevention will replace containment.

But genes are not destiny. Early evidence suggests that a single gene for anything is extremely rare, especially when it comes to behavior. If one parent has manic-depressive illness, the risk that the infant will develop the disability is still less than 20 percent. Natural selection weeds out the major genetic defects, and mania and depression are not among them. Attachment in infancy, the attention we receive, and what we learn from others will forever be important in shaping who we are. In untangling the web of exhilaration and despair, genetic screening will determine risk, not destiny.

Broad ethical issues remain. Will neuroscience and genetics help bring wisdom to our understanding of the disorders of mood and spur new treatments for those who suffer these painful afflictions? Or will some members of our society harness genetic insight to sharpen discrimination and drain compassion, advocating another eugenic surge under some pseudo-scientific banner? We must remain vigilant, but I am confident that our humanity will prevail, for all of us, in truth, have been touched by the disorders of the emotional self. Mania and melancholia are illnesses with a uniquely human face.

For the moment, I have hung the mask from the Dorsoduro upon my study wall. It's a daily reminder of the carnival, of moods apart, and of the privileged time in which I make my professional journey.

Appendix

Antidepressants and Other Medications for Mood Disorder

A Glossary and Guide

First, a note of caution: Antidepressants and mood-stabilizing drugs are powerful medicines that should be taken only under the guidance of a physician who is knowledgeable in psychopharmacology. The chemical operating systems of the brain are similar to those of other body organs and nerve centers, including the autonomic nervous system, the heart, and the gastrointestinal tract. Thus, drug interventions that modify mood may change blood pressure, heart rate, bowel habits, sexual performance, et cetera. A detailed understanding of the physiological effects of the medications, and interactions occurring among them, is essential to safe prescription and to an optimum therapeutic outcome.

This appendix lists the medications most commonly used in the treatment and prevention of mood disorders in Europe and the United States, circa 1998. It is not an exhaustive list. Some drugs prescribed in Europe, for example, are not available in the United States, and vice versa. Also, because the proprietary names differ among countries, to reduce confusion the tables include only the chemical names of the medicines. The dose range given is that within which clinical improvement usually occurs. However, the required dose must be carefully assessed for each individual and will vary with age, degree of illness, tolerance for side effects, and the speed of metabolism and excretion of the drug from the body. Also, if the upper limit of the dose range is exceeded, severe toxicity may occur and require emergency intervention.

ANTIDEPRESSANTS

Antidepressants are valuable in the treatment of major depression and in the prevention of episodes in those individuals who suffer recurrent depressive illness. In bipolar illness, the antidepressants must be used cautiously because some, particularly the tricyclics, can precipitate mania. Hence, to reduce this possibility in those with recurrent bipolar illness, antidepressants are invariably prescribed with a mood stabilizer, such as lithium, or one of the anticonvulsants.

Some antidepressant drugs are also effective in psychiatric syndromes other than depression, including anxiety, chronic pain, obsessive-compulsive behavior, and panic disorder. This is because, although different brain systems are disturbed in these syndromes, the operating chemistry is commonly similar to that dysregulated in mood disorder.

Tricyclic Antidepressants

Named tricyclics because of their three-ringed chemical structure, these drugs have become the standard by which other antidepressants are measured. Tricyclics perturb several neurochemical systems, including acetylcholine neurotransmission, which results in a drying of the mouth and eyes, constipation, and sometimes difficulty in passing urine. In their antidepressant action, the tricyclics can be divided into those acting predominantly on serotonin regulation and those acting on norepinephrine (NE). However, to a greater or lesser degree, each disturbs the neuronal re-uptake mechanisms of both systems and also has a minor influence upon the dopamine system. (See Table 1.)

Selective Serotonin Re-Uptake Inhibitors (SSRIs)

Because the action of the SSRIs is confined principally to the serotonin system, these drugs have few side effects compared to the tricyclic antidepressants. The molecular structure of the transporter system that recycles serotonin into the neuron, and the gene that is responsible for manufacture of the transporter, has been determined. The SSRIs discretely block the activity of this transporter system and the re-uptake of serotonin into the neuron is selectively prevented. Although these drugs also disturb the transporter systems that control norepinephrine and dopamine, such activity is minor in comparison. The practical advantage is an absence of anticholinergic side effects (no dry mouth, constipation, et cetera) and less effect upon the heart. This makes the drugs more patient-friendly. Adverse effects are not absent, however. Im-

Table 1
Tricyclic Antidepressants

Drugs	Therapeutic Dose Range	Predominant Action	Major Side Effects
Amitriptyline	100–300 mg	serotonin/NE	dry mouth, sedation
Clomipramine	100–300 mg	serotonin	dry mouth
Desipramine	100–250 mg	NE	rapid pulse
Doxepin	75–300 mg	NE/serotonin	dry mouth, sedation
Imipramine	100–300 mg	NE/serotonin	dry mouth, low blood pressure
Lofepramine	140–210 mg	NE	sweating, rapid pulse
Nortriptyline	50–200 mg	NE/serotonin	dry mouth
Protriptyline	20–60 mg	NE	insomnia, sweating
Tianeptine	35–40 mg	serotonin	GI distress
Trimipramine	100–300 mg	NE/serotonin/DA	sedation

potence, difficulty with orgasm, nausea, and gastrointestinal intolerance are common (up to 30 percent of individuals), and agitation and confusion—the serotonin syndrome—sometimes occurs secondary to excess serotonin in the body. But with these caveats, the SSRIs are safe and effective antidepressants that have become the preferred drugs in the initial treatment of most depressions. There is evidence that in severe melancholic depression the tricyclic antidepressants have an advantage, with ECT being more effective than either class of drug. (See Table 2.)

Other Antidepressants

These include the monoamine oxidase inhibitors (MAOIs). First discovered in the 1950s, these drugs remain valuable treatments for "atypical" depression where "bodily" disturbances (pain syndromes, low energy, decreased sleep without mood changes) predominate. The MAOIs inhibit the activity of enzymes that destroy both adrenergic and serotonergic neurotransmitters, and their antidepressant action is thought to occur through an increase in the levels of these messenger amines at the synapse.

Because this increase occurs in the body generally, not just in the brain, the MAOIs can be dangerous drugs unless a diet to control amine intake is strictly adhered to. Dramatic increases in blood pressure, with headache, nausea,

Table 2
Selective Serotonin Re-Uptake Inhibitors

Drugs	Therapeutic Dose Range	Predominant Action	Major Side Effects
Citalopram	20–60 mg	serotonin	GI distress
Fluoxetine	20–60 mg	serotonin	GI distress, sexual difficulties
Fluvoxamine	150–300 mg	serotonin	GI distress
Paroxetine	20–40 mg	serotonin	GI distress, sexual difficulties
Sertraline	50–200 mg	serotonin	GI distress, sexual difficulties

and sweating, can occur if incompatible foods (aged cheese, red wine, salted fish, sausage, chocolate) or drugs (cold medicines, amphetamines, and other antidepressants) are consumed. For these reasons, the diet and drug restrictions must be managed by a knowledgeable physician. The risk of such reactions has been reduced, however, with the manufacture of drugs that are "reversible" in their blockade of the monoamine oxidase enzyme. Moclobemide, available in both the USA and Europe, is an example of such a reversible MAOI.

Several other drugs with significant antidepressant properties are also available. These include the recently manufactured Venlafaxine, which targets both the serotonin and norepinephrine transporter systems but without impairment of cholinergic function. Venalfaxine has already established itself as an effective agent in a broad range of depressions. Mianserin, recently marketed in Great Britain and the United States, is another highly effective antidepressant which blocks several neuronal receptors to increase serotonin in the brain.

Several older antidepressants, for which the mechanism of action is unclear, are also available. A special mention here must be made of the ancient herbal remedy, Saint John's wort, which, in Germany, has recently returned as the most popular medication prescribed for depression. The active ingredients of this herb (with the botanical name Hypericum perforatum) probably inhibit the monoamine oxidase enzymes to produce a mild antidepressant effect, with few side effects. Extension clinical trials are now underway in several countries to compare the efficiency and safety of Hypericum with the manufactured antidepressants. (See Table 3.)

In selecting an antidepressant, the physician considers the symptoms that predominate in the depressive syndrome, their severity, and also how the side-effect profile of the drug will improve or worsen the symptoms experienced

Table 3
Other Antidepressants

Drugs	Therapeutic Dose Range	Predominant Action	Major Side Effects
Amoxapine	200–300 mg	mixed	stiffness of movement
Buproprion	250–350 mg	dopamine	restlessness, insomnia
Maprotiline	150–200 mg	NE	insomnia
Mianserin	60–90 mg	serotonin	minimal
Nefazodone	200–600 mg	mixed	headache
Reboxetine	8–10 mg	NE	insomnia, sweating
Trazadone	300–600 mg	mixed	drowsiness/sedation, penile erection
Venlafaxine	75–300 mg	serotonine/NE	increased blood pressure
Phenelzine	45–90 mg	MAOI	danger of hypertensive crisis
Tranylcypromine	30–60 mg	MAOI	danger of hypertensive crisis
Moclobemide	450–600 mg	reversible MAOI	reduced danger of hypertension

(for example, an individual with restlessness and insomnia may benefit from a sedating antidepressant).

The liver is the predominant site of destruction for most drugs. In particular, a family of enzymes (cytochrome P450) is responsible for the metabolism of antidepressants and many other drugs. Through competition for these enzymes, and other mechanisms, the speed of metabolism of a drug changes when other drugs (for example, antibiotics and other commonly prescribed drugs) are added to a treatment program. Understanding the implications of these interactions between drugs is very important and is another reason why psychotropic medications must be taken only through a physician's prescription and oversight.

ANTI-MANIC AGENTS AND MOOD STABILIZERS

In acute mania, the dopamine-blocking agents, (commonly known as neuroleptics and principally used in the treatment of schizophrenia) are widely employed to decrease excitement and to control racing thoughts. Chlorpro-

Table 4
Anti-Manic Agents and Mood Stabilizers

Drugs	Therapeutic Dose Range	Predominant Action	Major Side Effects
Lithium	600–1,800 mg	neuronal second messenger systems	thyroid dysfunction
Carbamazepine	600–1,200 mg	neuronal conduction	reduced blood count
Valproate	750–2,000 mg	neuronal conduction	liver dysfunction

mazine, Thioridazine, Trifluroperazine, and Haloperidol are among the most commonly used. All of these drugs are extremely potent and cause many side effects, including anticholinergic effects (dry mouth, constipation, et cetera), sedation, low blood pressure and a fast heart rate, stiff arms and legs and other disturbances of body movement (called "Parkinsonism"). Weight gain also occurs when these drugs are prescribed over a long period of time. Another long-term complication, emerging over months and years in some individuals, are abnormal motor movements of the face and tongue called tardive dyskinesia. Fortunately, in bipolar illness it is rare for the neuroleptics to be required for prolonged periods of time.

Other drugs effective in acute mania are lithium carbonate and the anti-convulsants valproate and Tegretol. Of these, only lithium carbonate and valproate are approved by the Federal Drug Administration in the United States for the treatment of acute mania. (See Table 4.)

The blood levels of lithium and the anticonvulsants must be monitored carefully—especially in the early stages of treatment when the dose is being stabilized—because the gap between a therapeutic dose and the dose when toxic side effects occur is relatively narrow. Once stabilized, the dose is monitored every two to four months. Interaction with other medications, fever, dehydration, et cetera, may alter the dose required, and thus all individuals who take these medications must do so under the regular guidance of a physician. The doses required to prevent recurrence of illness (prophylactic) are usually lower than those required in the treatment of acute mania. However, at the moment, only lithium is a proven prophylactic agent; the anticonvulsants appear to be so, but long-term clinical studies are incomplete.

In the long-term pharmacological management of bipolar disorder, mood stabilizing drugs are frequently used in combination. When lithium alone does not stabilize the illness, an anticonvulsant is added. Sometimes two anticonvulsants are used without lithium.

Anti-anxiety drugs, principally the benzodiazepines, are commonly prescribed to help stabilize mood on a short-term basis. There are many benzodiazepines and they have been put to various uses depending upon their speed of action and the length of time they stay in the body. Hence, Triazolam and Temazepam have been used as hypnotics because the speed at which they produce sedation is rapid; they are also quickly cleared from the body. Diazepam, although still sedating, is used to decrease anxiety, producing less drowsiness and lasting longer in the body.

The two benzodiazepines commonly used in bipolar illness to stabilize excitement and improve sleep are Clonazepam and Lorazepam. Within prescribed limits, they can be used at patient discretion to manage periods of excessive anxiety. However, benzodiazepines, like alcohol, can be addicting and they must be used cautiously.

Pharmacology, by itself, is rarely a complete treatment for mood disorder. Only with psychotherapeutic intervention and self-education, plus a careful recording of the mood shifts and their relationship with the environment, will most disturbances of mood be brought under optimum control. It is also important to emphasize that the antidepressants and mood stabilizers do not elevate normal mood. These are drugs that perturb and stabilize abnormal mood.

Notes

Prologue

Page xiv, " . . . LIFE AS A GREAT SPLENDOR": Diego Valeri, *A Sentimental Guide to Venice* (Florence: Guinti-Martello, 1978). The unabashed sentimentality of the several short essays in this little book capture the moods of Venice, alternately boisterous and expansive, sad and melancholy. Published originally in Italian, it was translated into English by Cecil C. Palmer.

Page xiv, THE VENETIAN CARNIVAL: *Carnival* means "to put away meat." Today, the carnival in Venice is approximately a ten-day festival occurring in February before Shrove Tuesday, which marks the beginning of Lent, but in the Middle Ages such revels sometimes lasted two months. In Venice, the carnival reached its peak in the eighteenth century and was famous for its excess and debauchery.

Page xiv, THE MASK: The carnival mask grew popular through the *commedia dell'arte,* which was the forerunner of the pantomime. The characters in this popular entertainment reflect the diversity of human temperament. The Harlequin, a witty, amorous clown with a great deal of energy, is the sanguine (hyperthymic or manic) temperament, while at the opposite pole Pagliaccio, the lovesick, melancholy character that developed into the modern circus clown, represents the depressive or melancholic temperament.

Page xv, IN JESTER'S GARB: The jester of the medieval court entertained with caustic wit and mercurial mood. Usually dressed in a multicolored doublet and tight breeches, he had bells on the costume and ass's ears—as in the MondoNovo mask found in Venice.

Page xv, MANIA IS TO DEPRESSION AS FIRE IS TO THE ASHES OF THE FIRE: For this fine metaphor I am indebted to Dr. Athanasio Koukopoulos, my friend and colleague from Rome.

Page xvii, FRENCH PHYSICIAN CLAUDE BERNARD: Bernard (1813–1878), after unsuccessfully trying his hand as a playwright, turned to medicine and was probably the first to recognize that the secret to life is the control of the immediate environment. His book, *An Introduction to the Study of the Experimental Method in Medicine*, trans. H. C. Green (New York: Macmillan Press, 1927), is worth reading as a classic in the history of medical science. See also note on homeostasis, chapter 1.

Page xviii, A NEW DIMENSION IN SELF-CARE: With the advances in pharmacology and general knowledge about the nature of mania and severe depression, it is possible for patients to educate themselves about their illness and to help manage it, much as diabetes or asthma can be managed.

Page xix, MANY . . . ACKNOWLEDGE THEIR STRUGGLES: An increasing number of celebrities, artists, and public leaders, especially in America, are willing to step forward regarding their own experience of depression and mood disorder. These include, for example, William Styron (author), Patty Duke (actress), Art Buchwald (humorist), Jules Feiffer (satirist), Joan Rivers (actress), Rod Steiger (actor), and Mike Wallace (television newsman). For personal descriptions of their experiences, plus the experiences of others, see the compendium *On the Edge of Darkness* (New York: Doubleday, 1994) by K. Cronkite, who herself has experienced depression. Biographical books and memoirs include W. Styron, *Darkness Visible* (New York: Random House, 1990); P. Knauth, *A Season in Hell* (New York: Harper and Row, 1975); P. Duke and G. Hochmann, *A Brilliant Madness* (New York: Bantam, 1992); M. Manning, *Undercurrents* (New York: HarperCollins, 1994); K. Redfield-Jamison, *An Unquiet Mind* (New York: Knopf, 1995).

CHAPTER ONE: A GLIMPSE OF MELANCHOLY

Page 1, "NO ONE EVER TOLD ME THAT GRIEF FELT SO LIKE FEAR": In the opening of his book *A Grief Observed*, C. S. Lewis (1898–1963) catches the essence of grief. As we have come to know more about the neurobiology of grief, it is recognized that disruption of attachment creates the same fearlike experience in adults that any challenging, uncontrolled stressor produces in the young. This primal emotional reaction of fear is a function of the brain's amygdala complex, which I take up in some detail in chapter 6. Clive Staples (C. S.) Lewis was born in Ireland and spent most of his adult life teaching on the English faculty at Oxford University, and later at Cambridge. A devout Christian, he was noted for his acid wit and scathing criticism. Many of his philosophical writings remain extremely popular, particularly *The Screwtape Letters* and *The Allegory of Love,* but it is his books for children that made him most famous—especially the series *The Chronicles of Narnia,* and its first book, *The Lion, the Witch, and the Wardrobe. A Grief Observed* was first published in 1961, after the death of Joy Gresham, to whom Lewis had been married for four years. See C. S. Lewis, *A Grief Observed* (San Francisco: HarperCollins, 1989).

Page 1, "AFTER THE MIND HAS SUFFERED FROM AN ACUTE PAROXYSM OF GRIEF, AND THE CAUSE STILL CONTINUES, WE FALL INTO A STATE OF LOW SPIRITS": These are the opening lines of the chapter on grief in Darwin's book *The Expression of Emotion in Man and Animals,* first published in 1872 (the American edition was published by Appleton & Co., New York, 1897). The work had grown out of his earlier, controversial, *The Descent of Man.* Darwin faced considerable opposition from the Anglican Church for his theories, including those of emotional expression, and the assertion that human beings and great apes had evolved from a common ancestor. The Church considered the human power of expression and emotion unique—the result of divine creation. Darwin's book created the first stepping stones to understanding emotion as a pre-verbal signaling system shared with other social mammals, particularly primates. For those interested in the details of these colorful debates, see A. Desmond and J. Moore, *Darwin* (New York: W. W. Norton, 1991); and W. Irvine, *Apes, Angels, and Victorians: Darwin, Huxley, and Evolution* (New York: McGraw-Hill, 1955).

Page 4, EMOTIONAL EXPRESSION EVOLVED: Evolutionary psychology has become a popular means of attributing a larger purpose to individual behavior; it is really an updated version of Darwin's interest in emotional expression as social communication. The basic thesis is that happiness and sadness survive because they are advantageous in certain situations, improving "Darwinian fit" and the continuation of the species. Individually positive or negative emotion must be considered part of a larger system that serves the vigor of the social group rather than the individual. For a review of this particular perspective, see R. Nesse, "Evolutionary Explanation of Emotions," *Human Nature* 1 (1990): 261–289.

Page 5, THE WORD *EMOTION:* has a double meaning in English, which is rather confusing. It is employed largely as I have placed it in context here, to denote expression of emotion, but it is also used to describe the subjective experience. Some individuals prefer to limit the word *emotion* to expression, and reserve *feeling* for subjective experience. Personally, I think such efforts confuse further because emotion, for most of us in adult life, is the association of memory of past experience with a feeling. Thus I use the term *emotion* in this book to refer to a combination of feeling and memory and reserve *feeling* for the raw subjective experience of being aroused prior to "emotional" designation being made. Emotion is therefore used here as the experience which is both recognized subjectively, and expressed.

Page 7, ELEVEN TO FIFTEEN MILLION PEOPLE ... ARE AFFLICTED WITH MOOD DISORDERS: see M. H. Miller, "Dark Days: The Staggering Cost of Depression," *The Wall Street Journal,* Thursday, December 2, 1993. See also for details of comparative prevalence with other psychiatric illnesses, R. Kessler, K. McGonagle, S. Zhao, et al., "Lifetime and Twelve Month Prevalence of DSM-III-R Psychiatric Disorders in the United States: Results from the National Co-morbidity Survey," Archives of General Psychiatry 51 (1994): 8–19.

Page 7, FEWER THAN ONE-THIRD OF THESE MILLIONS EVER RECEIVE TREATMENT: J. Fawcett, "Overview of mood disorders: diagnosis, classification, and management," *Clinical Chemistry* 40 (1994): 273–278; *Mental Illness Awareness Guide for Decision Makers,* American Psychiatric Association, 1994.

Page 9, HOMEOSTASIS: The concept of homeostasis in the body's function was proposed by Claude Bernard (1813–1878). He is known as the founding father of experimental medicine. Bernard attended medical school in Paris, where he trained under the physiologist François Magendie. He held many highly regarded posts at institutes in France, including the Sorbonne and the Museum of Natural History, and won the prestigious Academy of Sciences prize in experimental physiology four times. He is famous for his concept of the *"milieu intérieur,"* translated as the internal environment of an organism. His many discoveries focused principally on the physiology of digestive function, including the role of the pancreas, the body's buildup and breakdown of glycogen in maintaining a stable blood sugar level, and the ability of the sympathetic nervous system to control the dilation and constriction of blood vessels in response to temperature changes in the environment.

The concept of regulatory control in the arrangement of living beings was foreign to the thinking of the nineteenth century. Bernard was the first physiologist to recognize that life depended upon exerting control over the immediate environment; he hypothesized that we are born with an innate ability to maintain order within our bodies through constant change. The principles of homeostasis and an interior milieu led directly to research on the brain's control of bodily hormones.

Page 11, EVENTS WE REMEMBER WITH PARTICULAR CLARITY ARE THOSE ABOUT WHICH WE HAVE STRONG FEELING: Over thirty years ago, as a medical student, I was driving on a remote mountain road in Turkey when a large bee flew in through the car window. I still have a vivid mental image of every detail of the corner of the road at which I stopped in panic and alarm. Similarly, people remember the moment in which they heard of the assassination of John F. Kennedy and other "stunning" events. Neurobiology tells us that the brain is primed to remember such events of emotional importance by the release of the fear hormones, epinephrine and norepinephrine, which sensitize the brain memory banks to the event. This can be understood as a clever evolutionary device ensuring that we remember most clearly those events that are most dangerous; the amygdala and hippocampus appear to be the brain centers particularly involved. I discuss this further in chapters 6, 7, and 8. Interested readers should also review L. Cahill, B. Prins, M. Weber, L. McGaugh, "Beta—adrenergic activation and memory for emotional events," *Nature* 371 (1994): 702–4. The research is also discussed in "New kind of memory found to preserve moments of emotion," Science Times section of the *New York Times,* October 25, 1994.

Page 12, GREGORIO MARAÑON: see *"Contribution à l'étude de l'action émotive de l'adrénaline,"* Revue Française d'Endocrinologie 2 (1924): 301–325. Gregorio Marañon's report was an important signpost in the development of theories about emotion bridging the earlier theories of James and Lange, and the later work of Schachter and his colleagues at Columbia University in New York (see below).

Page 13, STANLEY SCHACHTER AND HIS COLLEAGUES: The great American psychologist William James had proposed that the crucial variable in subjectively experienced emotion is a person's awareness of his physical response to the situation that generates that "emotion." Danger triggers a bodily change which involves the movement of muscles used in fleeing or fighting, or the visceral reaction of a racing heart or adrenaline rush. The awareness of these physiological responses is the experience of emotion—in this case of fear—argued James. Carl Lange, a contemporary of James, proposed essentially the same hypothesis; hence this particular theory of emotions is called the James–Lange theory. However, the question arises: If emotions are indeed merely awareness of visceral states, how do different emotions emerge from similar states of arousal? For example, the autonomic nervous system is similarly activated in situations producing both rage and fear. Yet we experience these as different emotions even though our bodies react with the same physiological changes. Building on Marañon's work, Landis and Hunt in America evaluated the effects of injecting subjects with adrenaline to see the effects of autonomic arousal. Subjects reported palpitations, tremors, and sweaty palms when, according to the James–Lange theory, adrenaline should have produced a feeling of rage or fear in the subjects. In fact, the majority of subjects simply reported the physical symptoms without "emotional" connotation, while some individuals reported feeling as if they were angry or afraid, distinguishing the feeling from the "real" emotion.

To account for these differences, Schachter and Singer proposed another theory of emotion which included the role of cognitive factors. Their theory posits that physical arousal provides a general state of excitement which is then shaped into a specific emotion by the person's interpretation of the situation. If the hands are shaking and the heart is racing and we have just seen a bear, we will attribute the physical reaction to being afraid of bears. Therefore, argued Schachter, emotional experience is the *interpretation* of autonomic arousal, taking into account the circumstances of the moment. To test this hypothesis, the experimenters injected subjects with adrenaline, which the subjects thought was a vitamin supplement. Some subjects were told the true effects of the drug, while others were told that it produced unrelated symptoms. As each subject was waiting for the drug to take effect, he or she met an actor employed by the experimenters who posed as a fellow subject. In some cases the actor behaved in a manner that provoked anger in the subject, and in other instances behaved in a friendly, joyous manner. When asked to rate their mood after the waiting period, subjects who understood the effects of the drug reported less emotion. They attributed their arousal to the effects of the drug. Conversely, the uninformed subjects attributed their arousal to external factors and reported feeling happy or angry depending on the actor they had witnessed.

Schachter's theory assumes all emotions are based on a similar state of physical arousal, which is probably untrue. Investigators still debate the number of states of physiological arousal that may exist, but agree that the physiological experience is further defined by cognitive interpretation shaped by previous experience—that feeling and memory are linked together in the development of emotion.

Page 14, STUDIES SUGGEST THAT ONLY ABOUT FIVE OF EVERY HUNDRED PEOPLE WHO GRIEVE GO ON TO EXPERIENCE A MELANCHOLIC DEPRESSION: In the absence of a family history of depression, the incidence of severe depression in those who grieve does not appear to be increased above the general population. Grief is merely a ubiquitous form of challenge and a common form of lost attachment. For a discussion, see P. Clayton, "Bereavement," in *Handbook of Affective Disorders,* ed. E. S. Paykel (New York: Guilford Press, 1982), pp. 403–415. Thus grief seems to influence the timing, but not the prevalence, of the development of melancholic depression.

Page 18, DIAGNOSTIC CRITERIA OF THE AMERICAN PSYCHIATRIC ASSOCIATION: See more about this in chapter 5. The diagnostic and statistical manual (DSM) of mental disorders is now in its fourth edition. Serious classification of mental illness really began after World War II, when the World Health Organization (WHO) added mental illness to its International Classification of Disease (ICD). The American Psychiatric Association first published its diagnostic manual in 1952, and it has been regularly updated since. There is considerable overlap between WHO criteria of the ICD and the DSM, but they do stand separately, particularly in the details of diagnostic criteria. These are much more detailed in the American system. For further details, see *Diagnostic and Statistical Manual of Mental Disorders,* 4th ed., (Washington, D.C.: American Psychiatric Association Press, 1994).

Page 18, BY REASONABLE ESTIMATES, 12 TO 15 PERCENT OF WOMEN AND 8 TO 10 PERCENT OF MEN: See T. Helgason, "Epidemiological investigations concerning affective disorders", in *Origin, Prevention, and Treatment of Affective Disorders,* ed. M. Schou and E. Stromgren (New York: Academic Press, 1979), pp. 241–255. Also in D. A. Regier, W. E. Narrow, D. S. Rae, "The de-facto US mental and addictive disorders service system," *Archives of General Psychiatry,* 50 (1993): 85–94. This paper reports the most recent findings of the National Institute of Mental Health (NIMH) epidemiologic catchment area (ECA) study of prevalence rates of mental disorders in the United States. This study, which began in 1980, estimates that approximately 10 percent of individuals will have a diagnosable depressive illness during any one year—

approximately 17.5 million people. Slightly higher figures are reported by Ronald Kessler and his colleagues at the University of Michigan; see "Lifetime and 12 month prevalence of DSM-III-R psychiatric disorders in the United States," *Archives of General Psychiatry* 51 (1994): 8–19.

Page 18, ESTIMATED TO EXCEED FORTY BILLION DOLLARS EACH YEAR: These statistics come from a survey conducted for the National Mental Health Association and are quoted in the publication *Mental Illness Awareness Guide for Decision Makers* (American Psychiatric Association Division of Public Affairs, 1994). It is estimated "that depression costs society 40.7 billion annually of which 55 percent is paid directly by American business through lost income, due to depression related absenteeism and reduced productivity. An additional 17 percent is due to lost lifetime earnings, primarily through suicide, attributable to depression. Only 12 billion, or 28 percent of the total of 44 billion, spent on depression are accounted for by the direct cost of professional psychiatric and other medical services, rehabilitation, and counseling." The cost of depression to society is also discussed in "Dark Days: The Staggering Cost of Depression," *The Wall Street Journal,* Thursday, December 3, 1993.

CHAPTER TWO: DARKNESS VISIBLE

Page 21, DARKNESS VISIBLE: From William Styron, *Darkness Visible: A Memoir of Madness* (New York: Random House, 1990). One of the greatest living American writers, Styron is the author of such novels as *Lie Down in Darkness, The Confessions of Nat Turner,* and *Sophie's Choice*. He suffered a serious episode of melancholic depression in 1985. *Darkness Visible* chronicles that experience in compelling, illuminating prose, offering insights far beyond those of the usual biographical essays on this subject. This little book is one of the clearest descriptions of how the autonomous self can be slowly consumed by the illness until the only perceived avenue of escape is death itself.

Page 23, "THINGS HAVE DROPPED FROM ME . . . THE SHEER INABILITY TO CROSS THE STREET": Claire Dubois was thoroughly familiar with the writings of Virginia Woolf and considered *The Waves* to be the book that most eloquently expressed the mind of sadness. Apparently, the title is taken from Virginia Woolf's own experience with depression. In her writing she uses the metaphor where she describes depressive waves flowing over her; only by understanding them, she wrote, could she avoid being thrown down by them. In this particular quotation employed by Claire Dubois to describe her own loss of feeling, Bernard, one of the six characters in *The Waves* (who describe their personal experience in a series of monologues), underscores the paralysis and anhedonia of melancholia. First published in 1931, *The Waves* is now obtainable as a Harvest Book by Harcourt Brace Jovanovich: New York, 1978.

Page 23, HAMLET, STRICKEN BY THE BRUTAL MURDER OF HIS FATHER: Shakespeare's *Hamlet,* written some twenty years before Robert Burton's *Anatomy of Melancholy,* demonstrates the clear recognition in Elizabethan times of the effects of grief upon thought and emotion and its association with melancholy. Hamlet, in profound grief for his father, sees an image of him in his mind and goes on to describe the subjective mood of what is now called anhedonia.

Page 25, WILDER PENFIELD: Wilder Penfield (1891–1976) was a famous American-Canadian neurosurgeon who developed a surgical method for the successful treatment of severe epilepsy. As a Rhodes Scholar he was introduced to the study of neurophysiology by Sir Charles Sherrington. Penfield attended medical school at Johns Hopkins University and later practiced neurosurgery in New York City, before joining the faculty at McGill University in Canada. There he established the Montreal Neurological Institute. His discoveries in brain and behavioral relationships evolved from the application of neurophysiological techniques in the study of the living human brain. His surgical technique for removing offending scar tissue from the brains of epileptic patients employed preliminary electrical stimulation to the exposed areas of brain, with patients in the waking state to minimize surgical damage. In cooperation with his patients, this electrical stimulation enabled him to map out specific areas of brain function including the motor, sensory, and speech centers of the cortex. Throughout his career he was interested in the neural substrate of "consciousness." His theory focused upon the importance of the role of the upper brain stem in the integration of sophisticated human functions (a "centrencephalic" view) rather than the prevailing view, that the new cortex played the central role in the organization of human behavior. Although these ideas were quite controversial at the time, current neurophysiological views support his hierarchical theories of brain function, which have some similarity to the triune brain of MacLean. I offer a more detailed discussion of some of these issues in chapter 6 when I explore the anatomy of the emotional brain.

Page 28, . . . AND VIRGINIA WOOLF BECAME HER CLOSEST FRIENDS: The novelist, short story writer, and critic Virginia Woolf was born in London in 1882 and struggled with manic-depressive illness throughout her life. She died in 1941 by suicide. She is also recognized for her insight into her illness and its relationship to her creative process.

Virginia Woolf's mother died when she was 13 years old, followed by her father's death when she was 22, after which she and her siblings lived together in the Bloomsbury district of London. It was there, as a member of the Bloomsbury Group, that she first formulated the critical ideas and theories that were to form the basis of her creative work. In her unique method, the character and plot of traditional novels disappears, to be replaced by sensations of nature and individual consciousness. She progressively developed this method in each of her novels which included *The Voyage Out, Night and Day, Jacob's Room, Mrs. Dalloway, To the Lighthouse, Orlando,* and *The Waves,* among others. From the beginning she was concerned with describing "arrows of sensation," which she felt pierced the armor through which we shelter from others. Themes she introduces in *Mrs. Dalloway* include the conflict between the world of social graces and the world of thought and action; another theme was the reality and horror of living—as she experienced it through trying to cope with her mental illness. Both she and her husband, Leonard Woolf, with whom she started the Hogarth Press, kept extensive diaries chronicling the cycling of her moods and their effect on her writing. After her death, Leonard published much of this work, including critical essays and collections of short stories.

Page 31, MANY PEOPLE WHO HAVE MOOD DISORDERS BECOME DEPENDENT UPON ALCOHOL OR STIMULANTS: Claire Dubois's story is a common one, unfortunately, which compounds both the presentation of the illness and its treatment. There may be common genetic factors in alcohol addiction and mood disorder, as is represented in the case history by the probable alcoholism of Claire's mother. Several family studies have suggested an increased rate of alcoholism in relatives of those with depression compared to controls, particularly in relatives of unipolar patients. (See L. H. Price, J. C. Nelson, "Alcoholism and affective disorder," *American Journal of Psychiatry* 143 [1986]: 1067–1068.) There has been an extended debate regarding whether the depression causes the alcoholism or the other way round, but the most recent epidemiological studies support the conclusion that the alcoholism or drug addiction starts with the depression. For a review of the subject, see F. K. Goodwin, K. R. Jamison, "Alcohol and drug abuse in depressive illness," *Manic-Depressive Illness* (Oxford: Oxford University Press, 1990), pp. 210–226.

Page 32, "A TRUE WIMP OF A WORD FOR SUCH A MAJOR ILLNESS" and the "VERITABLE TEMPEST IN THE BRAIN": from William Styron, *Darkness Visible: A Memoir of Madness* (New York: Random House, 1990).

Page 32, EXIT EVENTS: A term coined by Professor Eugene Paykel. For more information about exit events and entrance events and how they effect mood disorders, see E. S. Paykel, "Life events and early environment," in *Handbook of Affective Disorder,* ed. E. S. Paykel (New York: Guilford Press, 1982), pp. 146–161; A. P. Schless, L. Schwartz, C. Goetz, J. Mendels, "How depressives view the significance of life events," *British Journal of Psychiatry* 125 (1974): 406–410; B. Glassner, C. V. Haldipur, "Life events and early and late onset of bipolar disorder," *American Journal of Psychiatry* 140 (1983): 215–217.

Page 33, WHEN CHILDREN LOSE A PARENT, OR PARENTS, IN CHILDHOOD THEY ARE AT INCREASED RISK FOR DEPRESSION, AND ALSO FOR SUICIDE: There is considerable scientific literature on this subject. For a sampling of that work, see B. Bron, M. Strack, G. Rudolph, "Childhood experience of loss and suicide attempts: significance in depressive states of major depressed and dysthymic or adjustment disordered patients," *Journal of Affective Disorders* 23 (1991): 165–172; C. Lloyd, "Life events and depressive disorder reviewed," *Archives of General Psychiatry* 37 (1980): 529–535; S. B. Patten, "The loss of parent during childhood as a risk factor for depression," *Canadian Journal of*

2

78 *Notes to Pages 33–39*

Psychiatry 36 (1991): 706–711; T. Crook, J. Eliot, "Parental death during childhood and adult depression: a critical review of the literature," *Psychological Bulletin* 87 (1980): 252–259.

Page 35, SEIZURES CAN OCCUR: Seizures—so-called "rum fits"—are infrequent in withdrawing alcoholics but when they occur are dangerous and usually of the "grand mal" type, where the whole body shakes and the tongue may be bitten, with loss of bladder control and consciousness. Probably the chances in Claire's case were small, but given her combined abuse of barbiturates (sleeping pills) and alcohol over several months, the risk was not one to be dismissed. Typically the seizures begin 48–72 hours after the last drink. The usual treatment, if seizures occur, is a benzodiazepine drug (see appendix) given intravenously, but a better strategy is to prevent their occurrence, as I describe in Claire's situation.

Mild to moderate withdrawal, usually with tremors, agitation, and abdominal distress, is very common in alcohol and barbiturate withdrawal and is associated with sweating and increased pulse rate—essentially a very severe "hangover." Approximately 5 percent of individuals go on to severe agitation, with panic and mental confusion (delirium tremens), when seizures are particularly likely to occur.

Page 36, ANTIDEPRESSANT DRUGS . . . PROD THE PENDULUM: Antidepressant drugs do not themselves replace something in the body, but rather encourage the body's natural adaptive response. I explore this way of thinking about specific treatments, and how they work in depression, in chapters 9 and 10.

Page 36, MAJOR DIFFERENCE BETWEEN TRICYCLICS AND . . . PROZAC: This question, together with the specific properties of amitriptyline, is taken up in detail in chapter 9 and in the appendix.

Page 38, "SLOWLY SINKING, WATERLOGGED, HER WILL INTO HIS": see V. Woolf, *Mrs. Dalloway* (New York: Harcourt, Brace, and Jovanovich, Modern Classics, 1990). A description of Lady Bradshaw, the wife of the consulting psychiatrist in the story, a woman bowed by her husband's search for dominion and power.

Page 38, HER OWN RESCUE OPERATION: For an elaboration of this idea, see my essay, P. C. Whybrow, "Adaptive Styles in the Etiology of Depression," in *Affective Disorders,* ed. F. Flach (New York: W. W. Norton, 1988).

Page 39, EXPLANATORY STYLE: A term used by Martin Seligman, psychologist and author of *Learned Optimism,* to describe the characteristic way in which individuals explain the relationship to their world. See M. E. P. Seligman, *Learned Optimism* (New York: Simon & Schuster, 1990).

Page 39, "STRUCTURED" PSYCHOTHERAPIES—PARTICULARLY COGNITIVE THERAPY: In the 1970s, out of discontent with the lack of objective criteria for implementation and change in psychotherapy, a group of manuals and methods for conducting psychotherapy emerged. Cognitive therapy, which I discuss in chapter 4 and in chapter 10, came to dominate this field. Cognitive

Therapy is the brainchild of Aaron T. Beck, Professor of Psychiatry at the University of Pennsylvania. His classic text, *Cognitive Therapy and the Emotional Disorders,* was first published in New York: International Universities Press, 1976. See also *Cognitive Therapy of Depression,* ed. A. T. Beck, J. Rush, B. Shaw, and G. Emery (New York: Guilford Press, 1979).

Chapter Three: A Different Drummer

Page 43, "There is a particular kind of pain": see K. R. Jamison, *An Unquiet Mind: A memoir of moods and madness* (New York: Knopf, 1995), page 67. This is a remarkable memoir of the personal experience of manic-depressive illness, recommended to all those who seek a better understanding of what the illness is like from the inside. Dr. Jamison, an American psychologist and a well-known authority on the illness, is the co-author with F. K. Goodwin of the definitive medical textbook on manic-depressive illness (F. K. Goodwin, K. R. Jamison, *Manic-Depressive Illness* (New York: Oxford University Press, 1990). *An Unquiet Mind* is an account of Dr. Jamison's personal struggles with the illness. The book documents particularly well the subjective mental changes of manic excitement and the complications of identifying the borderlands between a vibrant emotional "self" and the excesses of hypomania. Kay Jamison deals honestly and lucidly with this dilemma and many other issues pertinent to mania and its treatment in this excellent account.

Page 44, THE LAMB AND FLAG, AN OLD PUB IN THE COVENT GARDEN DISTRICT: This public house, just off Floral Street, at 33 Rose Street, is one of London's oldest. Close to Covent Garden (originally the convent garden for Westminster Abbey), the theaters and the bookstores of Charing Cross Road, it was a favorite haunt of mine as a medical student.

Page 45, INGMAR BERGMAN'S *THE SEVENTH SEAL:* Bergman's classic film is an allegory of man's search for meaning. The medieval knight, played by Max von Sydow, returns home from the Crusades through a plague-stricken Europe where he encounters Death. The knight prolongs his life by playing chess with Death, assisting along the way an innocent band of strolling players. Made in 1956, it was an instant success and became something of a "cult" film among university students at the time. For more information about the film see Melvyn Bragg, *The Seventh Seal* (London: British Film Institute, 1993), and *Four screenplays of Ingmar Bergman,* trans. L. M. Strom and D. Kushner (New York: Simon & Schuster, 1960). The film is also available on videotape from Embassy Home Entertainments, the Janus Collection.

Page 47, HARNESSED WITH A NATURAL TALENT . . . SUCH STATES CAN BECOME THE ENGINE OF ACHIEVEMENT: There is a continuing debate over whether the illness of manic depression contributes to creativity and leadership. To my mind, this is not the best way to pose the question. While the psychotic distortion and other disabling symptoms of the acute illness do not foster such human qualities (or only rarely, in extraordinary individuals), the experience of manic depression, especially the energy of early hypomania and the reflective insights of mild depression—the penumbra of the illness—certainly do so, in my experience. Those interested in this debate may wish to read D. J. Hershman and J. Lieb, *The Key to Genius* (New York: Prometheus Books, 1988). See also F. K Goodwin, K. R. Jamison, "Manic-Depressive Illness, Creativity, and Leadership," in *Manic-Depressive Illness* (New York: Oxford University Press, 1990), chapter 14, pp. 333–368; K. R. Jamison, *Touched with Fire: Manic-Depressive Illness and the Artistic Temperament* (New York: The Free Press, 1993).

Page 48, THE FLIGHT OF THE MIND: See Thomas Caramagno, *The Flight of the Mind: Virginia Woolf's Art and Manic-Depressive Illness* (Berkeley: University of California Press, 1992).

Page 48, "VIRGINIA WAS A LIFE ENHANCER": Nigel Nicolson's observations are quoted by T. Caramagno (Ibid., p. 49).

Page 49, "A VERY INTENSE AND TICKLISH BUSINESS": Virginia Woolf drew upon her own experience of mental illness in many of her novels. In *Mrs. Dalloway,* Septimus, a central character, experiences madness and commits suicide. Virginia describes in her diary just how difficult it was for her to approach this subject and develop the character without being either self-indulgent or triggering her own anxieties. This is discussed by Thomas Caramagno on page 211 of *The Flight of the Mind* (*op. cit.*), from where my quotation was taken.

Page 58, "LITHIUM . . . WAS STILL BEING TESTED FOR USE IN THE UNITED STATES: Ironically, only months after Emmett's death I was involved in some of the first clinical trials of lithium, at the University of North Carolina, and later in London at the Medical Research Council, where the value of lithium as a prophylactic agent in manic depression was under review. The first studies in England were conducted in 1963 (it had been introduced into Denmark a decade earlier by Mögens Schou). The first American study was conducted at the National Institute of Mental Health (NIMH) in 1968, and lithium was accepted by the Federal Drug Administration as a treatment for manic depression in the early 1970s.

Page 58, ADULT SITUATIONAL REACTION WITH PSYCHOTIC FEATURES: In a reaction to the rigid, "biological" diagnostic classification of Emil Kraepelin (see chapter 5), the American nomenclature of the '60s and '70s focused upon the contributing power of the environment in the development of mental illness. Hence the term "adult situational *reaction*" implied a set of overwhelming environmental stressors to which the victim's illness was a response. In defining specific vulnerability, the classification drew heavily on the psychodynamic theories of Freud and focused on those events in childhood thought to forge character. Emmett, classified as a narcissistic personality, was considered to have a character structure easily damaged by adverse circumstance.

Page 60, GRIEF CAN PRECIPITATE MANIA . . . ACUTE OR CHRONIC STRESS

MAY BOTH PRECEDE THE ONSET OF MANIA: Compared to depression, the precipitation of manic illness by stressful events has been neglected by English-speaking researchers. However, the French (and Italians) have long recognized "maniacal grief" (*deuil maniaque*) following the death of a loved person. In one study, specifically investigating stressful events and the onset of mania, 28 percent of patients developing mania, compared to 6 percent in a control group of surgical patients, reported adverse events prior to illness. Of the 28 percent, one-third had experienced bereavement and over half of the cases were first episodes of illness. This study was retrospective, from a review of case histories, and is therefore likely to be an underestimate. See A. Ambelas, "Psychologically Stressful Events in the Precipitation of Manic Episodes," *British Journal of Psychiatry* 135 (1979): 15–21. For a general discussion of the relationship between stress and the precipitation of illness, see also E. Paykel, "Recent life events in the development of depressive disorders," in *The Psychobiology of the Depressive Disorders,* ed. R. A. Depue (New York: Academic Press, 1979). Another point of interest here, which I take up in chapter 7, is that repeated stresses may increase vulnerability to mania—the concept of "kindling," first proposed by Robert Post of the National Institute of Mental Health in the United States. See R. M. Post, D. R. Rubinow, J. C. Ballenger, "Conditioning and Sensitization in the Longitudinal Course of Affective Illness," *British Journal of Psychiatry* 149 (1986): 191–201.

Page 61, *THE VOYAGE OUT:* London: Hogarth Press, 1975.

Page 61, THE GENETIC LEGACY . . . IN VIRGINIA WOOLF'S FAMILY: For a description of the family members and close relatives of Virginia Woolf who suffered mood disorder, see chapter 4 in T. Caramagno, *The Flight of the Mind* (*op. cit.*), pp. 97–113.

Page 61, MANIA AND MELANCHOLIA TRAVEL TOGETHER: Approximately 20 percent of those who experience serious depression also develop severe mania. See M. M. Weissman, J. K. Myers, "Affective disorders in a U.S. urban community," *Archives of General Psychiatry* 35 (1978): 1304–1311.

Page 61, THE TRAGEDY OF SUICIDE: National suicide rates vary from below 10 per 100,000 population each year in Ireland to over 35 per 100,000 in Hungary. The United States is around 12 per 100,000. Suicide increases with age, and males outnumber females at all ages. It also tends to run in families, suggesting a particular genetic vulnerability. For example, almost three-quarters of the suicides among the Amish in Lancaster County, Pennsylvania, a genetically homogeneous group over the past 200 years, occurred in just four families. See J. Egeland, J. Sussex, "Suicide and family loading for affective disorders," *Journal of the American Medical Association* 254 (1985): 915–918. The most common illnesses associated with suicide are depression and alcoholism; frequently they occur together. For a brief but comprehensive review of suicide, refer to A. Roy, "Suicide," in *Comprehensive Textbook of Psychiatry,* ed. H. I. Kaplan, B. I. Sadock (Baltimore: Williams and Wilkins, 1989), chapter 29.1, pp. 1414–1427.

Page 63, SUICIDE—THE "SAVAGE GOD" OF MOOD DISORDER: *The Savage God* was the title chosen by A. Alvarez, English poet and critic, for his book subtitled *A Study of Suicide*. Mr. Alvarez's book is remarkable for a careful integration of the experience of depression and alienation that leads to the suicide act, his extensive knowledge of literature, and the clinical studies of suicide. Particularly notable is his essay on Sylvia Plath, the American poet who suffered manic-depressive mood swings and killed herself in her thirty-first year. See A. Alvarez, *The Savage God, A Study of Suicide* (New York: W. W. Norton, 1990).

CHAPTER FOUR: A MIND OF ONE'S OWN

Page 67, "ALL LIFE DEPENDS ON A FLOW OF INFORMATION": J. Z. Young, *Programs of the Brain* (New York: Oxford University Press, 1978). John Zachary Young was the professor of anatomy at University College, London, during the 1950s and 1960s, when I studied medicine there. He taught that the body was best understood as a self-regulating machine, and his early textbook *The Life of Mammals* (Oxford: Oxford University Press, 1957), together with his dynamic lectures on anatomy and neuroscience, were very important to me in developing my own conceptual approach to the brain and behavior. Young considered the brain an organ of information-processing and the ordering of information as health. Disorder of information leads to illness, which is the essence of what happens in mood disorder.

Page 67, "IVAN ILYICH SAW THAT HE WAS DYING": L. Tolstoy, *The Death of Ivan Ilyich and Other Stories* (New York: The New American Library of World Literature, 1960). For me, Tolstoy's short story highlights the difficulty we all have in thinking objectively about the self. Over a lifetime of experience we have each developed a mind of our own, unique in memory and purpose. No abstraction can explain that, and to imagine that I—the self—can die or become diseased is the central conceptual difficulty. Disorders of mood, being disorders of the self, expose this difficulty.

Page 71, THESE MODELS, OR SCHEMA: *Schema* is a term used by cognitive psychologists to describe the brain's natural clustering of information into manageable units, subsequently filling in any missing information so that a decision can be made. The idea of schema was introduced into experimental psychology by Piaget, among others, in the 1920s and has persisted as an important neuropsychological concept in understanding how the brain handles information. Various other terms have been used that are similar—"patterns" and "templates" being the most common. Conceptually, when new information is acquired it is organized within the meaning of these structures, and their disturbance will distort information-processing. It is argued by some psychologists that distorted schema lie at the basis of specific learned psychopathologies, including depression. So, for example, the concept of attributional style (Martin Seligman) and Aaron Beck's cognitive triad (both of which are touched upon later in chapter 4) are examples of cognitive schema. For a discussion of these concepts, I recommend the book *Person Schemas and Maladaptive Interpersonal Patterns,* ed. M. Horowitz (Chicago: University of Chicago Press, 1991). Chapter 2, written by J. Singer and P. Salovey, "Organized knowledge, structures, and personality: person schemas, self schemas, prototypes and scripts," is particularly relevant.

Page 71, INFANTS HAVE A CORE REPERTOIRE OF EMOTIONAL BEHAVIORS: For an excellent discussion of the development of emotion in children see chapter 7, pages 121–130, and chapter 14, pages 203–218, in Vol. 1 (Fundamentals) of *Developmental Neuropsychiatry,* by James Harris (New York: Oxford University Press, 1995).

Page 72, SECONDARY EMOTIONS: Donald L. Nathanson, a psychoanalyst and student of Sylvan Tompkins, one of the first psychologists to explore the area of the secondary emotion, has written extensively on shame and pride and their importance to the development of "self" identity. For details, see Nathanson's book *Shame and Pride: Affect, Sex, and the Birth of the Self* (New York: W. W. Norton, 1992).

Page 72, CLIFF EXPERIMENT: James Harris discusses this experiment in chapter 14, Vol. 1 (Fundamentals) of *Developmental Neuropsychiatry*. The reference to the original work is R. D. Walk, E. J. Gibson, "A comparative and analytical study of visual depth perception," *Psychological Monographs* 75 (1961): 519.

Page 73, LANGUAGE ... IS SHAPED ACTIVELY BY THE ... WAY IN WHICH THE BRAIN ORGANIZES INFORMATION: Here I refer to Noam Chomsky's view of language and his rejection of the idea that language is merely a mimicry in the service of communication. Chomsky believes that there is a "genetically determined language faculty" and the structure of language can be reduced to its elements much the same way as chemistry or physics is so reduced. For a lucid discussion of some of these issues, see R. Salkie, "Basics," chapter 1 in *The Chomsky Update* (London: Unwin Hyman, 1990).

Page 74, SURVEYS SUGGEST: This information comes from the survey conducted for the National Mental Health Association in 1991 by Peter D. Hart Research Associates, Inc., Washington, D.C. In a telephone survey of 1,022 adults nationwide, 43 percent saw depression as a "sign of personal or emotional weakness."

Page 74, IT WAS ONCE FASHIONABLE TO BLAME DESCARTES: It is not my intention here to deprecate Descartes, who in his fifty-four years (1596–1650) provided the cornerstones of modern philosophy and science. Descartes was one of the greatest of all thinkers, and his development of a philosophical rationalization that suited the Church of the time is just one testimony to that. René Descartes was endowed with a great curiosity, and had a deep desire to dissect the brain, which he suspected had something to do with the nature of being. Recognizing that the Roman Catholic Church with its extraordinary power would oppose him, he proposed that the body was "nothing more than a . . . machine made of earth" of which the brain was a part, and offered in support of his position the famous metaphor of body and soul standing as two clocks, side by side, marking time together but each separate and divisible. One clock, Descartes proposed, belonged to God and the other to the anatomist. It was in the pursuit of science that Descartes sought intellectual justification for what we all tend to do anyway, to separate mind and subjective experience from the things of the objective world. For an excellent biography of Descartes, I refer the interested reader to S. Gaukroger, *Descartes: An Intellectual Biography* (Oxford: Clarendon Press, 1995).

Page 76, THE CONCEPT OF MIND: G. Ryle, *The Concept of Mind* (London: Hutchinson, 1949).

Page 78, SIGMUND FREUD BEGAN HIS CAREER AS A NEUROLOGIST: For a brief review of Freud's early professional career, and the psychoanalytic literature on depression and melancholia, see chapter 5, pages 81–93, in my book, written with Hagop Akiskal and William McKinney, *Mood Disorders, Toward a New Psychobiology,* (New York: Plenum Press, 1984). Freud's classical paper on depression, *Mourning and Melancholia,* was published in 1917; see *The Complete Psychological Works of Sigmund Freud,* Vol. 14, pages 243–258, (London: Hogarth Press, 1948). For an excellent discussion of "drive theory," its origin in Freud's work, and its relationship to the existential psychotherapies, read the introduction (pages 3–28) to Irvin Yalom's book, *Existential Psychotherapy* (New York: Basic Books, 1980).

Page 78, IVAN PAVLOV, THE GREAT RUSSIAN PHYSIOLOGIST: Ivan Pavlov, 1849–1936, earned the Nobel Prize for his work on digestion. His discovery of physiological conditioning, for which he is now remembered, was an extension of that work. In his original experiments with dogs, he found that the salivary response, normally triggered by food, could also be called forth by stimuli which the dogs had learned were associated with the *preparation* of the food. This association of the primary stimulus (food) and the secondary stimulus (a buzzer, in the case of Pavlov's experiment) was the foundation of his theory of classical conditioning. This later became the basis for Behaviorism, a school of psychology that had wide application to a variety of animal species, including human beings. Many psychologists helped develop the American movement in behaviorism, but probably the most famous was B. F. Skinner (1904–1990), who elaborated upon Pavlov's theories of classical conditioning, including the concept of behavioral reinforcement by reward or punishment.

Page 79, MARTIN SELIGMAN: For details of the original papers by Martin Seligman, see J. B. Overmier, M. E. Seligman, "Effects of inescapable shock upon subsequent escape and avoidance responding," *Journal of Comparative and Physiological Psychology* 63 (1967): 28–33; M. E. Seligman, "Failure to escape traumatic shock," *Journal of Experimental Psychology* 74 (1967): 1–9; M. E. Seligman, S. F. Maier, J. H. Gear, "Alleviation of learned helplessness in the dog," *Journal of Abnormal Psychology* 73 (1968): 256–62; C. S. Raps, C. Peterson, K. E. Reinhard, L. Y. Abramson, M. E. Seligman, "Attributional style among depressed patients," *Journal of Abnormal Psychology* 91 (1982): 102–8; and M. E. Seligman, L. Y. Abramson, A. Semmel, C. Von Baeyer, "Depressive attributional style," *Journal of Abnormal Psychology* 88 (1979): 242–7. Details of these early experiments and the evolution of Professor Seligman's concepts of learned helplessness and learned optimism will be found in pages 20–28 of his book *Learned Optimism* (New York: Simon & Schuster, 1990).

There is considerable debate over whether the original animal experiments support the anthropomorphic extension to "learned" helplessness. For example, in the original experiments in dogs, the helplessness had worn off after 72 hours, suggesting a neurophysiological exhaustion rather than learned cognitive change. Experimental psychologist Jay Weiss, then at Rockefeller University, further examined learned helplessness in rats and concluded that the passive state was secondary to a change in brain chemistry, probably norepinephrine depletion; again the behavioral effects of inescapable shock were found to be transient. For details of Professors Weiss's work, the reader is referred to "Stress Induced Depression: Critical Neurochemical and Electrophysiological Changes," in *Neurobiology of Learning, Emotion and Affect,* ed. J. Madden IV (New York: Raven Press, 1991).

Page 79, WHEN THE ANIMAL HAD CONTROL OVER THE SHOCK DEVICE: Control turns out to be a cardinal issue here, and in all stressful situations. The importance of control was clearly demonstrated in the early 1970s by Jay Weiss studying rats yoked together in pairs, where one could turn off the shock for both partners. In the experiment, the "executive" rat developed less stomach ulceration than the passive partner, who could do nothing about the random shock, even though (because the animals were yoked together) the total number of shocks received were the same. The issue of control, as I discuss in chapter 8, is important in the behavior and physiological consequences of human beings when under stress. The concept of control is similar to that of a "drive for mastery," which occupies most human beings throughout life.

Page 80, AARON T. BECK: Is known all over the world for his pioneering work in the development of "cognitive psychotherapy." Details of his work, as it applies particularly to depression, will be found in A. T. Beck, A. J. Rush, B. F. Shaw, and G. Emery, *Cognitive Therapy of Depression* (New York: Guilford, 1979).

Page 81, GALEN: Galen, the preeminent Greek physician of his day, served the emperor Marcus Aurelius (A.D. 130–200). He is notable for his work on human anatomy, although his observations were largely based on the dissection of pigs, dogs, and goats, and were sometimes incorrectly applied to humans. Galen described the sympathetic nervous system and the parts of the brain's anatomy that are visible to the naked eye. He was also the first physician to consider the pulse a diagnostic aid, recognizing that arteries carry blood, not air, throughout the body. His monumental work, *On the Use of the Parts of the Human Body,* served as a standard medical text for 1,400 years, and his treatment for melancholia—bleeding the patient to remove the black bile from the brain—was extensively, indeed excessively, used by Benjamin Rush, the "father" of American psychiatry, in the early nineteenth century. This practice of bleeding, or "letting," remains enshrined in the human anatomy classes taught even today. The name of the large, visibly prominent vein that lies in the hollow of your arm, just over the elbow joint, and the one still most frequently used when blood is drawn for examination, is the cephalic vein. *Kephalé* in Greek means "head," and for Galen, and Rush, this was the vein of choice when letting venous blood, laden with black bile, directly from the brain to relieve the oppressive shadow of melancholia. Those curious about Galen might read G. Sarton, *Galen of Pergamon* (Lawrence: University of Kansas Press, 1954); E. H. Eckernecht, *A Short History of Medicine,* revised edition (Baltimore: Johns Hopkins University Press, 1982).

Page 82, JAMES WATSON AND FRANCIS CRICK: For a gripping account of this Nobel Prize–winning piece of scientific sleuthing, read James Watson's book *The Double Helix* (New York: New American Library, 1969).

Page 82, GENES ... THE UNIQUE LIBRARY OF COMPREHENSIVE INSTRUCTIONS: For a comprehensive and lucid discussion of the importance of genetics in psychiatry and behavior, I highly recommend S. H. Barondes, *Molecules and Mental Illness* (New York: W. H. Freeman, Scientific American Library, 1993).

Page 82, HUMAN GENOME PROJECT: The ultimate goal of the Human Genome Project, beyond developing the comprehensive reference library, is to work out how the protein components of the cell are manufactured and regulated by individual genes, and how variations in the genetic code can lead to disease and to maladaptive behavior. The technologies are now available to screen the genome and achieve this goal. With this information we will be able to identify the variation in coding instructions responsible for those changes in the chemical regulation of the emotional brain that permit the phenotypic expression of behavioral illnesses, such as mania and melancholia. Once we understand the neuronal mechanisms involved we will be in a better position to design specific drugs and treatments to reverse or minimize the abnormality, and thus prevent the limbic dysregulation that drives the syndromes of mood disorder.

Page 83, PROFESSOR THOMAS BOUCHARD ... STUDIES OF IDENTICAL TWINS: An interesting review of twin studies and their importance in unraveling the nature-versus-nurture debate appeared in the *New Yorker,* August 7, 1995. Written by Lawrence Wright and entitled "Double Mystery," it goes into some detail about Professor Bouchard's studies. However, those who wish to explore the original studies, containing the suggestion that approximately 40 percent of human behavior is tied to the genome, should refer to Thomas Bouchard, "Genes, environment, and personality," *Science* 264 (1994): 1700–1701.

Page 84, A. A. MILNE'S CLASSIC BOOK *WINNIE THE POOH:* In a careful reading of A. A. Milne's stories about Christopher Robin and his friends, one can learn a great deal about temperament and mood. Eeyore, the old gray donkey, for example, is a wonderful portrait of dysthymia, and Pooh himself is something of an extrovert, perhaps a hyperthymic. The particular story in which Tigger is "unbounced" will be found in chapter 7 of *The House at Pooh Corner,* which has been republished numerous times by Dutton since its original publication in 1928.

Page 85, GALEN'S PROPHECY: Jerome Kagan's book *Galen's Prophecy* (New York: Basic Books, 1994) is well worth the investment for those interested in a more detailed review of temperament and its relation to emotional states and mood. Professor Kagan reviews his extensive studies over the fifteen years that he has been following the original cohorts of children. Professor Kagan also provides a detailed analysis of Galen's original theories, discussing some of the social implications of the nature-versus-nurture controversy. Other references include J. Kagan, "Behavior, biology, and the meanings of temperamental constructs," *Pediatrics* 90 (1992): 510–3; and J. Kagan, J. S. Reznick, J. Gibbons, "Inhibited and uninhibited types of children," *Journal of Child Development* 60 (1989): 838–45.

Page 88, SHYNESS AND INSTABILITY OF EMOTIONAL LIFE . . . ASSOCIATED WITH SYMPTOMS OF DEPRESSION . . . IN ADOLESCENT GIRLS: For details of these studies, see I. M. Goodyear, C. Ashby, P. M. E. Altman, C. Vize, P. J. Cooper, "Temperament and Major Depression in 11 and 16 Year Olds," *Journal of Child Psychology and Psychiatry* 34 (1993): 1409–1423; H. A. Klein, "Temperaments and Self Esteem in Late Adolescence," *Adolescence* 27 (1992): 689–694.

CHAPTER FIVE: UNIQUE AND SIMILAR TO OTHERS

Page 91, "MAN DESIRES A WORLD WHERE GOOD AND EVIL CAN BE CLEARLY DISTINGUISHED": In his essay on "The Depreciated Legacy of Cervantes," Kundera argues that mankind's discomfort with ambiguity lies at the base of the obsession with religion and science. The experience of being human, Kundera contends, is inherently ambiguous. This is portrayed in the allegory of Cervantes' novel *Don Quixote,* published in 1605, when Cervantes was fifty-eight years old. I believe the tangled social history of mental disorder also is explained by this complex tension between the ambiguity of being human and the certainty of religion and science that Kundera discusses in his essay. The pendulum of human opinion about these disorders has swung between the viewpoint that disturbed human passion (the emotions) is entirely the work of God and the reflection of a disturbed spirit (the seven deadly sins) and biological reductionism, where molecules explain it all. The same tension is still reflected in the changing fashions of psychiatric diagnosis. Human behavior lies upon a continuum and resists being sorted into categories, but distinguishing categories are essential in the diagnosis and treatment of disease. The essay on Cervantes appears in Milan Kundera's *The Art of the Novel* (New York: Harper and Row, Perennial Library, 1988).

Page 91, "ALL CASES ARE UNIQUE, AND VERY SIMILAR TO OTHERS": I first read T. S. Eliot's play *The Cocktail Party* when I was in medical school in 1961. The play is set in an ordinary London drawing room and follows the convention of a comedy of manners but has profound things to say about the nature of human behavior and its combination of unique and common forms. A mysterious visitor, later identified as a psychiatrist, Sir Henry Harcourt-Reilly, is the philosopher physician in the play who counsels Edward, the protagonist, during a period of confusion and crisis. Through Dr. Harcourt-Reilly, T. S. Eliot explores ideas about the complexity of identifying cause in abnormal behavior, about caring for the self, and about the nature of prejudice in human relationships. *The Cocktail Party* was first published in London by Faber and Faber, 1950.

Page 92, JOHN MOOREHEAD: Both John Moorehead and his sister Angela are composite characters, as is their extended family living in the North End of Boston. Each individual experience, however, is genuine. Thus, the hyperthymic personality (John) who develops depression in midlife, the young woman with winter depression (Angela), and the cyclothymia of Josephine, Angela's mother, are all vignettes taken from specific case histories and taped interviews. I have changed time and place to preserve anonymity, and have brought them together as one family for literary convenience.

Page 93, PERVASIVE ANXIETY: Anxiety is a common presenting symptom in depression and is frequently the reason why the diagnosis initially is missed. Out of context from the other disturbances of depression (sleep, appetite, concentration, and so on), it is frequently dismissed as evidence of minor stress. This commonly leads to prescriptions of diazepam and other anti-anxiety agents, which do little for depression.

Page 93, IT WAS CALLED ACEDIA IN THE MIDDLE AGES: Throughout this chapter I have woven together a brief social history of melancholia and mania with the medical evolution in diagnostic methods. In doing so, it has been difficult to do justice to the extraordinarily rich materials that are available in both subjects. The most comprehensive analysis of the history of the syndromes of mania and melancholia is to be found in S. Jackson, *Melancholia and Depression: From Hippocratic Times to Modern Times* (Connecticut: Yale University Press, 1986). An essay specifically on acedia by Dr. Jackson may be found in the bulletin *The History of Medicine* 55 (1981): 172–185, entitled "Acedia the sin and its relationship to sorrow and melancholia in medieval times."

Page 93, I'M A GAUDY-MINDED MAN: *Gaudium* is from the root *gauder,* the Latin word meaning "to rejoice." However, it also means an object of mockery—a showy ornament, an act of trickery—and it was in this context that John Moorehead used the term, based on his considerable Latin scholarship. His reference is to the song *Gaudeamus igitur,* implying the abandonment of his responsibilities. *Gaudy* is the name given to the annual college feast at some universities. Interestingly enough, *gaudium* is also a word associated with mania in some medieval texts.

Page 95, HIPPOCRATES: What we now consider the work of Hippocrates was probably the work of a school of scholars over many decades. The major treatise from the school, *The Nature of Man,* saw illness not as a visitation from some divine power, but as a battle between the disease process and the self-healing ability of the body. Treatment, therefore, centered upon assisting the patient through his particular nature (*physis*) to react in his own special way against the disease. These ideas led to the development of the dynamic concept of body humors. The works of Hippocrates are discussed in most texts of medical history, for example that by H. E. Sigerist, *The History of Medicine* (New York: Oxford University Press, 1961), two volumes.

Page 96, GALEN'S TIME: Galen was a Greek physician practicing in Rome in the second century (A.D. 131–201). He was the most skilled practitioner of his time and left a long list of miracle cures as an example of his art. He elaborated upon the theories of Hippocrates and wrote three books on the temperaments alone, including a chapter devoted to melancholia. He considered melancholia to be a group of diseases having as their root cause an abnormality of black bile in the body. Fear, despondency, darkness, and the color of dark venous blood were all linked in this theory. In fact, from it flowed the notion of bleeding the patient to remove the black bile that was casting the shadow of darkness over the brain. Bloodletting, a therapeutic intervention that lasted well into the nineteenth century, was used extensively by such physicians as Benjamin Rush. Rush was known to bleed his patients unmercifully, taking as much as 40 ounces (approximately 5 pints) at a time.

Page 97, HISTORY THAT REACHES BACK . . . TO THE TEACHING OF THE EARLY ROMAN CHURCH: The Church of Rome rose as the City of Rome fell; by the fifth century the city had been conquered by the Goths. Peter, who had been martyred by Nero, was the direct link with God and the first of the succession of Popes who carried forth the doctrine of the Christian church. Augustine (A.D. 354–430) lived in modern-day Algeria; he was raised by his mother as a Christian but lapsed, and then returned to Christianity in his thirties. He was the seminal thinker in developing the Church's authority and paving the way for Gregory (A.D. 540–604). A wealthy Roman who held many civil offices before becoming Pope in the latter years of his life, Gregory divided the spiritual from the civil and established once and for all the right of the church over spiritual matters, encouraging monasticism, the celibacy of the clergy, and its exemption from any civil proceedings. This established the Church's dominance over the emotional, or passionate, element of human experience.

Page 98, AN INFLUENTIAL MONK NAMED JOHN CASSIAN: John Cassian is the early Christian theologian credited with having brought Eastern practice to Rome and the Christian Church. He was particularly interested in prayer and the monastic life; his name was the first to be linked with acedia, one of the deadly sins. Cassian's contribution is well described in Jackson, *Melancholia and Depression* (op. cit.).

Page 98, THE ACT OF CONFESSION AND CATHARSIS: Reports of suffering similar to that experienced by John Moorehead can be found throughout the chronicles of human experience. Indeed, it is not his experience of melancholia that is remarkable but society's changing interpretation of what this suffering represents. For two thousand years, from monastic cells to the courtrooms of the Inquisition, countless individuals described similar pain, and through the misguided moral judgments of others, they commonly became the victims of extraordinary tragedy. Reginald Scot, an enlightened humanitarian and Justice of the Peace in England in 1584, drew attention to the plight of many melancholics then being burned for their morbid confessions: "The force which melancholy has and the effects that it workith in the body of a man or woman are almost incredible. These melancholic persons imagine they are witches and by witchcraft they . . . imagine they can transform their own bodies. If you read the executions done upon witches either in times past in other countries, or lately in this land, you will see such impossibilities confessed as none having his right wits will believe . . . (A) witch confessed at the time of her death, at execution, that she had raised all the tempests and procured all the frosts and hard weather that had happened in the winter of 1565. That madness, grave and wise men believed." Despite Reginald Scot's humanitarian concerns, it was to be another hundred years before such practices were to cease. The English burned their last witch in 1684. To learn more about Reginald Scot and other fascinating historical writings, the reader is referred to R. Hunter, I. Macalpine, *Three Hundred Years of Psychiatry: 1535–1860* (New York: Carlisle Publishing, 1982). Reginald Scot's writings will be found on pages 32–35.

Page 98, DEPRESSION IS MORAL WEAKNESS: The guilt and shame that go along with depression in Western society seem to have roots in this early Christian concept of suffering and personal responsibility. In talking with my colleagues from India, for example, it is clear that the core symptoms of melancholia are identical in the two cultures, but the explanation that individuals give for their illness is quite different. In the Hindu and Moslem religions, the suffering is designated by God, and thus while the same sense of pain and dejection is found as a cardinal symptom in the melancholic syndrome, the cause is assigned to outside the individual and the body—not perceived as some reprehensible weakness of will on the part of the sufferer.

Page 99, SCIENCE IS A METHOD: René Descartes was in his twenties when he put forward his discourse on method. Descartes offered four rules of logic with which to approach the questions of the universe: first, maintaining a posture of doubt; second, dividing every problem into as many parts as possible; third, reflecting upon these elements; and lastly, beginning the analytic task. This was in direct contrast to the prevailing philosophy at the time, stemming from Plato, that there was in the universe a set of independent realities embodied in ideas or forms. These were immutable archetypes, and insight could only be reached by the study of dialectic—a method of inquiry that proceeds by constant questioning of assumptions. The Platonic method, exactly opposite to that of Descartes, seeks to explain a particular idea always in terms of a more general condition. Descartes' method of reduction forms the basis of the Western scientific method of inquiry and is at the heart of today's scientific and biological revolution.

Page 99, EXTREMES OF EMOTIONAL EXPERIENCE: One of the complications in diagnostic classification is the obvious fact that while human behavior extends in infinitely small steps along a continuous variable, eventually we must agree upon a cutoff point in order to categorize illness. This point of transition, this Rubicon, is usually based upon individual tolerance for pain and society's agreement to accept the pain as the specific deviance of illness. Only recently has there emerged a willingness to consider depression as possible illness. But the point at which illness begins along the dimension of sadness is still debated. However, in the classification of serious mental disability, just as in the disability of other bodily systems, there comes a point when the pain is so obvious that ninety-nine individuals out of one hundred will agree to its aberrance. These are the behavioral disturbances found at the extreme poles of the continuum of emotional experience, and which we categorize as mania and melancholia.

Page 100, INTRUSIVE, RUMINATIVE THOUGHTS INCLUDING THOSE OF SUICIDE: Suicidal ideas can be understood as an extension of the rumination which frequently occurs in melancholia, especially when the ideas have the quality that Moorehead described as "intrusive." As a depression deepens, such thoughts are invariably present in one form or another. Although frequently denied by the sufferer as something shameful and frightening, they are an important measure of the depth of depressive illness, and must always be inquired about. At least 15 percent of individuals with manic-depressive disease kill themselves in the course of their illness, and reciprocally, of those who commit suicide, probably 50 or 60 percent have a serious depression.

Page 100, THE CLUSTERING OF SYMPTOMS COMMON TO THIS ILLNESS: This is another reference to the common pathway of illness, first referred to in chapter 1. This concept is very important, not only in understanding how these illnesses cluster, but how they can be treated and successfully managed. Although the origin of illness may be environmental, such as an overwhelming challenge, or physiological, such as an infection playing upon a basic vulnerability, there comes a time when a common pathway of behavioral disorganization is triggered, which then incorporates other activities of brain until the syndrome of melancholia or mania is evident. It is this recruitment and common constellation of physiological and behavioral events that I refer to as the final common pathway.

Page 102, THE FIRST STEPS IN DEVELOPING A DIAGNOSIS FROM ANY SYNDROME: The description of phenomena—that is, things observed—in science is called *phenomenology;* the classification of those phenomena is called *nosology.* These common terms in medicine form the basis of *The Diagnostic and*

Statistical Manual (DSM) of the American Psychiatric Association, now in its fourth edition (Washington, D.C.: American Psychiatric Association Press, 1994). The categorization I have used in *A Mood Apart* derives principally from the DSM-IV, utilizing the concept of core symptoms around which additional symptomatology must be clustered if the definitive diagnosis of mania or depression is to be made. The diagnostic and statistical manual uses a multi-axial system of assessment. The first axis or dimension is that which describes the clinical syndrome itself; Axis Two is designated to include ongoing behavioral characteristics of the person, which might include temperament, or severe central nervous system disease such as mental retardation. Axis Three includes any general medical conditions that may influence the psychiatric disorder, and Axis Four includes environmental and psychosocial problems such as housing, poverty, single parenthood, and so on, that will influence the illness. Finally, Axis Five is reserved for a general global assessment of how all the previous axes come together to specify the general functioning of the individual and the disability he or she suffers.

Page 105, ROBERT BURTON IN . . . *THE ANATOMY OF MELANCHOLY:* A comparatively recent edition of this delightful volume was published in 1990 by University of Oxford Press.

Page 106, TEMPERAMENT: Temperament, or emotional style, is sometimes confused with the diagnostic categories of mood disorder. They are distinct. A diagnosis is the name given to a cluster of symptoms which wax and wane with time, and when present incapacitate normal function. Temperament is an enduring emotional style—a way of relating to the world—that is consistent over a lifetime. While it may compromise social interaction, and some behavioral characteristics may be similar to those seen in severe mood disorder, temperament is not an illness. The specific temperaments of dysthymia, cyclothymia, and hyperthymia, however, do predispose to illnesses of mood, particularly to bipolar disorder, as I have described in the Moorehead family. The diagram on the following page will help to clarify the interrelationship.

Page 107, PAINFULLY SHY, SHE PREFERRED THE SAFETY OF HER MOTHER'S SIDE: One characteristic of the thymic temperaments of the bipolar spectrum is that they are seen early on in childhood. In that regard, they may reflect something similar to what Jerome Kagan has described in his extremes of the shy child versus the uninhibited child. In my example, John, the uninhibited child, is the one who grows into the hyperthymic man, and Angela, the shy, retiring girl, into the dysthymic woman who later develops what American psychiatrists call bipolar II disorder—ongoing depressive symptoms with occasional episodes of hypomania, which in Angela's case were in rhythm with the seasons.

Page 107, JOSEPHINE MOOREHEAD . . . WAS . . . OF CYCLOTHYMIC TEMPERAMENT: Just as in mania, the mood is not always euphoric, but frequently irritable and negative. Those with cyclothymia, which seems to mirror manic-depressive disease with a lower amplitude, also have episodes of irri-

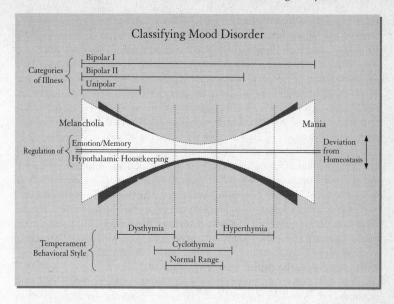

Classifying Mood Disorder

Categories of Illness
- Bipolar I
- Bipolar II
- Unipolar

Melancholia — Mania

Regulation of
- Emotion/Memory
- Hypothalamic Housekeeping

Deviation from Homeostasis

Temperament Behavioral Style
- Dysthymia
- Hyperthymia
- Cyclothymia
- Normal Range

tability associated with "high" periods. Thus it is in keeping with the common profile of cyclothymia for Josephine Moorehead to have periods where she was excited, and then irritated, before falling into a depressive state. This may be associated with periods of excitement and apparent happiness, or arise suddenly out of a normal mood. The most important thing in the identification of cyclothymia is the lability of mood and the consistency of that lability over time.

Page 108, THROUGH THE WRITINGS OF PROFESSOR HAGOP AKISKAL ... AND DOCTOR ATHANASIO KOUKOPOULOS: One problem in the study of predisposing personality and temperament in mood disorder is that the surveys are frequently conducted after the individual has had the first depressive illness. Such backward glances at behavior, observed after an individual has become ill, cannot fail to be tainted by hindsight and circumstances. But the alternative, to follow children who may be at risk because of depression in their families, takes not only a considerable investment of time and patience on the part of the researcher—Jerome Kagan has spent two decades of his career following his cohort of shy and uninhibited children—but is also very expensive.

A compromise that has been adopted by Hagop Akiskal is to search for depressed individuals with large families and many relatives, who can together provide rich and varied descriptions of the family tapestry from which the depression emerged. Given enough relatives (because only the most certain facts are ever agreed upon in large families, argues Akiskal), reasonable accuracy of information and agreement regarding the characteristic behavior of the sufferer *before* the onset of illness can be obtained. Dr. Akiskal and his colleagues believe they have found such families in the villages of rural Italy. Based on these and other observations, Akiskal has described predisposing "affective constitutions"—temperaments—in those who have later developed mood disorder. It should perhaps be no surprise that the temperaments Akiskal defines are similar to those outlined by Galen and others some 2,000 years ago. Take, for example, some of the words and phrases used to characterize those whom Akiskal calls the "extroverted individual of sanguine temperament." These are words reflective of manic states: irritable, cheerful, overoptimistic and exuberant, naive, overconfident, self-assured, boastful, bombastic, grandiose, full of plans, impudent, imprudent, carried away, restless, impulsive, overtalkative, warm, people-seeking, overinvolved, meddlesome, uninhibited, stimulus-seeking, promiscuous. Similarly, those of depressive temperament, roughly equivalent to the melancholic, are described as characteristically gloomy, pessimistic, humorless, incapable of fun, quiet, passive, indecisive, skeptical, hypercritical, complaining, brooding, given to worry, conscientious, self-disciplining, self-critical, self-reproaching, self-derogatory, preoccupied with inadequacy and failure.

Page 111, CONFIRM THE WORK OF EMIL KRAEPELIN: The principal temperaments of Akiskal's classification are also very similar to descriptions by Kraepelin in his book *Manic-Depressive Insanity and Paranoia* (Edinburgh: E. and S. Livingstone, 1921). For the dysthymic temperament, Kraepelin uses the term *depressive;* what Akiskal calls hyperthymic, Kraepelin first termed *constitutional excitement:* later he changed it to *manic temperament.* The cyclothymic temperament, regular fluctuations between manic and depressive behaviors, is similar in Kraepelin's classification to the irritable temperament. For Kraepelin, the irritable temperament was more common than the cyclothymic temperament; he put them respectively at about 12 percent and 4 percent of the patients that he saw in Munich. From my own practice, I believe the cyclothymic temperaments and irritable temperaments are one and the same. I have rarely seen a consistency in either that does not eventually overlap into the other.

For further reading regarding Dr. Akiskal's concepts of the thymic temperaments, see "The Importance of Measures of Affective Temperaments in Genetic Studies of Mood Disorders," *Journal of Psychiatric Research* 26 (1992): 257–268; Akiskal and Akiskal, "Reassessing the Prevalence of Bipolar Disorders: Clinical Significance and Artistic Creativity," *Psychiatry and Psychobiology* 3 (1988): 29–37.

Page 111, A HUMANITARIAN MOVEMENT: Otherwise called the "moral treatment" of the insane. This Quaker movement, which began in England under the leadership of William Tuke, spread rapidly across Europe and to the American States. The Quakers in America established several hospitals for the insane in the early 1800s. Friends' Hospital in Philadelphia was the first and is still in operation. The Brattleboro Retreat, the Hartford Retreat (later changed to the Institute for Living), and the Shepherd Pratt Hospital in Baltimore all began from this tradition. This supplanted the care of the insane in general hospitals such as the Pennsylvania Hospital in Philadelphia, and also in Colonial Williamsburg, where a reproduction of the hospital stands on the original site.

Parenthetically, it is a tribute to the flexibility of the Christian doctrine that most of the nineteenth-century asylums considered religious devotion integral to the care of their patients. A feature of the large English mental hospitals built in that period are chapels, many of which were quite magnificent in their architectural form. Thus the moral treatment of the insane replaced moral persecution.

Page 112, DEMENTIA PRAECOX—WHAT WE NOW CALL SCHIZOPHRENIA: Kraepelin's formulation of dementia praecox hinged on the typically chronic and progressive course of the disabling disease, thus distinguishing it from manic-depressive psychosis. His nomenclature was accepted by most European psychiatrists, except in Switzerland where Manfred Breuler first coined the term *schizophrenia*. Over decades, especially in America, these diagnostic distinctions eroded until the term *schizophrenia* became associated with any psychosis. This led to considerable confusion between manic psychosis and the psychosis of schizophrenia, so that mania was very rarely diagnosed in America until the introduction of Lithium in the 1970s. This is one reason why many people with manic-depressive psychosis who are now in their fifties and sixties were initially diagnosed as having schizophrenia and received inadequate treatment.

Page 113, A STRIKING DIFFERENCE BETWEEN THE NUMBER OF PEOPLE WHO SUFFER DEPRESSION ALONE (UNIPOLAR ILLNESS) AND THE NUMBER OF THOSE WHO EXPERIENCE DEPRESSION IN ASSOCIATION WITH MANIA: There are several outcome studies which support these differences. Here I have drawn on those of George Winokur. See M. Tsuang, G. Winokur, R. Crowe, "Psychiatric disorders among relatives of surgical controls," *Journal of Clinical Psychiatry* 45 (1984): 420–422; L. Robbins, Z. Helzer, M. Weissman, et al., "Lifetime prevalence of specific psychiatric disorders in three sites," *Archives of General Psychiatry* 41 (1984): 949–958; G. Winokur, R. Wesner, "From unipolar depression a bipolar illness: 29 who changed," *Acta Psychiatrica Scandinavica* 76 (1987): 59–63; G. Winokur, M. Tsuang, R. Crowe, "The Iowa 500: affective disorder in relatives of manic and depressed patients," *American Journal of Psychiatry* 139 (1982): 209–212. In making this distinction, I deviate from Winokur's classification, which distinguishes individuals with depression secondary to some medical illness from those with unipolar depression in general. Winokur's distinction may be useful, however, in that it potentially identifies a group of people with unipolar depression who are genetically predisposed, sorting them from those who develop the same symptoms from the general stress and strains of life. Here, in the interests of simplification, I have clustered all unipolar patients together.

Page 113, GENETICALLY IDENTICAL TWINS: Here I refer to a Danish study of manic-depressive disorder in twins by A. Bertelsen, B. Harvald, M. Hauge, "A Danish twin study of manic-depressive disorder," *British Journal of Psychiatry* 130 (1977): 330–351.

Page 114, RECENT SURVEY OF FIVE HUNDRED BIPOLAR MEMBERS OF THE NATIONAL DEPRESSIVE AND MANIC-DEPRESSIVE ASSOCIATION: J. Lish, S. Dime-Meenan, P. C. Whybrow, et al., *Journal of Affective Disorders* 31 (1994): 281–294.

Page 118, AS ANGELA REACHED MATURITY HER PERIODS OF SEVERE DEPRESSION DEVELOPED A MARKEDLY SEASONAL PATTERN: There are many individuals like Angela Moorehead for whom episodes of depression begin in adolescence, later to emerge as bipolar illness. The several studies conducted in such individuals suggest that mood lability is the subtle sign that identifies bipolarity in these early years. Thus, an individual who appears to have unipolar depression, but who is emotionally labile as was Angela Moorehead, is very likely to develop bipolar illness later in life. A dominance of the depression in adolescence, therefore, may be misleading, especially if sleep and weight changes during the depressive episodes are not readily apparent, or are dismissed. For further details on the longitudinal course of the two illnesses, I recommend H. S. Akiskal, J. D. Maser, P. J Zeller, et al., "Switching from unipolar to bipolar II: An 11-year prospective study of clinical and temperamental predictors in 559 patients," *Archives of General Psychiatry* 52 (1995): 114–123; G. Winokur, W. Coryell, H. S. Akiskal, J. Endicott, M. Keller, T. Mueller, "Manic-depressive (bipolar) disorder: the course in light of a prospective ten-year follow-up of 131 patients," *Acta Psychiatrica Scandinavica* 89 (1994): 102–110.

Page 118, PERIODS OF DEPRESSION ARE DISTINGUISHED BY THEIR SENSITIVITY TO SEASONAL CHANGES: This topic has become extremely popular and is often the subject of articles in newspapers and magazines, for example the *New York Times,* Health section, Wednesday, December 29, 1993, "Scientists find ways to reset biological clocks in dim winter"; the *New York Times,* Science section, Tuesday, March 14, 1995, "Modern life suppresses an ancient body rhythm." For a detailed and scholarly review of the subject, see D. Avery, K. Dahl, "Bright light therapy and circadian neuroendocrine functions in seasonal affective disorder," in *Hormonally Induced Changes in Mind and Brain,* ed. J. Schulkin (San Diego: Academic Press, 1993), chapter 11.

CHAPTER SIX: THE LEGACY OF THE LIZARD

Page 121, "THE HUMAN BRAIN HAS EVOLVED": The quotation is from *The Triune Brain in Evolution* (New York: Plenum Press, 1990), a comprehensive review by Paul MacLean of his research on behavior and the evolution of the brain, work upon which I draw heavily in this chapter. It was Maclean's research, spanning thirty years, that led to the general acceptance of the limbic system as the anatomical home of emotion. For those interested in the evolution of the human brain, it is a text worth exploring, being full of anatomical and research details while adding a philosophical, sometimes whimsical, view of the complexity of human development. For a brief essay by Paul MacLean on his basic concepts and ideas, see pp. 33–57, *Human Evolution: Biosocial Perspectives,* ed. S. L. Washburn and E. R. McCown (Menlo Park, Calif.: Benjamin Cummings Publishing Co., 1978).

Page 123, PICTURES OF [THE BRAIN'S] ACTIVITY: For a comprehensive review of brain imaging, I recommend *Images of Mind* by Michael Posner, a cognitive psychologist from the University of Oregon, and Marcus Raichle, who is professor of neurology and radiology at Washington University School of Medicine in Missouri and a pioneer in brain imaging (New York: W. H. Freeman Co./Scientific American Library, 1994).

Page 123, THE NUMBER OF SYNAPSES ... VARIES RELATIVE TO AN ANIMAL'S ACTIVE ENGAGEMENT WITH ITS ENVIRONMENT: The first reports that the complexity of the environment can change brain development appeared during the 1970s. Littermate rats, reared under different environmental conditions, had marked variation in their brains, especially in the cortex. Those housed in empty cages had a thinner brain cortex and a lower brain weight, by as much as 6 percent, compared to their siblings reared in cages with interesting toys. See M. R. Rosensweig and E. L. Bennett, "Experimental influences on brain anatomy and brain chemistry in rodents," in G. Gottlieb (Ed.), *Studies on the Development of Behavior and the Nervous System* (New York: Academic Press, 1978). Recent research has shown that the effects are particularly pronounced early in development and appear within a few days in young rats, showing the intimate relationship between early experience and the micro-anatomy of the brain. C. S. Wallace, V. A. Kilman, G. S. Withers, W. T. Greenough. "Increases in dendritic length in the occipital cortex after four days of differential housing in weanling rats." *Behavioral and Neural Biology* 58 (1992): 64–68.

Page 124, THE DEVELOPMENT OF THE INDIVIDUAL HUMAN BRAIN: For a concise but comprehensive review of this complicated subject, see chapter 3, "Life-Span Development of the Brain," in *Brain, Mind and Behavior* by Floyd Bloom and Arlyne Lazerson (New York: W. H. Freeman, 1988).

Page 127, LIZARDS EXHIBIT PRIMATIVE SOCIAL BEHAVIORS: The source of information here, beyond my own daily observations at the Center, is chapter 6 of Paul MacLean's *Triune Brain in Evolution,* page 99, "Reptilian Behavior as Typified by Lizards." Much of the original information that Maclean presents came from field studies of the Mexican black iguana, conducted by Llewellyn Evans.

Page 128, THE LIMBIC SYSTEM: MacLean was the first to use the term *limbic system,* but in his formulation he drew heavily upon the work of James Papez. In 1937, in an essay in the *Archives of Neurology and Psychiatry* 38: 725–743, Papez proposed that the hypothalamus, the hippocampus, and the cingulate gyrus (part of the limbic lobe) and their interconnections were the anatomical basis for emotion. He did not include the amygdala; the important role that these nuclei play in emotion was not recognized and integrated until much later.

Page 130, GRASPING A VISUAL IMAGE OF THE BRAIN:

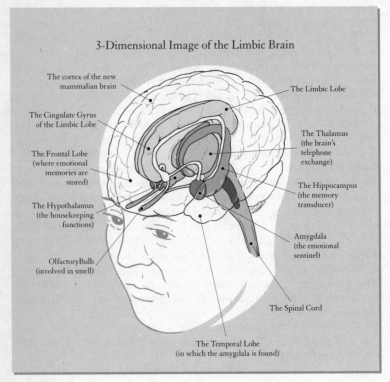

Page 130, STEPHAN SZABO . . . THE LESSER ANTILLES: Stephan Szabo had a gift for making things amusing in medical school. In the early 1960s, studying brain anatomy *was* something akin to ancient cartography. Just the shapes and relative position of the islands of gray matter, in a sea of white brain substance, seemed to be important. There was little understanding of how the brain nuclei connected together or what function they served. The Lesser Antilles are the chain of old, volcanic islands that curve their way through the Caribbean Sea from Puerto Rico to the coast of Venezuela, from the Virgin Islands to the Grenadines.

Page 132, EACH AMYGDALA IS ALSO INTIMATELY CONNECTED TO THE HYPOTHALAMUS: The activities of the limbic alliance that sustain emotional behavior also profoundly influence the general homeostasis of the body through neuronal projections that link the amygdala and hypothalamus to the endocrine system. It is through these connections that chemical changes in the body can be detected during the stress response, something which I elaborate upon in chapters 7 and 8. The higher cortical activities of decision-making and control, which are so uniquely human, profoundly influence the stress response through these pathways, interacting with the limbic emotion of fear and anxiety. Thus, perceived threat and abstract fear are equally as powerful in generating the stress response as is physical injury.

Page 134, THE STORY OF PHINEAS GAGE: This fascinating story of a young man, a foreman for the Rutland and Billington Railway in Vermont, has been carefully reevaluated by Hannah and Antonio Damasio from the University of Iowa. Using three-dimensional computer techniques and accurate measurements from Gage's skull, which has been preserved in the Anatomy Museum at Harvard University, the Damasios have reconstructed the injury done to Phineas's brain. Drawing upon a knowledge of modern neuropsychology and the descriptions of his behavior at the time, they have integrated their findings and proposed specific areas of neuroanatomical damage that may explain his change in behavior. Details of the Damasios' reconstruction are found in their *Science* article, 264 (1994): 1102-1110. See also A. Damasio, *Descartes' Error: Emotion, reason and the human brain* (New York: Putnam and Sons, 1994).

Page 134, WILDER PENFIELD, THE CANADIAN NEUROSURGEON: Dr. Penfield published several books describing his work with epileptic patients and the mapping of the functions of the living brain. The most comprehensive among them is W. Penfield, T. Rasmussen, *The Cerebral Cortex of Man: A clinical study of localization* (New York: Macmillan, 1950). Chapter 9 deals particularly with the temporal cortex. *The Mystery of the Mind* (Princeton, N.J.: Princeton University Press, 1975), is a summary of Penfield's work and his integration of it into a comprehensive theory of brain activity. The individual who served as the main source of clinical information for the sufferings of Claire Dubois did meet Wilder Penfield. I understand that Dr. Penfield delighted in walking around the medical center with his patients, discussing their experience in illness and their recovery, and expanding on his general philosophy of life. In passing the Montreal Neurological Institute, he liked to point out that he had physically located it between Psychiatry and the General Hospital, thus establishing a bridge between the mind and the body.

Page 134, THE SUBSTANCE OF THE BRAIN ITSELF IS INSENSITIVE TO PAIN: The trigeminal nerve, the nerve to the face and scalp, also innervates the dura, which is the tough membrane that covers the brain. However, once the surgeon is inside that covering, the brain tissue is insensitive to the probe.

Page 135, OVER FIVE HUNDRED INDIVIDUALS WITH TEMPORAL LOBE EPILEPSY: Dr. Penfield studied well over 1,000 patients altogether, of whom approximately one-half had temporal lobe epilepsy. Surgery for the treatment of epilepsy is undertaken only after efforts to obtain reasonable control with medication have failed. Before surgery proceeds, the specific site of the epileptic discharge must be identified to ensure that only that part of the brain from which the seizure originates is removed. As an aid to this very delicate task, especially before the availability of modern imaging procedures, Penfield conducted careful exploration of the region of the brain thought to be abnormal, using mild electrical stimulation. This made possible an ongoing discussion between the conscious patient and the surgeon throughout the operation. From these explorations, Penfield located, again and again, those brain areas that produced movements of the limbs, sensations in various parts of the body, and so on. These studies confirmed what physicians had learned over centuries about the effects on behavior of brain injury, strokes, and localized tumors. Easily identified were those areas controlling the arms and legs—the motor cortex, which lies at the very back of the frontal lobe in the middle of the brain. Immediately opposite, across the central sulcus, is the sensory cortex, which registers incoming stimulation from the body surface.

Penfield discovered some other interesting things with his electrodes. One was that the movements he induced were very crude—mere clumsy movements of the limbs and hands. The smacking of the lips and grimacing of the face that were induced did not have the fine, coordinated movements we associate with an adult individual, but rather with a young infant. Similarly, the sensations described when the sensory cortex was stimulated were crude representations of the usual experience of touch or movement. Stimulation of certain areas of the sensory brain gave the patient the impression that a limb was moving, even when it was not. Stimulating the back of the brain, where sight is organized, resulted in the patient seeing only color and light, and in the auditory area he heard only simple qualities of sound, such as tone and volume. Penfield suggested that the reason for this was that his stimulation engaged only a small part of a larger neural net, or "system," that was responsible for the functions of seeing, hearing, moving, or sensing the position of a limb. Subsequent research has borne out his hypothesis.

Page 135, THE MYSTERY OF THE MIND: This quotation is from page 21 of Dr. Penfield's book (Princeton, N.J.: Princeton University Press, 1975), which also contains many other case histories.

Page 136, DURING HIS MANIAS . . . STEPHAN HAD REMOVED HIS CLOTHES: Episodes of automatic behavior are a fascinating link between epilepsy and manic-depressive psychosis. In some epilepsy, a patient may appear unconscious but, in fact, continues to perform stereotyped habitual patterns of behavior, such as walking and talking, and apparently purposeful behavior, such as opening or closing windows. Some form of this automatic behavior occurs in 50 to 75 percent of individuals who have temporal lobe epilepsy. MacLean considers these automatic behaviors in epilepsy and in manic excitement to be early, perhaps genetically programmed, activities that are triggered by the seizure or manic irritability of the brain. He suggests they are equivalent to chest beating, which occurs in primates and is also sometimes seen in manic excitement.

Page 136, EACH SIDE OF THE BRAIN CONDUCTS BUSINESS PRINCIPALLY WITH THE OPPOSITE SIDE OF THE BODY: I refer here to the motor and sensory systems of the body. Approximately 90 percent of the population is right-handed, and this appears to have been the distribution since the beginning—judging by cave painting, early tools, and other such points of reference. Manifestly it is preferable to have one side of the brain in charge when one is wielding an ax. However, there are exceptions to this lateralization, one being in how we hear, which involves both sides of the brain. Smell, in contrast, is handled on the same side of the brain and is not crossed over, as are motor and sensory function. Smell is very important in the emotional expression of some lower mammals, and this may be of significance as we learn more from research about the lateralization of emotional function in mood disorder. Also, nerves from both sides of the brain are involved in the control of some of the facial muscles, which is why the upper part of the face is frequently spared after a stroke.

Page 136, THE CORPUS CALLOSUM . . . STUDIES IN INDIVIDUALS WHERE THIS BRIDGE HAS BEEN SEVERED: In some rare instances an individual is born without a corpus callosum, but those who have been studied are largely individuals who have had the bridge divided to treat intractable epilepsy. Roger Sperry was a pioneer in the study of hemispheric specialization, and his work is well summarized in F. Bloom, A. Lazerson, *Brain, Mind, and Behavior* (New York: W. H. Freeman, 1988), chapter 9. This is a beautifully illustrated book, which is well worth the investment for those interested in a general overview of brain and behavior.

Page 137, WHERE . . . AN INTRACTABLE MELANCHOLIA DEVELOPS (WITH) BRAIN INJURY: A large number of clinical reports describe changes in emotional behavior following brain injury. However, as one might imagine from the diffuse nature of the damage that occurs in such illness, there are very few systematic reviews or comparative studies in the literature. Probably the best summary of this very complicated area is the paper by D. V. Jeste, J. B. Lohr, and F. K. Goodwin, "Neuroanatomical studies of major affective disorders. A review and suggestions for further research," *British Journal of Psychiatry* 53 (1988): 444–459.

Page 138, WITH SUCH MAPS THE ANATOMICAL INTEGRITY OF DIFFERENT BRAIN REGIONS CAN BE DETERMINED, BUT NOTHING CAN BE SAID ABOUT FUNCTION: The informed reader may be concerned here that I have forgotten about magnetic resonance imaging (MRI), which has the potential to determine both structure and function. I have left MRI out of the general text to avoid confusion in an already complicated chapter.

The technology of MRI is dependent upon the physics of atoms exposed to a powerful magnetic field, when many of them behave like little compass needles and can be made to line up together by artfully manipulating the magnet. Then, by prodding the atoms with radio waves and briefly shifting their alignment, a measurable signal is produced as they move within the magnetic field. Computerized integration of these signals yields a detailed anatomical picture, far more detailed than X-ray CT of any body organ, and in the brain these MRI pictures are detailed enough to identify the major nuclear centers of the limbic system. Thus, MRI has the capacity to "photograph" structure. However, extraordinary potential also lies in using it to measure function—the physiological changes of the working brain. The physics of this opportunity lies in a discovery made by Linus Pauling in 1935, when he found that the amount of oxygen carried by hemoglobin—which gives blood its red color—changes the magnetic properties of the hemoglobin. This has been exploited by scientists at Bell Laboratories, who have now demonstrated that MRI can detect these small changes in the brain. The really interesting part, however, is that when the brain cells are working hard, they do not actually use more oxygen, although more blood rushes to the site, presumably carrying the other elements required (such as glucose) for metabolism. This means that there is an increased amount of oxygen at the sites of activity which can be measured through the MRI, thus providing the potential of a functional image of blood flow.

This rather complicated piece of neurobiology is lucidly explained, together with the physics of PET and SPECT, in an excellent article by Marcus Raichle, "Visualizing the Mind," *Scientific American* magazine 270 (1994): 58–64. Other modalities of brain imaging that have caught public attention were discussed in a *Newsweek* magazine cover story entitled, "The Brain: Science Opens New Windows on the Mind," April 20, 1992, pp. 66–72.

Page 139, WAYNE DREVETS: I chose this study, among several, of the PET imaging of mood disorder because it is one of the most complete and easily understood, although it is confined to individuals with unipolar depression. The group of investigators at Washington University School of Medicine in St. Louis, Missouri, has carefully developed the techniques of PET to investigate behavior over several years, and this particular study, which was described in the *Journal of Neuroscience* 12 (1992): 3628–3641, is a good example of their work.

Page 140, PREFRONTAL CORTEX: Some individuals may argue that the whole prefrontal cortex is a part of the limbic circuitry. It is difficult to say precisely where one ends and the other begins. What we do know is that this is the part of the brain that deals with working memory and with emotional memory in particular.

Page 140, RECURRENT THOUGHTS OF SUICIDE: Here I refer to the repetitive, obsessive ideas that plague depressed individuals and frequently persist even after recovery in individuals with severe manic-depressive disease. They can be obsessive thoughts or even obsessive acts, like checking for facial blemishes or other physical concerns. This suggests an anatomical overlap between depression and obsessive-compulsive disorder. There is evidence that in obsessive-compulsive illness, disturbance of the frontal lobe and the limbic cortex is present, particularly disturbance of the cingulate gyrus.

Pages 140–141, CONSISTENT FINDING . . . CONFIRMED IN SEVERAL RESEARCH CENTERS: The specific studies referred to here appeared in *Psychiatry Research* 52 (1994): 215–236. Some of Dr. Gur's work has also been published in *Science* 267 (1995): 528–531. Other studies from the National Institute of Mental Health have also implicated the same areas of brain in the induction of sad thoughts, which are found activated in individuals with depression. M. S. George, T. A. Ketter, P. I. Parekh, B. Horwitz, P. Herscovitch, R. M. Post, "Brain Activity During Transient Sadness and Happiness in Healthy Women," *American Journal of Psychiatry* 152 (1995): 341–351, and from St. Louis, J. V. Pardo, P. J. Pardo, and M. E. Raichle, "Neural Correlates of Self-Induced Dysphoria," *American Journal of Psychiatry* 150 (1993): 713–719.

Page 141, RESEARCHERS AT THE UNIVERSITY OF IOWA: Researchers at the University of Iowa, and particularly Dr. Robert Robinson, were some of the first to systematically correlate the relationship of post-stroke behavior and depression with the site of the lesion. The paper I refer to here regarding mania is one of the few in the literature. See R. Migliorelli, S. E. Starkstein, A. Teson, et al., "SPECT findings in patients with primary mania," *The Journal of Neuropsychiatry and Clinical Neurosciences* 5 (1993): 359–383.

Page 141, AT THE UNIVERSITY OF PENNSYLVANIA: Here I refer to two papers by colleagues at the University of Pennsylvania. See J. D. Amsterdam and D. P. Mozley, "Temporal Lobe Asymmetry With Iofetamine (IMP) SPECT Imaging in Patients with Major Depression," *Journal of Affective Disorders* 24 (1991): 43–53; L. Gyulai, A. Alavi, K. Broich, J. Reilley, W. Ball, P. C. Whybrow, "Iofetamine (IMP) Single Photon Computed Emission Tomography in Rapid Cycling Bipolar Disorder—A Clinical Study," to be published in *Biological Psychiatry*.

Page 142, THE AMYGDALA REGION . . . STILL DISPLAYED A PATTERN OF INCREASED BLOOD FLOW: In psychiatry, efforts are made to distinguish behaviors that come and go with the episode of illness and those that persist even after the behaviors of illness have departed. The former are considered state dependent, that is, dependent upon the state of depression itself, and the latter, referred to as *traits,* are thought to reflect a persistent biological vulnerability. In the findings of this experiment Dr. Drevets speculates that the increased activity of the amygdala, in the absence of depressive symptoms, is a trait of vulnerability. Other potential traits of illness in depression are disturbances of sleep and some changes in the chemistry of serotonin (see chapter 9) that persist in the cerebrospinal fluid. Finding trait "abnormalities" is considered by some investigators to be telltale evidence that the vulnerability is genetic.

Page 143, THE KLÜVER–BUCY SYNDROME: For those interested in a detailed account of this syndrome I suggest J. P. Aggleton, R. E. Passingham, "Syndrome Produced by Lesions of the Amygdala in Monkeys," *Journal of Comparative and Physiological Psychology* 95 (1981): 961–977. A review article on the same subject by Joseph Ladue is found in *The Amygdala: Neurobiological*

Aspects of Emotion, Memory and Mental Dysfunction, ed. J. P. Aggleton (New York: William Wiley-Liss, 1992), pp. 339–351.

Page 143, YOU WILL RECALL MY EARLIER DESCRIPTION OF THE CHILD HESITATING: Here I refer to my earlier discussion, in chapter 4, of the cliff experiment (see page 72).

Page 143, DR. DAMASIO DESCRIBED A YOUNG WOMAN: see R. Adolphs, D. Tranel, H. Damasio, A. Damasio. "Impaired Recognition of Emotion in Facial Expressions Following Bilateral Damage to the Human Amygdala," *Nature* 372 (1994): 669–672. Also recommended for a concise summary of Prof. Damasio's ideas is, "Toward a Neurobiology of Emotion and Feeling: Operational Concepts and Hypotheses," *Neuroscientist* 1: (1995) 19–25.

Page 144, SECONDARY . . . SELF-CONSCIOUS EMOTIONS: For an interesting review of secondary, or self-conscious, emotions, see M. Lewis, "Self-Conscious Emotions," *American Scientist* 83 (1995): 68–78.

Page 144, DAVID CLARK: See D. M. Clark, J. D. Teasdale, "Diurnal Variation in Clinical Depression and Accessibility of Memories of Positive and Negative Experience," *Journal of Abnormal Psychology* 91 (1982): 87–95. Any clinician (or any individual who has lived with a depressed person) knows the distortion which mood places upon the recall of memories. In this paper, Professor Clark compares the retrieval of memories during different mood disorder states in the same patient, confirming clinical and anecdotal impression.

CHAPTER SEVEN: THE VITAL BALANCE

Page 149, CLAUDE BERNARD: From *Leçons sur les phénomènes de la vie communs aux animaux et aux végétaux* (Paris: Ballière, 1878–1879), quoted by Irving Kupfermann in "Hypothalamus and Limbic System: Peptide Neurons, Homeostasis, and Emotional Behavior," chapter 47 in *Principles of Neuroscience,* by E. R. Kandell, J. H. Schwartz, and T. M. Jessell (Eds.) (East Norwalk, Conn.: Appleton and Lange, 1991).

Page 150, HANS SELYE, THE HUNGARIAN-CANADIAN RESEARCHER AND PHYSICIAN: Selye (1907–1986), in his autobiographical book *The Stress of My Life* (New York: Van Nostrand Reinhold, 1979), describes how his complex family roots in Europe helped make him aware of stress and its effects upon the body. He was born of a Hungarian father, who served as a surgeon in the Austro-Hungarian army of World War I, and an Austrian mother. He lived in territory that was initially Hungarian and then became part of Czechoslovakia. He became interested in "the syndrome of just being sick" as a medical student of nineteen. He was struck by the number of people who, regardless of their diagnosis, complained of being tired, unable to sleep, having little appetite, some loss in weight—many of the symptoms seen in depressive syndromes. However, throughout his work Selye emphasized that contrary to widespread public opinion, stress is not synonymous with nervous depression, tension, fatigue, or discouragement; he preferred to characterize stress as a nonspecific response of the body to challenge. Selye received his Ph.D. in organic chemistry and his M.D. from the University of Prague during the 1930s, later studied in the United States at Johns Hopkins University, and then transfered to McGill University to work with Professor Collip, one of Canada's most distinguished biochemists at the time. After briefly returning to Prague, he was offered a permanent position at McGill and spent the rest of his professional life at that university.

For those interested in the development of the concept of stress and Hans Selye's contribution to it, his autobiography is recommended, as is his fundamental text, *The Stress of Life* (New York: McGraw-Hill, 1956).

Page 152, QUALITIES CHARACTERISTIC OF AN "OPEN" SYSTEM: Appreciating the dynamic qualities of living creatures aids in the understanding of adaptation and disease. The dynamic concepts of "open systems," which draw upon mechanical and mathematical models, can be generically applied across all living things. For a discussion of greater depth on this subject, see chapter 8, "Theoretical Aspects of Living Systems," in my earlier book, written with colleagues Hagop Akiskal and William McKinney, entitled *Mood Disorders: Toward a New Psychobiology* (New York: Plenum Press, 1984), pp. 153–172. And for those interested in the mathematics underlying dynamic behavior and biological time, try A. Winfrey, *The Geometry of Biological Time* (New York: Springer-Verlag, 1980). Dr. Winfrey explores the dynamic behavior of living systems from the heart to brain clocks. He draws analogies from nonliving examples using the abstract concepts of systems theory and mathematics. Also, see *Order out of Chaos,* co-authored by Ilya Prigogine, who won the Nobel Prize in 1977 for work on the thermodynamics of nonequilibrium systems. This is also an excellent book for those interested in systems theory and the general language of dynamics (Ilya Prigogine and Isabelle Stengers, *Order out of Chaos* [New York: Bantam Books, 1984]).

Page 152, THE ECONOMY OF AN OPEN SYSTEM: The metaphor of a commercial economy is very useful in understanding living systems and, in fact, played a part in the early development of Charles Darwin's theories of evolution. Adrian Desmond and James Moore in their book *Darwin* (New York: W. W. Norton, 1991) describe Darwin's heavy financial investment in the Industrial Revolution and his interest in the economics of manufacturing. The concept of specialized labor practices to provide a market niche in industrial development was extrapolated by Darwin to survival in nature—how populations and species sort their own "unpressured nook" to protect themselves and maintain a dynamic economy with their surroundings. For a detailed discussion of this, see pages 420–421 in Desmond and Moore, op. cit.

Page 153, THE ... AUTONOMIC ... NERVOUS SYSTEM ... QUIETLY AND CONTINUOUSLY TUNES THE BODY'S PHYSIOLOGY: While I have not detailed it here, not wishing to distract the reader, there is a very interesting older literature on the neurophysiology of depression which supports the clinical experience of the physician that the autonomic nervous system is highly aroused during melancholia and also in mania. This supports the endocrine evidence that increased arousal and stress are a linking biological mechanism that triggers vulnerability to mood disorder. Research in depression that shows changes in muscle tone, in the arousal responses to stimuli, and in sleep patterns, and studies of sodium metabolism that may be pertinent to the excitability of the neuron, all point to an increased irritability and sensitivity of autonomic nervous system function. Many years ago I reviewed this early literature, and the paper remains pertinent. See P. Whybrow and J. Mendels, "Toward a biology of depression: Some suggestions from neurophysiology," *American Journal of Psychiatry* 125 (1969): 1491–1500.

Page 153, THE BODY'S IMMUNE SYSTEM ... IS INTIMATELY ENTWINED WITH THE HORMONES OF THE ENDOCRINE SYSTEM: Stimulated partly by the AIDS epidemic in the 1980s and 1990s, there has been a rapid increase in our knowledge about the regulation and function of the immune system. It has become clear that an intimate relationship exists among stress, immune defense, and neuroendocrine function, with the steroid axis playing a key role. The cells of the immune system that react to invasion—including macrophages, neutrophils, lymphocytes, and natural killer cells—all appear to be involved in these regulatory processes. Immune defense of the body includes the mobilization of these cells and also hormonelike molecules that the immune system cells produce in response to challenge. A large number of these molecules are messengers, collections of amino-acids called peptides. Of these, interleukin 1 (IL–1), a peptide produced by macrophages, seems to be the important messenger between the immune system and the adrenal axis. The presence of a foreign protein in the body will quickly lead to increased production of interleukin 1, which is fed back via the bloodstream to alert the hypothalamus that the body is under attack. Exactly how interleukin 1 works in the hypothalamus is not yet clear, but the possibility is that it sensitizes neurons containing corticotrophin-releasing factor (CRF) to the effects of norepinephrine and leads to an outpouring of ACTH from the pituitary gland, and then subsequently to the release of the stress hormone cortisol from the adrenal cortex. This increased production of steroids dampens the immunological reaction of the body, thus completing the classic feedback loop of a homeostatic system. Hence the limbic brain and adrenal cortex are interconnected in the regulation of immune activity. While the details of this enormously complex intertwining are not completely understood, the implications for those who suffer repeated depressions is clear. In a situation of chronic psychological stress, perhaps due to social disruption or bereavement, or a recurrent episode of depression triggered by environmental stress, high levels of circulating steroids will dampen the immune response and increase the chances of infection. Just as disturbances of general health can trigger depression, so can depression trigger disturbances of general health. For a recent review and research exploring these consequences of depression for immunity and general body metabolism, see E. M. Sternberg, G. P. Chrousos, R. L.

Wilder, P. W. Gold, "The stress response and the regulation of inflammatory disease," *Annals of Internal Medicine* 117 (1992): 854–866; and D. Michelson, C. Stratakis, L. Hill, J. Reynolds, E. Galliven, G. P. Chrousos, P. W. Gold, "Bone mineral density in women with depression," *New England Journal of Medicine* 335 (1996) 1176–1181.

Page 154, REACTIVE HOMEOSTASIS . . . PREDICTIVE HOMEOSTASIS: I believe this useful distinction was first suggested by Dr. Moore-Ede, of Harvard University. See M. Moore-Ede, "Physiology of the circadian timing system: Predictive versus reactive homeostasis," *American Journal of Physiology* 250 (1986): R737–R752.

Page 154, WE ARE NOT AT OUR BEST IN THE EARLY HOURS OF THE MORNING: Numerous studies have shown that human performance deteriorates during the night hours, on any number of tasks. One investigation, for example, testing the ability to operate a flight simulator, found that performance in the early hours of the morning was similar to that after moderate alcohol consumption. Major accidents, such as the Three Mile Island and Chernobyl nuclear plant accidents, frequently occur in the early morning hours. Indeed, industrial performance as mundane as meter readings over a twenty-year period in a Swedish gasworks shows pronounced deficits during the night shift. For an interesting discussion of these circadian issues, the reader is referred to D. F. Dinges, "An overview of sleepiness and accidents," *Journal of Sleep Research* 4 (Suppl 2) (1984): 4–14.

Page 156, IN A STUDY THAT I CONDUCTED IN NORTHERN NEW ENGLAND: The details of this study, and other details of seasonal behavior in animals and in humans, will be found in *The Hibernation Response,* a book I wrote with Robert Bahr (New York: Arbor House, 1988); and P. C. Whybrow, "Where there's mud there's momentum: A psychobiologist looks at the New England springtime," *Yankee* magazine, April 1979.

Page 157, SEASONAL AFFECTIVE DISORDER (SAD): For one of the original descriptions of this syndrome, see N. E. Rosenthal, D. A. Sack, J. C. Gillin, et al., "Seasonal affective disorder: A description of the syndrome and preliminary findings with light therapy," *Archives of General Psychiatry* 41 (1984): 72–80.

Page 157, SAD IS MORE COMMON IN ALASKA THAN IN SAN DIEGO: The source paper for this intriguing finding is L. Rosen, S. Targum, M. Bryant, et al., "Prevalence of seasonal affective disorder at four latitudes," *Psychiatry Research* 31 (1990): 131–144.

Page 157, LIGHT THERAPIES FOR THE TREATMENT OF WINTER DEPRESSION: For a review of phototherapy in the treatment of winter depression, see D. Avery, K. Dahl, "Bright light therapy and circadian neuroendocrine function in seasonal affective disorder," in *Hormonally Induced Changes in Mind and Brain,* ed. J. Schulkin (San Diego: Academic Press, 1993), pp. 357–390. The first report of successful light therapy in an individual with manic-depressive disease appeared in 1982. For this reference, see A. Lewy, H. Kern,

N. Rosenthal, T. Wehr, "Bright artificial light treatment in a manic-depressive patient with a seasonal mood cycle," *American Journal of Psychiatry* 139 (1982): 1496–1498. This is a very popular subject for magazine writers and has a seasonal cycle of its own, with December, January, and February being the peak times. A particularly fine article appeared in the *Chicago Tribune* on Sunday, December 24, 1995, entitled "The Science of the Seasons: Winter, A Long Hard Trek to Light."

Page 157, MASTER CLOCKS . . . IN THE SUPRACHIASMATIC NUCLEI (SCN) OF THE HYPOTHALAMUS: These little clusters of nerve cells received their forbidding name because they sit above the area in the hypothalamus where the optic nerves from the eyes partially cross each other (the chiasma) on the way to the brain. It was first recognized in 1972 by two scientists working independently—Robert Moore and Irving Zucker—that destruction of these nuclei in the rat brain prevented the rhythmic coordination of behavior with the light–dark cycle. Furthermore, the eating, sleeping, and waking activities and the hormonal rhythm of cortisol were disorganized. For the original papers, see R. Y. Moore, Z. B. Eichler, "Loss of circadian adrenal corticosterone rhythm following suprachiasmatic lesions in the rat," *Brain Research* 42 (1972): 201–206; and F. K. Stephan, I. Zucker, "Circadian rhythms in drinking behavior and locomotive behavior of rats are eliminated by hypothalamic lesions," *Proclamation to the National Academy of Science (U.S.A.)* 69 (1972): 1583–1586. The activity of these brain clocks is genetically prescribed and the "speed" of the clock varies among species. The human clock, for example, runs with a period of approximately 24.6 hours, whereas the mouse clock is faster at 23.6 hours, while the hamster's is almost exactly 24 hours. It is now recognized that the phasic manufacture of proteins, driven by gene instruction, provides the oscillating mechanisms of the SCN. For those interested in the details of molecular research, I recommend the review article by M. Hastings, "Circadian rhythms: Peering into the molecular clock," *Journal of Neuroendocrinolgy* 7 (1995): 331–341.

Page 158, SOCIAL HABITS . . . HAVE SUBTLE ZEITGEBER EFFECTS: Most of us are more dependent upon the environment and the cues of our daily routine than we would like to believe in the maintenance of rhythmic behavior. The morning cup of coffee, the timing of meals, exercise, and our comfortable habits, are all zeitgebers in the general sense. That they support the regimentation of our physiology is recognized when they are taken away for one reason or another, such as moving to a new location or to a new job. We are creatures of habit. The brain has developed its own habits to manage predictable planetary change by encoding the information into the genome. Although light is the most powerful zeitgeber in synchronizing these brain rhythms with the day–night cycle and maintaining a stable existence, many of the social zeitgebers emerge as of great practical importance when one is working to reestablish a daily rhythm in manic-depressive disease. Individuals with bipolar disorder frequently are driven by an internal cycle that disregards external cues, and they find organizing themselves in synchrony with the day very difficult.

Page 158, SIGNALS FROM THE BRAIN CLOCKS TO THE PINEAL: The pathway that the fibers take from the eye to the pineal is rather complicated. After passing through the SCN, where the impulses set the timing of the clock, they exit the brain via the hypothalamus and pass down through the nerves of the neck to the superior cervical ganglion. From there they loop back to the pineal gland. For those interested in these details, I would recommend A. J. Lewy, "Human melatonin secretion: A marker for the circadian system and the effects of light," in *Neurobiology of Mood Disorders,* ed. R. M. Post, J. C. Ballenger (Baltimore: Wilkins and Wilkins, 1986).

Page 159, A SPRING AND SUMMER PATTERN OF BIRTH: Herd animals, such as sheep and deer, are sexually most active in the fall of the year, when the photo-period—the balance between light and dark—is declining. This ensures that the offspring will be born predominantly in the spring and early summer. Testosterone levels in the males—including the growth of antlers in the bucks of the deer herd, which is testosterone dependent—are tied to this changing rhythm of melatonin secretion. In humans there is evidence of a seasonal variation in testosterone production too. See M. Lagoguey, A. Reinberg, *Journal of Physiology* 257 (1976): 19–20. There is also evidence that the number of summertime births increases in higher latitudes. I confirmed this to be the case in northern New England when I calculated the probable month of conception from the date of live births, over a two-year period, in Vermont and New Hampshire. The number of conceptions showed a relative increase in the late summer and early fall. Cowgill reviewed this seasonal phenomenon in a paper during the 1960s entitled "Season of Birth in Man," which will be found in *Ecology* 47 (1966): 614–623.

Page 159, THE RESEARCHERS DISCOVERED A REMARKABLE THING: The influence of light on the daily and seasonal behavior of animals has been recognized for a long time, but in the 1970s it was thought that human behavior was relatively immune to the effects of light. Artificial light in the home for many years had extended "daytime" well beyond the seasonal allotment of sunlight. Then, in 1980, Alfred Lewy and his associates, in an article published in *Science* magazine (*Science,* 210 [1980]: 1267–1269), demonstrated that light does suppress human melatonin secretion. In their original experiments they used a light source of approximately 3,000 lux. One lux, the standard measure of illumination, is the illumination provided upon a surface one meter from a point of light equivalent to the power of one candle flame. A bright, sunny day is approximately 50,000–100,000 lux, and a cloudy day somewhere between 1,000 and 10,000 lux. However, even the best-lit offices rise only to about 500 lux and the average lamp is probably somewhere between 100 and 200 lux. This means that when one is reading by such a lamp the room is in darkness as far as the pineal gland is concerned, and the internal rhythm of melatonin secretion proceeds without impediment.

Page 160, THE RESULT IS A SLEEPY BRAIN: The sleep–wake cycle is commonly disturbed in all forms of mood disorder and in many other illnesses, and is considered by most researchers as additional evidence of dysregulated limbic system function. Studies of these changes using the electroencephalogram (EEG) have been important in understanding the limbic disturbances occurring in depression for two reasons. First, the amount of sleep–wake activity each day is a measure of arousal, and secondly, sleep studies represent the most consistent and extensive body of information about circadian rhythms in both depression and mania. While the EEG is only a crude reflection of brain function, it is a reliable physiological measure and reveals the level of brain activity through the night. Studies conducted in the 1960s and 1970s defined a specific architecture of a night's sleep in normal people. From being awake, the individual proceeds slowly through a series of sleep stages to the deepest (slow-wave) sleep. In those depressed, it takes longer to fall asleep and the time to the first dreaming period is decreased. The number of spontaneous awakenings during the night is increased, the levels of sleep change frequently through the night, and the deepest level of sleep (slow-wave sleep) is rarely achieved. David Kupfer and his colleagues at the University of Pittsburgh have drawn attention to the shortened time between sleep onset and the first dreaming episode (marked by a rapid rolling of the eyes without waking, which usually occurs within ninety minutes in normal people) as a hallmark of the depressed state.

While in unipolar depression the total time asleep is usually reduced, in those with bipolar illness the time asleep frequently increases during depressive episodes. This is particularly true for the markedly seasonal forms of bipolar illness, such as Angela Moorehead suffered, when in the winter months sleep lengthens dramatically, and then shortens in the spring. For those interested in reading more about the subject of sleep and dreaming, I recommend the book by J. A. Hobson, *The Dreaming Brain* (New York: Basic Books, 1988).

Page 161, INDEED A NEW INDUSTRY . . . SUCCESSFULLY MANAGING HUMAN HIBERNATION AND SAD: Many small companies have been established following the observations of the beneficial effects of light in winter depression. A variety of desk and standing lamps, for home and business use, has been produced, in addition to some intriguing light visors that can be worn rather like a baseball cap. The length of time needed for these devices to reverse the depression depends upon the intensity of the light they produce. They range in strength from approximately 2,000–10,000 lux, and the time required varies proportionally, from approximately two hours to about thirty minutes a day. Such light systems are increasingly recognized as valuable for synchronizing the body rhythms even in those without seasonal affective disorder. Industrial society has relegated many people to an indoor habitat twelve months of the year. Personally, I find a bright light system very valuable for waking up on a winter's morning.

Among those companies in the United States and Canada marketing lighting units for domestic use during the winter months are:

The SunBox Company
1132 Taft Street
Rockville, MD 20850

Northern Technologies,
8971 Henri Bourassa West
Montreal, Canada H4S 1P7

and

Helios Light Technologies
4445-B Breton SE, Suite 268
Kentwood, MI 49508

SunRiser Enterprises, 10756, Exeter Avenue N.E., Seattle, WA 98125, makes control systems to program home lighting to simulate an early spring dawn.

Page 161, THERE IS LITTLE EVIDENCE THAT USE OF MELATONIN ALONE CAN ALLEVIATE WINTER DEPRESSION: Evidence for an absolute change in melatonin production during winter depression is weak. In those free of depression, melatonin is capable of influencing the sleep circadian rhythm depending upon the time of the day at which it is given. (Upon this evidence, proponents have suggested melatonin as a treatment for jet lag.) There is also debate over the relationship between the disturbed cycle of melatonin excretion found in seasonal affective disorder and the disorder itself. Some researchers, including Alfred Lewy, now at the University of Oregon, believe that the timing of the melatonin rise in the 24-hour cycle is a barometer of circadian disturbance. Personally, I believe this interpretation of the evidence to be correct. Others are not convinced, arguing that it is the absolute amount of light falling upon the retina of the eye during any 24-hour period that is important in the treatment of winter depression. However, the latter opinion does not explain why bright light therapy in the early morning is most effective in reversing the symptoms of winter disorder and in the general organization of circadian rhythms. For a concise and interesting discussion of this whole area of research, I recommend D. Avery, K. Dahl, "Bright Light Therapy and Circadian Neuroendocrine Function in Seasonal Affective Disorder," in *Hormonally Induced Changes in Brain and Mind,* ed. J. Schulkin (San Diego: Academic Press, 1990), pp. 357–390.

Page 162, ESPECIALLY SENSITIVE TO CHANGES IN LIGHT INTENSITY AND ... ANY ADAPTIVE DEMAND THAT REQUIRES RESETTING THE BRAIN CLOCKS: See A. J. Lewy, T. Wehr, P. Goodwin, et al., "Manic-depressive patients may be super sensitive to light," *Lancet* (1981): 383–384.

Page 162, MANIA IS MOST COMMON IN THE LATE SUMMER: See P. A. Carney, C. T. Fitzgerald, C. E. Monaghan, "Influence of climate on the prevalence of mania," *British Journal of Psychiatry* 152 (1988): 820–823.

Page 162, CHRONIC SLEEP DEPRIVATION: Probably many have noticed that after a night without sleep, during the next day one feels reasonably well, indeed quite happy and energetic, but on the second day there is a letdown and great tiredness. In those of bipolar predisposition this experience is magnified. There is no second-day letdown, but continued acceleration and then manic behavior. There are many clinical reports of this triggering of mania by sleep deprivation, and I have seen it many times in my own practice. A controlled study by Thomas Wehr and his colleagues in 1982 confirmed that manic-depressive individuals, especially those prone to switching from one mood state to another (rapid cycling patients), will develop mania if kept awake all night. Some individuals switched back into depression after sleeping, but in others the mania persisted for days or weeks. For the discussion of the original studies see T. Wehr, D. A. Sack, N. E. Rosenthal, "Sleep Production as a Final Common Pathway in the Genesis of Mania," *American Journal of Psychiatry* 144 (1987): 201–204.

Page 165, FLOOD OF STRESS HORMONES . . . GROWS DAMAGING: While the hormonal arousal of reactive homeostasis is a defensive system designed to protect the body, if the same hormones are chronically released into the bloodstream, negative changes due to their presence may occur. Robert Sapolsky of Stanford University has presented evidence that feedback of steroid hormones to the hippocampus may impair memory and also the regulation of corticosteroid production, creating a vicious cycle. In depressed people, and in older individuals with memory disorders such as Alzheimer's disease, Sapolsky speculates that this may lead to progressive failure of hippocampal function and memory. For further discussion of this interesting theory, see L. Jacobson, R. Sapolsky, "The role of the hippocampus in feedback regulation of the hypothalamic-pituitary-adrenocortical axis," *Endocrine Reviews* 12 (1991): 118–134.

Page 166, BRAIN KINDLING: For the details of Dr. Post's work and a discussion of kindling, see R. M. Post, D. R. Rubinow, J. C. Ballenger, "Conditioning and sensitization in the longitudinal course of affective illness," *British Journal of Psychiatry* 149 (1986): 191–201.

CHAPTER EIGHT: OF HUMAN BONDAGE

Page 170, AN . . . EXPERT IN COLD STRESS AND SURVIVAL FOUND CROSSING THE POLAR ICE CAP CONSIDERABLY LESS DEMANDING THAN GIVING AN IMPORTANT LECTURE: Details of this unusual study will be found in H. W. Simpson, "Field studies of human stress in polar regions," *British Medical Journal* 1 (1967): 530–533. Another volume which describes stress in various combat situations is *The Psychology and Physiology of Stress: With Reference to*

Special Studies of the Vietnam War, ed. P. G. Borne (New York: Academic Press, 1969).

Page 171, NEURONS OF THE LOCUS CERULEUS: The locus ceruleus received its unusual name because when the brain is freshly sliced, the nucleus has a slightly bluish tinge. The neurons from this major arousal system utilize the messenger norepinephrine and function as the vigilance system of the limbic alliance. They modulate particularly the hippocampus by increasing neuronal receptor sensitivity to chemical messengers. I describe these messenger systems in more detail in chapter 9. For the technical details, see G. Aston-Jones, S. L. Foote, F. E. Bloom, "Anatomy and physiology of locus ceruleus neurons: Functional implications," in *Norepinephrine,* ed. M. G. Zeitler and C. R. Lake (Baltimore: Williams and Wilkins, 1984), pp. 92–116.

Page 171, RATS ARE FORCED TO SWIM: When a rat is placed in a bowl of water from which it cannot escape, for a while it will swim around the bowl vigorously until, finding that there is no escape, it gives up the struggle. Named the Persolt Swim Test after the psychologist who first used it as a stress paradigm, this simple experiment has an extraordinarily high correlation with the effectiveness of antidepressant drugs. Drugs that relieve depression in human beings increase the length of time the rat will struggle. In other screening tests, antidepressants will also reverse the learned helplessness model of Seligman in animals exposed to inescapable shock, although this is not as consistent, nor as simple, as the Persolt test. Similarly, the depressive reaction that occurs in primates upon separation is reduced by drugs that have an antidepressant action in human beings.

Page 172, THE ACTIONS OF HYDROXYCORTICOSTEROIDS, OF WHICH CORTISOL IS A FAMILY MEMBER, ARE ESSENTIAL TO LIFE: For a comprehensive review of the effects of adrenal steroids on the brain, see B. S. McEwen, R. S. Sakai, R. L. Spencer, "Adrenal steroid effects on the brain: versatile hormones with good and bad effects," in *Hormonally Induced Changes in Mind and Brain,* ed. J. Schulkin (San Diego: Academic Press, 1993), pp. 157–189; and P. C. Whybrow, "Neuroendocrinology," in *Biological Bases of Brain Function and Disease,* ed. A. Fraser, P. Molinoff, A. Winokur (New York: Raven Press, 1993).

Page 173, REGULATED FEEDBACK BECOMES DISORGANIZED AND IMPRECISE DURING PERIODS OF MELANCHOLIA: For a concise review of the endocrine disturbances that occur in mood disorders, the reader is referred to F. Holsboer, "Neuroendocrinology of mood disorders," *Psychopharmacology: The Fourth Generation of Progress,* ed. F. E. Bloom, D. J. Kupfer (New York: Raven Press, 1995), pp. 957–969.

Page 174, METABOLISM OF THYROID HORMONES . . . CHANGED . . . DURING DEPRESSION: Thyroid hormone metabolism, together with the steroid hormones, are the endocrine systems most consistently disturbed in depression. For a detailed review, see M. S. Bauer, P. C. Whybrow, "Thyroid hormones and the central nervous system in affective illness: Interactions which may have clinical significance." *Integrative Psychiatry* 6 (1988): 75–100; and P. C. Whybrow, "Sex differences in thyroid axis function: Treatment implications for affective disorders," *Depression* 3 (1995): 33–42.

Page 179, HOIST WITH HIS OWN PETARD: A petard was a small bomb, first used in the seventeenth century. A box made of tin or wood was filled with gunpowder and was used to blow a hole in a door or wall to give access to a building. To be *hoist by one's own petard* was to be injured by one's own bomb. Here, of course, John Moorehead's reference was to the suffering he had endured as a direct result of his efforts to change the nature of the faculty organization at the college where he taught.

Page 180, THE PHYSIOLOGY AND BEHAVIOR OF . . . RHESUS MONKEYS [IN] RESPONSE TO SEPARATION AND . . . SOCIAL STRESS: Many studies of primate behavior under stress have been conducted by numerous investigators. For a useful summary of this work, see S. J. Suomi, "Primate separation models of affective disorders," in *Neurobiology of Learning, Emotion and Affect,* ed. J. Madden IV (New York: Raven Press, 1991), pp. 195–214; and S. V. Vellucci, "Primate social behavior—anxiety or depression?" *Pharmacology and Therapeutics* 47 (1990): 167–180.

Page 182, SOME 5 OR 10 PERCENT OF INDIVIDUALS NEVER EXPERIENCE OBVIOUS SADNESS: This is counterintuitive. How can a mood disorder not be associated with a disturbance of mood? However, it is an accurate observation which highlights the problems inherent in giving specific names to broad diagnostic categories. Mood is just one of the disturbances of behavior that occurs when the limbic system disorganizes, as I have emphasized throughout this book. It is not always mood that is primarily disturbed. Indeed, in some patients mood is never disturbed. The syndrome of melancholia may appear "masked" as another disorder, where physical complaints of insomnia, an aching body, little energy, and so on, result in a different diagnosis. Many visits to physicians and a large number of expensive diagnostic procedures may be undertaken before the correct diagnosis is arrived at.

Page 182, PHILIP GOLD . . . HAS POINTED OUT . . . THAT . . . DEPRESSIVE BEHAVIORS MIMIC . . . CHRONIC STRESS: Dr. Gold has demonstrated, in a meticulous series of clinical studies, that depression is a state of chronic endocrine arousal—a syndrome where the stress response fails to turn off after the challenge that provoked it has declined, or disappeared. The chronic increase in circulating stress hormones, particularly cortisol and the hydroxycorticosteroids, may directly explain many of the behavioral disturbances seen in depression. For reviews of Dr. Gold's research, see P. W. Gold, F. K. Goodwin, G. P. Chrousos, "Clinical and biochemical manifestations of depression: Relation to the neurobiology of stress: Part I," *New England Journal of Medicine* 319 (1988): 348–353; and P. W. Gold, F. K. Goodwin, G. P. Chrousos, "Clinical and biochemical manifestations of depression: Relation to the neurobiology of stress: Part II," *New England Journal of Medicine* 319 (1988): 378–374.

Page 182, CRH ... HAS DIRECT EFFECTS UPON THE BEHAVIORS OF THE LIMBIC BRAIN: For a discussion of the research in animals of the effects of the corticotrophin-releasing hormone in the brain, I refer the reader to the work of George Koob. Two representative papers are G. F. Koob, "Behavioral actions of corticotrophin releasing factor in the central nervous system," *Journal of Cell Biochemistry* Suppl. 12D (1988): 299–309; and G. F. Koob, "Behavioral responses to stress," in *Stress, Neurobiology, and Neuroendocrinology,* eds. M. R. Brown, G. F. Koob, C. Rivier (New York: Marcel Bekker, 1991), pp. 255–271.

Page 183, THE RESEARCH WE WERE CONDUCTING: For convenience, two actual studies are combined in this vignette. One of them, the comparison of thyroid hormones with placebo as a method of enhancing the effectiveness of antidepressants, was conducted in the early 1970s, rather than the 1980s. That study is described in two papers: A. J. Prange, A. Coppen, P. C. Whybrow, R. Noguera, R. Maggs, "The comparative anti-depressant value of L-tryptophan and imipramine with and without attempted potentiation by L-iodothyronine," *Archives of General Psychiatry* 26 (1972): 474–478; and P. C. Whybrow, A. J. Prange, R. Noguera, J. E. Bailey, "Thyroid function and the response to L-iodothyronine in depression," *Archives of General Psychiatry* 26 (1972): 242–245. There is nothing contrived, however, about the evaluation and treatment of John Moorehead in a clinical research unit. These units, which have been established in many academic health centers of the United States, have been vital in sustaining the evolution of clinical research. The facilities provide excellent patient care in collaboration with the other hospital units, and at the same time develop further knowledge of the pathology and effective treatment of serious illness.

Pages 183–184, HAMILTON SCALE FOR DEPRESSION: The Hamilton Scale is one of the oldest rating scales developed in the assessment of depression. Carefully employed by trained personnel, it provides an objective measurement of the severity of depression, which is essential in comparing response to treatment by different patients and different patient groups. Such scales also permit comparison among different research centers. The idea is a simple one: Each symptom of depression is evaluated carefully against agreed-upon criteria, by trained observers, and given a numerical score. The total score can be compared across time by repeated examination, and thus the response to treatment, or the natural progression of the illness, can be followed. Behavioral rating scales are therefore no different in concept, or practice, from taking repeated blood pressures or measuring an individual's pulse. Rating scales have been very important in the development of objective clinical assessment in psychiatry, starting in the 1960s.

Page 185, "THE HYPERSECRETION OF CORTISOL ... ": This is a quotation from the paper by Dr. Sachar entitled "Disrupted cortisol secretion in psychotic depression," *Archives of General Psychiatry* 28 (1973): 19–24. In a series of studies, Dr. Sachar built the first evidence for a disturbance of cortisol regulation being linked to the behavioral changes of severe depression. In his paper "Corticosteroid responses to the psychotherapy of depressions: Evaluation during confrontation of loss," *Archives of General Psychiatry* 16 (1967): 461–470, several patient case histories are described. I have drawn upon them for that part of John Moorehead's story where changes in steroid response to a stressful event are detected during his hospitalization.

Pages 185–86, THE LONDON PSYCHOANALYST RENÉ SPITZ: The original publication of this work is R. A. Spitz, "Anaclitic depression: An inquiry into the genesis of psychiatric conditions in early childhood," *Psychoanalytic Study of the Child* 2 (1942): 313–347.

Page 186, RESEARCH WITH YOUNG MONKEYS PARALLELS THESE FINDINGS IN HUMANS: Here again I have drawn upon the work of Stephen Suomi, among others. The specific papers are: S. Suomi, "Early stress and adult emotional reactivity in rhesus monkeys," *The Childhood Environment and Adult Disease, Ciba Foundation Symposium* 156 (1991), 171–188; S. Suomi, "Uptight and Laid-Back Monkeys: Individual Differences in Response to Social Challenges," in *Plasticity of Development,* ed. S. Branch, W. Hall, E. Dooling (Cambridge, Mass.: M.I.T. Press, 1991), pp. 27–55; S. J. Suomi, H. F. Harlow, C. J. Domek, "Effect of repetitive infant-infant separation of young monkeys," *Journal of Abnormal Psychology* 76 (1970): 161–172; and C. L. Coe, S. P. Mendoza, W. P. Smotherman, S. Levine, "Mother-infant attachment in the Squirrel Monkey: Adrenal response to separation," *Behavioral Biology* 22 (1978): 256–263.

Page 186, THE HORMONE OXYTOCIN: Oxytocin is the hormone of love and attachment. It has profound effects upon the behavior of parents and infants, but also is important in the development of attachment in adult life. A fascinating story is that of the prairie vole, a small rodent that lives in the grasslands of North America. Once paired, the male and female prairie vole commonly stay together for life. In fact, three quarters of these parings persist until one member dies. The social organization of these creatures has recently been described in L. Getz, S. Carter, "Prairie-vole partnerships," *American Scientist* 84 (1996): 56–62. Thomas Insel has described a series of experiments with the prairie vole and has compared the prairie vole with the asocial mountain vole, which does not form the same affiliative links. Differences in the oxytocin receptor distribution in the brain are apparent. From these and studies in primates, it appears that the limbic structures have oxytocin receptors that are largely unchanged across the mammalian phylogeny. For details, see T. R. Insel, "Oxytocin; a neuropeptide for affiliation: Evidence from the behavioral, receptor, autoradiographic, and comparative studies," *Psychoneuroendocrinology* 17 (1992): 3–35. See also T. R. Insel, L. Shapiro, "Oxytocin receptor distribution reflects social organization in monogamous and polygamous voles," *Proceedings of the National Academy of Sciences* 89 (1992): 5981–5985; C. S. Carter, L. Getz, "Monogamy and the Prairie Vole," *Scientific American* 268 (1993): 100–106.

Page 187, SEYMOUR LEVINE: For a detailed review of Professor Levine's work on social buffering and infant separation, see S. Levine, "Psychobiologic consequences of disruption in mother-infant relationships," in *Perinatal Development: A Psychobiologic Perspective,* ed. N. A. Krasnoger, E. M. Blass, M. A. Hofer, W. P. Smotherman (Orlando, Fl.: Academic Press, 1987), pp. 359–376.

Page 188, RAT PUPS DEPRIVED OF MATERNAL CARE . . . HAVE INCREASED SENSITIVITY TO STRESS: Professor Charles Nemeroff and his colleagues at Emory University in Atlanta isolated male rats from their mothers, and littermates, for six hours daily during infancy (before weaning, between the age of two days and twenty days), and found that in adulthood they reacted in a biochemically distinct way to the mild stress of foot shock, compared to control animals. The corticosteroid stress hormones in the brain were increased, together with their brain receptors. See C. O. Ladd, M. J. Owens, C. J. Nemeroff, "Persistent changes in corticotrophin-releasing factor neuronal systems induced by maternal deprivation," *Endocrinology* 137 (1996): 1212–1218.

Page 188, JEROME KAGAN'S RESEARCH . . . STRONGLY SUGGESTS THAT VARIATIONS IN INDIVIDUAL TEMPERAMENT ARE ENCODED IN THE GENOME: For a discussion of this particular aspect of Professor Kagan's research, see "Continuity and discontinuity in development," in *Plasticity of Development,* ed. S. Brauth, W. S. Hall, R. J. Dooling (Cambridge, Mass.: M.I.T. Press, 1991), pp. 11–26.

Page 192, WAS THIS MAN BECOMING HYPOMANIC?: Switching a patient from depression to hypomania through the use of antidepressant drugs is a possibility that every clinician must keep in mind and guard against when treating individuals with a vulnerability to bipolar illness. The switch commonly occurs rapidly, within the first week of using the antidepressants, so in John Moorehead's case the change in behavior was late for this to be the explanation. However, given the history of bipolar illness in his family, and his sudden "flight into health," I was concerned that I had precipitated in him a hypomanic state beyond his usual hyperthymic temperament. His wish to visit his sister, his energy and ebullience in my office were all of concern to me, which is why I gave him the names of psychiatrists to contact in both San Diego and Boston. Thyroid hormones can also switch people into mania, and thus the combination of antidepressants with thyroid hormones in bipolar persons can be particularly dangerous. Had his symptoms grown any worse during the two days that he was with us, I would have broken the research code to find out whether he was on this combination. In fact, John did not escalate into mania, but returned to the emotional set point characteristic of his temperament prior to illness.

CHAPTER NINE: PILLS TO PURGE MELANCHOLY

Page 195, PILLS TO PURGE MELANCHOLY: The chapter title is taken from the collection of ballads entitled *Wit and Mirth, or Pills to Purge Melancholy,* published in 1719 by Thomas D'Urfey, a popular playwright and songwriter of the time. D'Urfey was much given to singing his own compositions in public and was particularly known for his humor and the performances he gave to "make merry the vacant hours" of the royal court. I first came across some of his ballads at Colonial Williamsburg, where a group of singers and instrumentalists reenacted one of D'Urfey's concerts. In the seventeenth and eighteenth centuries, many herbal remedies were promoted for the relief of melancholy, but other ways of stirring the passions were also prominent. Music, in particular, was considered an antidote to melancholia. Robert Burton in *The Anatomy of Melancholy* comments that "of the means which philosophers and physicians have prescribed to exhilarate a sorrowful heart . . . in my judgment none [is] so present, none so powerful, none so apposite as a cup of strong drink, mirth, music, and merry company." Music, says Burton, "affects not only the ears but the arteries. It erects the mind and makes it nimble . . . mirth and merry company cannot be separated from music." Those interested in learning further should read "Music, a remedy" in *The Anatomy of Melancholy,* R. Burton (New York: Vintage Books, 1977), p. 115. For music as a therapeutic tool in modern times, see S. B. Hanser, L. B. Thompson, "Effects of a music therapy strategy on depressed older adults," *Psychological Sciences* 49 (1994): 265–269. Thomas D'Urfey in the title of his collection intended a double entendre, and it is as a reminder that pills and good companions are still both essential to optimum recovery from depression that I use it here.

Page 195, SUSAN DIME-MEENAN: Susan Panico (as she is now) is the executive director of the National Depressive and Manic-Depressive Association and a leading public advocate for those with mood disorder. Ms. Panico, herself, has overcome periods of extreme illness, which have served to temper her leadership. The highlights of Susan's story were reported by Dianne Hales in the news article "My Journey Back to Sanity," *McCall's* magazine, October 1994.

Page 195, MIKE WALLACE . . . *LARRY KING LIVE,* CNN: This hour-long dedication to improving the general knowledge about depression through the stories of public figures emphasizes the cultural shift that is now occurring in America. Increasingly, it is becoming accepted that mood disorders are illnesses of the emotional brain which require understanding and appropriate treatment, rather than to be stigmatized as a mark of moral deficit. Joining Larry King and Mike Wallace, the journalists on this broadcast, were Naomi Judd (entertainer), Kay Jamison (psychologist and author), and Art Buchwald (syndicated columnist and professional humorist). All of the guests have personally suffered mood disorder and now successfully manage the illness, leading unusually productive lives.

Page 197, I HAD WORKED ONE SUMMER AT A MENTAL HOSPITAL: This was the summer of 1956, when I was seventeen. Chlorpromazine, the first drug of any significant therapeutic value in schizophrenia, had been introduced in 1952 but had not yet penetrated the country lanes of Hertfordshire, where I grew up. It was during this decade that psychopharmacology was born, and it became widely accepted that the study of neurochemistry held promise for the treatment of mental disorder. Various scientific societies were formed in the 1950s that were to lead the field forward. Most notable were the Collegium Internationale Neuropsychopharmalogicum in 1957, and the International Brain Research Foundation in 1960. The American College of Neuropsychopharmacology was also founded in 1960. For details of these interesting years, see *Discoveries in Biological Psychiatry,* ed. F. Ayd Jr., B. Blackwell (Philadelphia: J. B. Lippincott, 1970).

Page 197, ONE OF THE HOSPITAL'S SEVERAL UNITS FOR CHRONICALLY ILL MEN: In the 1950s the treatments for those with chronic psychoses, including manic-depressive disease, had changed little from the late 1800s. Apart from electroconvulsive therapy, the care afforded was principally an asylum, with good nursing care and an opportunity to participate in the community life of the hospital when remission of the illness occurred. Although the North Wing unit for the chronically ill was frightening to me, the patients were well cared for and the staff dedicated in their responsibility. Units such as the one I remember emptied rapidly as the new psychotropic medications became available and those individuals who responded to them left the hospital. While the lives of many individuals were greatly improved by these changes, others suffered because of a lack of organization and financial support for the community programs that are required to substitute for hospital care. Over subsequent decades, especially in America, this has resulted in a large number of those with chronic mental illness, including mood disorder, being poorly cared for in the community.

Page 197, GIVEN OPIATES AND EXTRA DOSES OF PARALDEHYDE: Opium had been used for the treatment of manic excitement since the late eighteenth century; ether and chloroform were popular in the nineteenth; and bromides, barbiturates, and paraldehyde during the early part of the twentieth century. Nothing worked very well. Most of these medications were general sedatives rather than having any selective effect upon the mania per se. In the 1950s, when I was working in the English mental hospital, paraldehyde was the backbone of the pharmacology. A liquid not unlike vodka to look at, it has a strong aromatic odor and a burning, disagreeable taste. Dispensing it was an elaborate ritual because, when exposed to the light, paraldehyde decomposes to the toxic substances of acetaldehyde and acetic acid (used on salads and called vinegar). To protect against such disasters, I was dispatched each day to collect thirty-six amber-colored bottles, one for each patient, in a large basket. After the doses were dispensed, the unit lapsed into the uneasy quiet of men breathing heavily as paraldehyde's strange odor pervaded the air—an experience not easily forgotten.

Page 199, TOO EXCITED TO SLEEP: Another example of how chronic sleep deprivation induces mania. In the first few days postpartum, most mothers are sleep deprived, and where the vulnerability to bipolar disorder exists, rapid induction of mania can occur. For discussion, see T. A. Wehr, D. A.

Sack, N. E. Rosenthal, "Sleep production as a final common pathway in the genesis of mania," *American Journal of Psychiatry* 144 (1987): 201–204.

Page 201, A TWOFOLD INCREASE ABOVE THE USUAL, NONPREGNANT LEVELS OF SEX HORMONE PRODUCTION IS COMMONPLACE: See A. J. Connely, J. I. Mason, "Placental steroid hormones," *Clinical Endocrinology and Metabolism* 4 (1990): 249–272.

Page 201, POSTPARTUM "BLUES": Also called the baby blues, this common syndrome is best understood as that loss of emotional homeostasis which occurs in women *without* other vulnerability for mood disorder. In a large prospective study where women were followed from the second trimester of pregnancy until nine weeks after the birth of their child, Michael O'Hara and colleagues found that a previous history of depression, a family history of depression, plus the absolute drop in the level of circulating estrogens, were the most significant variables predicting depression. See M. W. O'Hara, J. A. Schlechte, B. A. Lewis, E. J. Wright, "Prospective study of post partum blues: Biological and psychosocial factors," *Archives of General Psychiatry* 48 (1991): 801–806.

Page 202, PSYCHOSIS . . . OCCURS EARLY IN THE PUERPERIUM: Here I refer to a study by Dean and Kendall at the University of Edinburgh, who studied seventy-one women who had developed puerperal illness within ninety days of delivery and compared them to carefully matched controls. Forty percent of the psychoses occurred within a week of delivery, and *all* the manic episodes (as Melanie Branch suffered) were within nineteen days of delivery. See C. Dean, R. E. Kendall, "The symptomatology of puerperal illnesses," *British Journal of Psychiatry* 139 (1981): 128–133. In another study, Meltzer and Kumar reviewed 142 mother and baby admissions to psychiatric hospitals in southeast England. Again the authors found that "almost without exception mothers with manic symptomatology have an onset of illness within the first two weeks" after the delivery. See E. S. Meltzer, R. Kumar, "Puerperal mental illness, clinical features and classifications: A study of 142 mother and baby admissions," *British Journal of Psychiatry* 147 (1985): 647–654.

Page 202, DELIRIOUS MANIA: Kraepelin describes this syndrome as "the profound clouding of consciousness and extraordinary and confused hallucinations and delusions. The attack usually begins very suddenly, only sleeplessness, restlessness, or anxious moodiness may already be conspicuous one or two days beforehand. Consciousness rapidly becomes clouded and the patient becomes confused and bewildered, completely losing orientation to time and place. Everything appears to have been changed, they may think they are in heaven, in Herod's palace, mistakes are made about people and their surroundings. Numerous hallucinations appear and the mood changes rapidly from being timid and lacrimous . . . to unrestrainedly merry, erotic, or ecstatic." See E. Kraepelin, *Manic Depressive Insanity and Paranoia* (Edinburgh: Livingston, 1921), p. 70.

Page 204, HALOPERIDOL LACTATE: Haloperidol was the first butyrophenone drug effective in schizophrenia, and was introduced in 1958 by Janssen

Pharmaceuticals. It has the advantage of being effective when administered as a tablet, liquid, or by injection. Haloperidol lactate is the liquid, and when placed in juice it is an effective method of tranquilizing somebody who is confused and suspicious. Of course, it is always important to administer such powerful medications with the informed consent of the patient, or the relatives if the patient is psychotic and under a commitment order. Melanie, in her psychotic state, had been committed to the hospital as a person dangerous to herself and to her baby, and under these circumstances I judged haloperidol to be the most appropriate dopamine-blocking agent that would also provide sedation. Haloperidol binds to postsynaptic dopamine receptors, blocking the activity of the dopamine produced by the brain. This blockade reduces the neurotransmission and rapidly dampens the excessive energy, sleeplessness, and agitation which are characteristic of mania. For details of its discovery, see P. A. J. Janssen, "The Butyrophenome Story," in *Discoveries in Biological Psychiatry,* ed. F. Ayd Jr., B. Blackwell (Philadelphia: J. B. Lippincott, 1970), pp. 165–179.

Page 205, THE SYNAPSES . . . ARE THE CONNECTING LINKS BETWEEN INDIVIDUAL NEURONS:

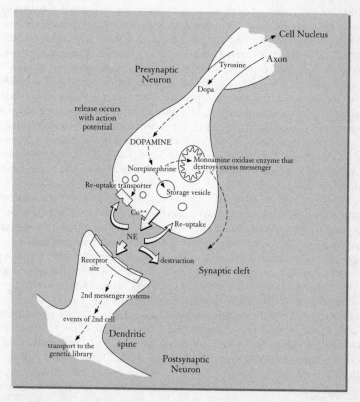

The synapse, the junction between two nerve cells, was named by Charles Sherrington in 1897 from the Greek word meaning "to clasp." The synapse is the key to information-processing in the brain. In the human cortex it is estimated there are 10^{14} synaptic contacts, or 30,000 times the number of people in the world. The events occurring at the synapse are tightly regulated to ensure accurate transmission of information, and are supported by the continuous production of messengers, their release, and ultimate recycling or destruction. The metabolic cycle of dopamine and norepinephrine is illustrated here, but the principles of regulation of the other "mood" neurotransmitters are similar. Drugs that change behavior interfere with these regulatory mechanisms and disturb the information passing among the neurons. Depending upon the action of the drug, mood will either be stabilized, as with lithium in mania or an antidepressant in depression, or disordered, as with cocaine. As a reference text for this complicated subject, see J. R. Cooper, F. E. Bloom, R. H. Bloom, *The Biochemical Basis of Neuropharmacology* (New York: Oxford University Press, 1996).

Page 206, SYNAPTIC TRANSMISSION (IS) NEUROCHEMICAL RATHER THAN ELECTRICAL: Camillo Golgi (1843–1926), the great Italian neuroanatomist, using his silver impregnation stain, had demonstrated the presence of neurons in the late 1800s. Using the Golgi method, a Spanish scientist, Santiago Ramón y Cajal, demonstrated a few years later that the brain was not wired together as a neural network, as Golgi had suggested, but was a series of individual neurons. This idea was scientifically accepted by 1906, but it would be another fifty years before it was agreed that the transfer of information from one neuron to another was chemical rather than electrical, emphasizing just how rapidly research advanced in the last decades of the twentieth century. For an interesting discussion of this scientific disagreement, read A. Oliverio, G. M. Shepherd, "Neurons: The Cells of Thought," in *The Enchanted Loom: Chapters in the History of Neuroscience* (New York: Oxford University Press, 1991), pp. 240–245.

Page 207, DRUGS INHIBITING THE MONOAMINE-OXIDASE ENZYMES: Monoamine Oxidase Inhibitors (MAOIs) remain in use and are effective antidepressants, especially in some atypical depressions where physical complaints predominate. The monoamine enzymes destroy the amine neurotransmitters and also other amine substances in the body. Because many MAOIs block these enzymes permanently, a nasty side-effect of the drugs, should a large dose of amine substances be consumed in the diet, is high blood pressure. This makes the drugs dangerous unless prescribed carefully and in association with a strict diet, without aged cheese, red wine, and other amine-containing foods. For the early development of these drugs, see N. S. Kline, "Monoamine Oxidase Inhibitors: An Unfinished Picaresque Tale," in *Discoveries in Biological Psychiatry,* ed. F. Ayd, Jr., B. Blackwell (Philadelphia: J. B. Lippincott, 1970), pp. 194–204.

Page 207, *THE DOORS OF PERCEPTION:* See Aldous Huxley, *The Doors of*

Perception, and Heaven and Hell (New York: Harper and Row, 1963). Aldous Huxley (1894–1963), the English novelist and essayist, came from a scientific family. His grandfather Thomas Henry Huxley had been a great champion of Darwin and evolution. Known for his criticism of Western civilization— evident in his most famous book, *Brave New World*—Aldous was distressed by the rise of technology and a lack of spirituality in modern culture. In rebellion, he turned to mysticism and the use of hallucinatory drugs, particularly mescaline, which he describes in *The Doors of Perception*.

Page 208, CONNECTIONS BETWEEN THE VEGETABLE WORLD, THE CHEMICAL MESSENGERS OF THE BRAIN, AND THE EXPERIENCE OF MIND: There are many fascinating connections between the drugs used today in medicine and the plant world. Digitalis, for example, an extract from the foxglove, is still widely used in the treatment of heart disease. It has a chemical nucleus similar to that of the steroid stress hormones. Opium, from the juice of the poppy, has relieved painful suffering and calmed psychic agitation, including that of mania, for over two thousand years. Then came the discovery that natural opioid peptides are released as messengers in the brain under stressful conditions and physical injury. Appropriately named *endorphins* (meaning "internal morphine"), these "internal pain pills" help explain opiate addiction. Morphine and heroin, drugs of the opiate family, have a very high affinity for the brain receptors through which the endorphins operate.

Page 209, A MONOAMINE THEORY OF AFFECTIVE DISORDER: Several different research groups proposed monoamine theories of affective illness during the 1960s. In America, of predominant interest were the catecholamines, norepinephrine specifically; in Europe, serotonin was the messenger most investigated. Later it was recognized that both played important roles in mood. For the original papers, see A. J. Prange, "The pharmacology and biochemistry of depression," *Diseases of the Nervous System* 25 (1964): 217–221; J. Schildkraut, "Catecholamine hypothesis of affective disorders," *American Journal of Psychiatry* 122 (1965): 509–522; W. E. Bunney, M. Davis, "Norepinephrine in depressive reactions," *Archives of General Psychiatry* 13 (1965): 483–494; A. Coppen, "Depressed states and the indolealkylamines," *Advances in Pharmacology* 6 (1968): 238–291.

Page 209, MOLECULAR BIOLOGY: An excellent introduction to molecular biology and its application to psychiatry is the book by S. H. Barondes, *Molecules and Mental Illness* (New York: W. H. Freeman/Scientific American Library, 1993). A more detailed text, and probably hard going for the reader without a scientific background, is S. E. Hyman, E. J. Nestler, *The Molecular Foundations of Psychiatry* (Washington, D.C.: American Psychiatric Press, 1993).

Page 210, ONCE THE NEUROTRANSMITTERS ENGAGE THESE RECEPTORS: The recognition of specific receptors for the individual neurotransmitters and how these receptors interact with drugs active in changing behavior has been a cornerstone of the advance in the treatment of mood disorder. Once the drug structures were established, it was possible to build new molecules with a similar configuration and to test their ability to interact with the receptor directly by a technique called a *binding assay*. By tagging the drug with a radioactive compound, its affinity for the receptor could be determined and, slowly, through investigation, drugs with a better fit for the receptor could be developed. Similar techniques can be used to determine the activity of the transporter systems that recycle the neurotransmitters at the synapse. In the laboratory, brain tissue is homogenized and then through separation techniques fragments of the neuron containing synaptic receptors are isolated for study. Through these techniques, the relative potency of antidepressants and other drugs can be compared. Most drugs will bind to several different receptors, and this makes it possible to develop a profile of a drug's activity. Depending upon its clinical effect, one can therefore also deduce possible abnormalities in receptor function that may explain the change in neuronal functioning that determines the vulnerability to depression or mania. Investigation has been further advanced through molecular biology. Once the receptor structure has been determined by molecular analysis of the genetic material that manufactures them, the actual proteins that are the building blocks of the receptors now can be manufactured by molecular techniques.

Page 211, DISTINCT COMMUNICATION SYSTEMS WHICH RELY UPON SPECIFIC NEUROTRANSMITTERS: The brain contains anatomically distinct neuronal systems that employ the catecholamines and serotonin as their messengers. Each of the systems serves separate but similar roles in the limbic system, which is probably important in the fine tuning of the brain's communication among neurons. The dopamine, norepinephrine, and serotonin systems each serve local networks in the brain, but of greater interest are the long connecting fibers that start deep in the limbic brain and extend upward into the centers of the limbic alliance and on to the emotional cortex of the frontal lobe. These are the superhighways of the limbic system. The dopamine, neurepinephrine, and serotonin messengers all have multiple receptors with which they interact on the neuronal surface. Some of these receptors open and close ion channels and change the excitability of the receiving neuron; others stimulate the messenger systems and take information into the nucleus. Still others are responsible for the regulation and recycling of the messenger itself. Exactly how all these receptors interact and intertwine in the maintenance of balanced neuronal communication is unclear, but some of the receptor systems seem to be more important in emotions than others.

It would be a mistake to think that there are only the amine messenger systems in the brain. There probably are at least one hundred others that have already been discovered. Some of them involve smaller molecules than the amine transmitters; some of them utilize much larger peptide molecules. Each neuron can employ multiple sets of these messengers, although the evidence still suggests that as far as the monoamines are concerned, the neuronal systems remain fairly exclusive. Amine messengers seem to be very tightly regulated and rapidly produced and destroyed. This may be one reason why they have emerged as the fulcrum around which we can manipulate mood with the use of drugs.

Page 212, NEURONS ARE VERY SENSITIVE TO RAPID CHANGES IN THEIR HORMONAL ENVIRONMENT: Hormones are messengers just like the neurotransmitters, except that when they were initially discovered it was in association with a particular function or body organ. Hence they have names which no longer truly describe their action.

When too little or too much hormone is produced by the endocrine glands of the body, distinct clinical syndromes develop, many of which have profoundly disturbed mental states as part of their presentation. Thyroxin and the steroids are particularly powerful in modifying the general activity of the neuron and can profoundly modify behavior. Thus, mild disturbance of the endocrine glands must always be thought of in the assessment of somebody with depression. Thyroid hormones—and also, less commonly, changes in the steroid system—can compound depressive syndromes. For details, see P. C. Whybrow, "Neuroendocrinology," in *Biological Bases of Brain Function and Disease,* ed. A. Fraser, P. Molinoff, A. Winokur (New York: Raven Press, 1993).

Page 212, CHANGES IN THE MONTHLY OVARIAN CYCLE: These disturb other endocrine functions, and can profoundly change mood in some women. Of all age groups, women between the menarche and the menopause are at the highest risk for mood disorder. A general discussion of the links between the sex hormones and behavior will be found in my essay "Neuroendocrinology," referenced above. For greater detail regarding the physiology, and an extensive list of references, see E. Liebenluft, P. L. Fiero, D. R. Rubinow, "Effects of the menstrual cycle on dependent variables in mood disorder research," *Archives of General Psychiatry* 51 (1994): 761–781.

Page 212, THE LIMBIC BRAIN CENTERS OF THE AMYGDALA, HIPPOCAMPUS, AND HYPOTHALAMUS: See D. Greenstein, "Steroid hormone receptors in the brain," in *Neuroendocrinology,* ed. S. L. Lightman, D. J. Everitt (Oxford: Blackwell Scientific, 1986), pp. 32–48.

Page 213, COCAINE IS ONE OF THE MOST POWERFUL DOPAMINE-ENHANCING DRUGS: Cocaine was first isolated in 1880, by German scientists. When rubbed into the skin it deadens the sensation of pain, and it was rapidly introduced into medicine as a local anesthetic. However, it also has many stimulant properties that are powerfully addicting. Chewing the leaves of the Erythroxylon coca tree, from which cocaine is derived, is a habit favored by natives of Peru and Bolivia to increase endurance during physical labor, and in the nineteenth century, shortly after it was isolated from the coca leaf, cocaine assumed something of a romantic image. In those of normal mood it causes immediate stimulation with restlessness, excitement, and a subjective sense of increased intellectual power. Sir Arthur Conan-Doyle, the physician-author, had his character Sherlock Holmes take advantage of it, much to Dr. Watson's consternation, and Sigmund Freud explored its mental properties on himself, and others, during his early years as a neurologist. In contrast to many other drugs of addiction, where tolerance occurs, the craving for cocaine increases with chronic use, making it one of the most dangerous of the addictions. Like imipramine, cocaine blocks the re-uptake of the catecholamines, principally dopamine, by binding with the transporter that recycles the messenger. However, it also stimulates the release of dopamine, which probably accounts for its immediate euphoric effect. Imipramine, by contrast, in the absence of depression, produces little subjective change—except perhaps a slightly woolen feeling in the head and a certain dryness of the mouth. It is only when given in depression that imipramine is a mood-altering drug, and then only after two or three weeks of taking it every day. If cocaine is taken by those depressed, followed sometimes by an initial uplift in energy it will invariably deepen a melancholic mood. Thus, although both drugs block the recycling of catecholamines, in their behavioral effects cocaine and imipramine are dramatically different. Imipramine leads to improvement in the symptoms of depression, without addiction, while cocaine increases depressive symptoms, producing mania in those predisposed, and is powerfully addicting.

Page 213, THE SUDDEN WITHDRAWAL OF HIGH DOSES OF ESTROGEN CAUSES AN INCREASED NEURONAL SENSITIVITY TO DOPAMINE: For an interesting review of the relationship between estrogen and dopamine metabolism and the development of psychosis in the postpartum period, see S. A. Checkley, A. Wiecke, J. A. Bearn, M. N. Marks, I. C. Campbell, R. Kumar, "Hormones, genes, and the triggering of bipolar illness in the puerperium," in *Psychopharmacology of Depression,* ed. S. A. Montgomery, T. H. Corn (Oxford: Oxford University Press, 1994), pp. 57–65.

Page 214, FREDERICK K. GOODWIN: Dr. Goodwin, Professor of Psychiatry and Director of the Center on Neuroscience, Medical Progress, and Society at George Washington University, and formerly Director of the National Institute of Mental Health, is one of the world's leading authorities on manic-depressive illness. Together with Dr. Kay Jamison, he is the author of the definitive medical text on the subject. (F. K. Goodwin, K. R. Jamison, *Manic-Depressive Illness* [New York: Oxford University Press, 1990].) For details of his research into dopamine and mania, see F. K. Goodwin, R. L. Sack, "Central dopamine function in affective illness: Evidence from precursors, enzyme inhibitors, studies of central dopamine turnover," in E. Usdin, *The Neuropsychopharmacology of Monoamines and Their Regulatory Enzymes* (New York: Raven Press, 1974), pp. 261–279.

Page 214, DOPAMINE HELPS DRIVE THE MOTOR SYSTEMS OF THE BODY: The highest concentration of dopamine neurons is in the substantia nigra (*nigra* meaning "black") of the lizard brain. These dopamine neurons connect with the cells of the basal ganglia, the nerve center that helps coordinate movement. If these cells degenerate or are damaged—as happens in the illness of Parkinson's disease—the usual fluid movement of the arms and legs takes on a wooden, ratchetlike quality with a tremor in the hands and a shuffling walk. When high doses of dopamine-blocking drugs are prescribed for the treatment of mania, this same stiffness sometimes will appear.

Page 215, ONE FLEW OVER THE CUCKOO'S NEST: This classic book by Ken Kesey undoubtedly was intended as an allegory of good and evil but unfortunately has been adopted by a naive public as an accurate picture of life in a mental institution. ECT is characterized as a barbaric assault upon the dignity of victimized and wrongly confined patients, rather than a therapeutic intervention. K. Kesey, *One Flew Over the Cuckoo's Nest* (New York: Viking Press, 1964).

Page 216, ELECTROCONVULSIVE THERAPY IS ALSO EFFECTIVE IN BREAKING THE FLIGHT OF MANIA: For a general review of electroconvulsive therapy, see R. Abrams, *Electro-convulsive-therapy* (New York: Oxford University Press, 1988). Its effectiveness in mania is well documented in the review by S. Mukherjee, H. A. Sackeim, D. B. Schnur, "Electro-convulsive-therapy of acute manic episodes; A review of fifty years' experience," *American Journal of Psychiatry* 151 (1994): 169–176.

ECT is also highly effective for individuals who have melancholia, but who have not responded to antidepressant drugs. This is particularly true in those individuals who are delusional and psychotic. Only about 35 percent of such individuals will respond to antidepressants, whereas 85 percent recover with ECT. See O. Bratfos, J. O. Haug, "Electro-convulsive therapy and antidepressant drugs in manic-depressive disease: Treatment results at discharge three months later," *Acta Psychiatria Scandinavica* 41 (1965): 588–596. For a moving personal account of severe depression and recovery, including the experience of receiving ECT, read Martha Manning's unique memoir, *Undercurrents: A Therapist's Reckoning with Her Own Depression* (San Francisco: HarperCollins, 1995).

Page 217, MELANIE CONTINUED TO BE CONFUSED AT TIMES: Much has been written about the confusion associated with ECT, and confusion does occur. However, as was true during Melanie's illness, confusion is also an integral part of severe psychosis, both in mania and in melancholia. Thus one cannot easily determine the source of the confused thinking under such circumstances. In Melanie's case her confusion was lessening with the treatments. Research has also shown that if the pulse of electricity is passed only across the nondominant hemisphere—the right in most individuals—avoiding the speech and the memory centers of the dominant hemisphere, then confusion resulting from the ECT is considerably reduced. Hence it is this mode of treatment that is now most frequently used.

There is no evidence that ECT permanently impairs memory. Studies of performance six to nine months after treatment on a variety of memory tests have been found normal (see L. R. Squire, "Memory functions as affected by electroconvulsive therapy," *Annuls of the New York Academy of Science* 462 [1986]: 307–314). Similarly, forty-three patients examined before, at one week after, and seven months after a course of bilateral ECT showed that personal and public memories disrupted during the course of treatment had been completely recovered by the seventh month, in the overwhelming majority of patients (L. R. Squire, P. C. Slater, P. L. Miller, "Retrograde amnesia and bilateral electroconvulsive therapy: Long term follow-up," *Archives of General Psychiatry* 38 [1981]: 89–95).

Page 218, RAPID CYCLING: This variation of manic-depressive disease is said to be present when more than four episodes of illness occur in one year. An episode is either mania or depression; a cycle is mania and depression occuring together. In some individuals, periods of rapid cycling come and go through the course of the illness and usually are triggered by changes in medication, the hormones of the body, and particularly stressful periods of life. The frequency of cycling varies, and in most individuals is greater than four times a year. In the clinic that we developed at the University of Pennsylvania especially to care for such individuals, the average number of episodes each year was about sixteen, a degree of illness which dramatically interferes with daily living. The risk of developing rapid cycling is increased in women, in people with hypothyroidism, and in those receiving tricyclic antidepressant drugs. However, it will sometimes occur after brain trauma. It responds poorly to lithium, but can usually be controlled by a combination of medications, particularly the anticonvulsants, and sometimes by adding high doses of thyroid hormone. For a review, see M. S. Bauer, P. C. Whybrow, "Rapid cycling bipolar disorder: Features, treatment and etiology," in *Refractory Depression: Advances in Neuropsychiatry and Psychopharmacology,* Vol. 2, ed. J. D. Amsterdam (New York: Raven Press, 1991). For those interested in the particular relationship between being a woman and the risk of developing rapid cycling illness, see E. Leibenluft, "Women with bipolar illness: Clinical and research issues," *American Journal of Psychiatry* 153 (1996): 163–173. Details of tricyclic antidepressants precipitating rapid cycling will be found in T. Wehr, F. K. Goodwin, "Tricyclics modulate frequency of mood cycles," *Chronobiologia* 6 (1979): 377–385.

Page 218, MADNESS IS MORE RIGID AND INFLEXIBLE IN ITS ORGANIZATION THAN IS HEALTHY BEHAVIOR: Comparing seven rapidly cycling patients with twenty-eight normal controls who rated their mood daily for over a year, we found that those with manic depression are not truly cyclic for extended periods of time, but have nonetheless a more organized structure to their mood cycles than do normal individuals. For details of this study, see A. Gottschalk, M. S. Bauer, P. C. Whybrow, "Evidence of chaotic mood variation in bipolar disorder," *Archives of General Psychiatry* 52 (1995): 947–959; and the commentary that follows, C. Ehlers, "Chaos and complexity. Can it help us understand mood and behavior?" *Archives of General Psychiatry* 52 (1995): 960–964.

Page 220, LITHIUM IS PROVEN . . . TO DEFEND AGAINST RETURNING

EPISODES OF ILLNESS IN APPROXIMATELY 70 PERCENT OF PATIENTS: The ability of lithium to prevent episodes of illness has profoundly changed the lives of those with manic depression. The first hints of its prophylactic abilities were recorded in the 1950s, but not confirmed in extensive trials until the late 1960s. See A. Coppen, et al., "Prophylactic lithium in affective disorders: Controlled trial," *Lancet* 2 (1971): 275–279. The optimal blood level for maintenance lithium treatment is generally considered to be between 0.5 and 1 mEq/liter. Should the blood level drop too low, the prophylaxis is compromised, especially for manic episodes, and if it rises too high, then toxic symptoms to kidney, heart, and the brain itself may result. Maintenance therapy must always be under the guidance of a physician and frequent blood tests are required, especially in the early phases of the treatment when the correct dose necessary to maintain an appropriate blood level is being determined. The actual dose required will vary from patient to patient. Most pharmacology textbooks discuss these complex issues. However, for a particularly comprehensive review, see chapter 23, "Maintenance and Medical Treatment in Manic-Depressive Illness," in F. K. Goodwin, K. R. Jamison, *Manic-Depressive Illness* (Oxford: Oxford University Press, 1990), pp. 665–724.

Page 220, CARBAMAZEPINE: The first trials of carbamazepine in mania were conducted in Japan in the 1970s and later at the National Institutes of Mental Health in the United States. See T. Okuma, et al., "Comparison of the anti-manic efficacy of carbamazepine and chlorpromazine: A double blind controlled study," *Psychopharmacology* (Berlin) 66 (1979): 211–217; and J. C. Ballenger, R. M. Post, "Carbamazepine in manic-depressive illness: A new treatment," *American Journal of Psychiatry* 137 (1980): 782–790. In direct comparison, carbamazepine has been found to be comparable to lithium in the treatment of mania (see G. F. Placidi, et al., "The comparative efficacy and safety of carbamazepine versus lithium: A randomized double blind three year trial in 83 patients," *Journal of Clinical Psychiatry* 47 [1986]: 490–494). Carbamazepine is particularly valuable in individuals who suffer a mixture of depression and mania. Lithium, on the other hand, is superior in individuals who have clear-cut episodes of mania followed by depression with a free interval in between.

Page 221, DIVALPROEX SODIUM: This anticonvulsant is a slightly modified version of valproic acid, which like carbamazepine reduces kindling and is a powerful anti-epileptic drug. Used as an anticonvulsant since 1983, divalproex was approved in 1995 as an anti-manic agent by the Food and Drug Administration in the United States. In a double-blind comparative trial it has been shown to be equally effective to lithium in acute mania (see C. L. Bowden, et al., "Efficacy of divalproex versus lithium and placebo in the treatment of mania," *Journal of the American Medical Association* 271 [1994]: 918–924). As with lithium, the dose range of divalproex must be carefully monitored and usually adjusted individually for each patient. Doses range from 750–2,000 milligrams per day and higher. The drug is particularly useful in individuals who have mixed manic states, and can be combined effectively with lithium in the long-term maintenance of bipolar disorder. Whether divalproex alone is a prophylactic agent is being studied. In open trials it seems to be effective and may also be useful in cyclothymia and premenstrual syndrome. See F. M. Jacobson, "Low dose valproate: A new treatment for cyclothymia, mild rapid cycling disorders, and premenstrual syndrome," *Journal of Clinical Psychiatry* 54 (1993): 229–234.

Page 221, JOHN CADE, AN AUSTRALIAN PSYCHIATRIST: Dr. Cade has described his discovery of the anti-manic effects of lithium in "The story of Lithium," in *Discoveries in Biological Psychiatry,* ed. F. J, Ayd, B. Blackwell, (Philadelphia: J. B. Lippincott, 1970), pp. 218–229. His original paper was J. F. J. Cade, "Lithium salts and the treatment of psychotic excitement," *Medical Journal of Australia* 1 (1949):195–198. He describes several of the case histories in these papers, including the first patient who after three weeks was, as I quoted in the text, "enjoying the unaccustomed and quite unexpected amenities of a convalescent ward . . . "

Page 221, THE MEDICAL WORLD WAS AT FIRST SUSPICIOUS OF THIS MIRACLE: The reason was that, as Dr. Cade describes in the article cited above, lithium has had a rather erratic history. It was used as a treatment for gout in the mid-1800s and as a laxative in the early 1900s. Lithium bromide was used as a sleeping draft in the early 1920s (*and*, I learned on the good authority of my father, to reduce the libido of troops in the British Army!). Then, in the 1940s, when being used as a salt substitute, lithium chloride was found to cause heart failure and kidney failure, resulting in several deaths. These are just some of the reasons why the medical world was initially skeptical. Despite this colorful history, lithium is a safe drug when used appropriately and carefully monitored. However, there was another reason, and perhaps one of more practical importance, why there was no great rush to market lithium as a treatment for manic-depressive illness: Being merely a common salt it could not be patented, and therefore few pharmaceutical companies were interested in developing its use in psychiatry.

Page 222, LITHIUM . . . MODIFIES THE SIGNALING OF THE SECOND MESSENGER SYSTEMS: See J. M. Baraban, P. F. Worley, S. H. Snyder, "The second messenger systems and psychoactive drug action: Focus on the phosphoinositide system and lithium," *American Journal of Psychiatry* 146 (1989): 1251–1260.

Page 224, THE CHOICE OF AN APPROPRIATE ANTIDEPRESSANT: Details of the classes of antidepressants available will be found in the appendix on psychopharmacology. In addition, interested readers are referred to the guidelines published by the American Psychiatric Association on bipolar disorder ("Practice guideline for the treatment with patients with bipolar disorder," *American Journal of Psychiatry* 151, December 1994 Supplement) and on unipolar depression ("American Psychiatric Association Practice guideline for major depressive disorder in adults," *American Journal of Psychiatry* 150, April 1993 Supplement). Copies of the guidelines may be obtained from the

American Psychiatric Association Press, 1400 K Street, N. W., Washington, D.C. 20005.

Page 224, IMIPRAMINE . . . *PERTURBS* THE SET POINT OF MELANCHOLIC HOMEOSTASIS: There is evidence that ECT and antidepressants both perturb the brain and produce their effect by inducing adaptation. Studies in animals show that at the molecular level, a reduction in the sensitivity of adrenergic receptors occurs over two to three weeks, regardless of the specific molecular action of the antidepressant that is used in the experiment. ECT produces a similar reduction in sensitivity (called "down regulation") of receptors over approximately the same time frame. This suggests that in both instances it is the brain's adaptation that corresponds with the behavioral changes seen in depressed patients, not the acute intervention itself.

Page 225, PROFESSOR ARTHUR PRANGE . . . SEROTONIN ACTIVITY GIVES "PERMISSION" FOR MOOD DISORDER TO DEVELOP: Details of the Prange hypothesis and supporting papers will be found in A. J. Prange, I. C. Wilson, C. L. Lynn, L. B. Alltop, R. A. Stikeleather, "L-tryptophan in mania: Contribution to a permissive hypothesis of affective disorders," *Archives of General Psychiatry* 30 (1974): 56–65; and A. Coppen, A. J. Prange, P. C. Whybrow, R. Noguera, "Abnormalities of indoleamines in affective disorders," *Archives of General Psychiatry* 26 (1972): 474–482.

Page 225, TRYPTOPHAN IN THE DIET: Tryptophan has been used as a supplement to the treatment of depression together with ECT and with monoamine oxidase inhibiting drugs. In both instances, it increases the speed of response. See A. Coppen, P. C. Whybrow, R. Noguera, R. Maggs, A. J. Prange, "The comparative antidepressant value of L-tryptophan and imipramine with and without attempted potentiation by liothyronine," *Archives of General Psychiatry* 26 (1972): 234–241. Alternatively, low tryptophan diets, where patients drink an amino acid complement with tryptophan removed, can rapidly reduce the advantages gained from the treatment of depression by antidepressants. For details, see T. L. Delgado, D. S. Charney, L. H. Price, et al., "Serotonin function and the mechanism of antidepressant action: Reversal of antidepressant by rapid depletion of plasma tryptophan," *Archives of General Psychiatry* 47 (1990): 411–418.

Page 225, SEVERAL POSTMORTEM STUDIES OF THE BRAINS: The association between suicide and low serotonin in the brain is supported by postmortem studies of the brains of suicide victims showing increased numbers of synaptic serotonin receptors, presumably because the neurons have become more sensitive secondary to low serotonin during life. A reduction in serotonin is generally associated with aggressive behavior both to the self—that is, suicide—and also with homicidal behavior. For a review, see H. M. Van Praag, "Depression, suicide, and serotonin metabolism in the brain," in *Neurobiology of Mood Disorders,* ed. R. M. Post, J. C. Ballenger (Baltimore: Williams and Wilkins, 1984), pp. 601–618.

Page 226, PROZAC . . . AN ELIXIR FROM THE GODS: Prozac, the first drug marketed to specifically inhibit the re-uptake of serotonin, was introduced by the Eli Lilly company in 1988. It has a long half-life, and therefore a standard dose can be employed for almost all patients, usually a dose of 20 milligrams. (With other antidepressants—where doses are variable depending upon the time the drug stays in the body, the side effects produced, and the clinical response—the doctor and patient must work together closely in the first few weeks to establish the optimum dose.) This simplicity of prescribing, coupled with minimal initial side effects, rapidly established Prozac as a "wonder drug." Shortly thereafter, reports occurred that it increased suicidal and perhaps homicidal behavior. In 1993, Peter Kramer, a psychiatrist practicing in Rhode Island, published *Listening to Prozac,* in which he explored the effects of serotonin on the temperament and personality of some of the patients he had cared for in his practice. The book was an instant success, and Prozac's reputation rose again. In exceptional circumstances a large proportion of the population of a small town was placed on the medication (see "Depressed town finds happiness on Prozac: Psychologist puts 600 residents on the happy pill," *San Francisco Sunday Examiner and Chronicle,* January 30, 1994). The scientific story behind Prozac is that of serotonin, as was pointed out by Sherwin Nuland in a review article ("Talking back to Prozac," *New York Review of Books,* June 9, 1994). Nuland in his article turns the coin and cautions that serotonin can also cause adverse effects, as in the rare but deadly serotonin syndrome (with severe agitation, muscle twitches and seizures, similar to some reactions with the MAOI drugs), impotence, and other long-term side effects, some yet unknown, that are always discovered after a drug becomes established (see also "Singing the Prozac blues," *U.S. News and World Report,* November 8, 1993). The broader issue which is really stimulated by Peter Kramer's excellent book is that for the first time the public is responding to the idea that a nonaddicting, medically endorsed drug can constructively change behavior for some patients over the long term. And perhaps, suggests Kramer, even change behavior in individuals who were not considered to be clinically abnormal, but just worried or miserable. This watershed in public opinion is perhaps of more interest than Prozac itself, for it suggests that the pendulum is swinging back toward Greek times and a general recognition that biology and experience go hand in hand in the development of behavior.

Page 228, DEPRESSIVE AND MANIC-DEPRESSIVE ASSOCIATION: The Depressive and Manic-Depressive Association began in Chicago, Illinois, in the early 1980s and was incorporated as a national organization in 1986 (NDMDA). A decade later it had 275 chapters across America and Canada, and several associated chapters in Europe. The NDMDA is a nonprofit "support, education, and advocacy organization" that has been developed and led by individuals with depressive illnesses to specifically provide help for others in similar distress. In addition to the many educational services they provide, the NDMDA is a powerful national lobby. It has influenced public policy in many ways and establishes standards for medical practice and patient care from the patient advocacy standpoint. The national office has developed a strong working relationship with the scientific community, including a National Scientific Advisory Board, of which I am a vice-chairman. Similar organizations exist in Europe, including the Manic-Depressive Fellowship in England. The NDMDA address in Chicago is 730 North Franklin Street, Suite 501, Chicago, IL 60610-3526.

Chapter Ten: Thoughtful Reconstruction

Page 231, "Learn the ABC of science . . . ": This quotation comes from Ivan Pavlov's *Bequest to the Academic Youth of Soviet Russia,* written just before his death at the age of eighty-seven, on February 27, 1936. The complete text will be found in *The Practical Cogitator,* edited by C. P. Curtis and F. Greenslet, (Boston: Houghton Mifflin, 1962), pp. 98–99.

Page 231, ". . . it's only the passions that make you think": The Marquise Marie de Vichy-Chamrond du Deffand (1697–1780) was a French intellectual of the Enlightenment, known for her brilliant, witty correspondence. This line is from one of her love letters to Horace Walpole, the English author.

Page 233, many steps upon the scientific staircase: For an interesting discussion of the various levels of scientific investigation and their importance, especially when it comes to complex adaptive systems, read chapter 9, "What is fundamental?" in *The Quark and the Jaguar: Adventures in the Simple and the Complex,* by Murray Gell-Mann (New York: W. H. Freeman, 1994).

Page 236, I explained to Stephan how my thinking had evolved: I first wrote in detail about these ideas in 1984 in the book *Mood Disorders: Toward a New Psychobiology,* co-authored with Hagop Akiskal and William McKinney (New York: Plenum Press, 1984). Of particular relevance are chapters 8, 9, and 10 (pp. 153–216).

Page 236, the "ducking-stool": Some of these devices were very elaborate. I recall seeing an illustration of a gazebo built in the middle of a bridge over a small pond. The whole gazebo could be lowered into the pond for the water therapy. An interesting anecdote is the part "ducking" played in the founding of the Brattleboro Retreat, a Quaker asylum in southern Vermont which was established in the early 1800s as part of the movement for the moral treatment of the insane. Apparently a preacher living in the area had become mentally ill, with many delusions and ravings, and a physician had ordered his submersion. This improved the preacher's mental state, but gave him pneumonia from which he subsequently died. Shortly thereafter the physician died also. The physician's wife, alarmed by the trend, provided her life's savings for the establishment of a hospital for the insane, which later became the Brattleboro Retreat.

Page 236, the "tranquilizer" chair of Benjamin Rush: Details of this fiendish device will be found in various texts of the history of psychiatry. The original description, and a picture of the contraption, will be found on page 671 of *Three Hundred Years of Psychiatry,* edited by Richard Hunter and Ida Macalpine (New York: Carlisle Publishing, 1982). A variation upon the tranquilizing chair was one that could be rotated on a vertical axis, essentially centrifuging the patient who was restrained. Apparently many of these devices returned people to their senses, at least temporarily.

Page 237, THYROXIN . . . USED AS A THERAPEUTIC DRUG: For a discussion of the use of thyroid hormones in mood disorder, see P. C. Whybrow, "The Therapeutic Use of Triiodothyronine and High-Dose Thyroxin in Psychiatric Disorder," in *Medica Austriaca* 21 (1994): 47–52. For details regarding the predominance of thyroid disorder in women, see P. C. Whybrow, "Sex Differences in Thyroid Axis Function: Treatment Implications for Affective Disorders," *Depression* 3 (1995): 33–42.

Page 239, THORAZINE: In the United States, Thorazine™ is the trade name of chlorpromazine, manufactured by SmithKline Beecham. It is a dopamine blocking agent that reduces manic excitement.

Page 239, A COMMON EXPERIENCE FOR FAMILY MEMBERS: The major mood disorders, particularly manic depression and recurrent depression, cause significant disruption of family function and considerable anguish among friends and loved ones. A valuable book is *We Heard the Angels of Madness: A Family Guide to Coping with Manic Depression,* by Diane and Lisa Berger (New York: Quill, William Morrow, 1991). Written by the mother and sister of a young man who developed manic-depressive disease in his late teens, this is a powerful book that describes the difficulties families must overcome in finding the right doctor, insurance, and the resources required to successfully manage the disorder. The autobiography of Leonard Woolf, husband of the writer Virginia Woolf, also provides fascinating insight into living with manic-depressive disease. The volume of particular pertinence is the second, 1911 to 1969. See Leonard Woolf, *An Autobiography* (Oxford: Oxford University Press, 1980).

Page 240, THE STATISTICS ABOUT THOSE WHO SUFFER: See W. Coryell, J. Endicott, M. Keller, "Outcome of Patients with Chronic Affective Disorder: A Five-Year Follow-up," *American Journal of Psychiatry* 147 (1990): 1627–1633; J. F. Goldberg, M. Harrow, L. S. Grossman, "Course and Outcome in Affective Bipolar Disorder: A Longitudinal Follow-up Study," *American Journal of Psychiatry* 152 (1995): 379–384; and H. S. Akiskal, T. Walker, V. R. Puzantian, D. King, T. L. Rosenthal, M. Dranon, "Bipolar Outcome in the Course of Depressive Illness: Phenomenologic, Familial, and Pharmacologic Predictors," *Journal of Affective Disorders* 5 (1983): 115–128.

Page 240, POOR JOB PERFORMANCE . . . EVEN CRIMINAL ACTS: Articles appearing in the general press suggest that the public is becoming more aware of, and sympathetic to, the complexities of mood disorders as social illnesses. For example, on Sunday, April 21, 1996, in the National Report section of the *New York Times,* the case history of Alice Faye Redd was reported in considerable detail. Entitled "This Way Madness Lies: A Fall from Grace to Prison," the article chronicles Mrs. Redd's financial dealings, which her family attributed to manic-depressive disease. A detailed family history in the article shows clearly that many individuals in Mrs. Redd's family have suffered manic-depressive disease. In the "pyramid scheme" that Mrs. Redd devised, she was at one point paying 90 percent interest per month to those who invested money. Approximately $3.6 million was "lost in a cycle of loans, exorbitant interest payments and lavish spending that inevitably collapsed." See also "What To Do When a Star Executive Breaks Down? Managing a Manic-Depressive," by Julia Lieblich (*Harvard Business Review,* May–June 1994), and "Back from Hell," the story of Margot Kidder, "Once Superman's glamorous costar . . . she slept in a box . . . among the homeless." Cover story in *People,* September 23, 1996.

Page 242, THE CORE OF ANY PSYCHOTHERAPY: The brief discussion of psychotherapy provided in this chapter, within the context of a conversation with Stephan Szabo, cannot do justice to the richness of the subject. Nor can it document in detail the many positive results that psychotherapy can yield for those who suffer mood disorders, including prevention. Objective demonstration of the value of psychotherapy has been difficult and quantitative research on the subject is a recent phenomenon. For decades it was commonly expressed that psychotherapy was beyond research (many practitioners viewed it this way); how psychotherapy worked was considered to be a mystery hardly worth investigating. Fortunately, in the department of psychiatry at the University of Pennsylvania, Lester Luborsky and Aaron (Tim) Beck did not take this view and established a dynasty of psychotherapy research that is unparalleled in the United States. I have been fortunate to work with these preeminent men and have learned from them that psychotherapeutic change can be measured, given the right tools and the wit to investigate it. Lester Luborsky and Tim Beck established manuals for the conduct of dynamic and cognitive psychotherapy, the "how-to" books, that now are the backbone of objective psychotherapy research and many clinical programs.

Many short-term variants of dynamic psychotherapy now fall under the general description of supportive-expressive psychotherapy. An important trend, led by Lester Luborsky and his colleagues, has been the development of methods to "dissect" apart what happens between patient and therapist during these dynamic psychotherapies and objectively evaluate how they work. Dr. Luborsky developed the first reliable method of estimating the transference pattern in psychotherapy, a method he called the Core Conflictual Relationship Theme. See L. Luborsky, *The Symptom-Context Method* (Washington, D.C.: The American Psychological Association, 1996).

Martin Seligman, also a professor at the University of Pennsylvania, in the department of psychology, starting from the perspective of the experimental psychologist, has made many contributions to our understanding and to the treatment of mood disorder from the cognitive perspective.

For further reading in the general area of psychotherapy and psychotherapy research, I recommend A. T. Beck, A. J. Rush, B. F. Shaw, G. Emery, *Cognitive Therapy of Depression* (New York: The Guilford Press, 1979); L. Luborsky, P. Crits-Christoph, J. Mintz, A. Auerbach, *Who Will Benefit from Psychotherapy? Predicting Therapeutic Outcomes* (New York: Basic Books, 1988); and M. E. P. Seligman, *Learned Optimism: How to Change Your Mind and Your Life* (New York: Pocket Books, 1990). For an excellent introduction to the art of psychotherapy, see *First Steps in Psychotherapy: Teaching Psychotherapy to Medical Students and General Practitioners,* edited by H. H. Wolff, W. Knauss, W. Bräutigam (Berlin: Springer-Verlag, 1985). Finally, for an extraordinary volume that places psychotherapy in a human construct, a volume filled with history and insight, read Irvin Yalom's *Existential Psychotherapy* (New York: Basic Books, 1980).

Page 243, CLASSICAL PSYCHOANALYSIS: In uncovering the conflicts of the unconscious mind, psychoanalysts traditionally look back in time. Freud believed that early sexual strivings are critical in determining what happens later in life, and that these, and other "base" instincts, lie largely outside consciousness. (Today we might consider these instincts equivalent to the homeostatic systems that control the appetites for sex, food, and aggression, and locate them within the hypothalamus of the lizard brain.) Freud postulated that it is what we learn—the restraints of a disciplined upbringing and well-defined social rules—that keeps these primitive drives in check. In Freud's theoretical model (which became the language of clinical practice), the learned constraints, once internalized, become the superego, and the dark drives of the unconscious are the id. If the superego fails and the id prevails then the ego (the self), serving as a balance between the two, is in danger of being crushed in the battle. In guarding against such a catastrophe, Freudian therapists conceptualize an ego with adaptive mechanisms—similar in concept to the adaptive mechanisms of the physiological stress response—that are called the mechanisms of ego defense. These serve to reduce anxiety and are expressed through distinct behaviors, some of which promote healthy adaptation (the mature defenses, such as humor and altruism), while others compromise it. The *denial* of conflict or illness; the *projection* of suspicion and blame onto others; *sublimation* of desire into a socially acceptable form, as when an individual with a desperate need to be cared for spends a lifetime helping others—all these behaviors are examples of mechanisms that sustain the ego under stressful circumstances. Anxiety or depression emerge when immature ego defenses are overwhelmed or break down. At other times, when stress is minimal and the defenses are adequate, although the unconcious conflict remains, no illness is apparent to the untrained observer. For this reason a "neurosis" appears to ebb and flow throughout the life of an individual. For a discussion of the psychological mechanisms of defense, see Anna Freud's classic little book, *The Ego and the Mechanisms of Defense* (New York: International Universities Press, 1966).

Page 244, THE PROFILE . . . THAT RESPONDS WELL TO LITHIUM: See A. Koukopoulos, D. Reginaldi, G. Minnai, G. Serra, L. Pani, F. N. Johnson, "The long-term prophylaxis of affective disorders," *Advances in Biochemical Psychopharmacology,* edited by G. Gessa, W. Fratta, L. Pani, and G. Serra (New York: Raven Press, 1995).

Page 245, WINSTON CHURCHILL: There is significant evidence of manic-depressive illness in Winston Churchill's family, dating back to John Churchill, the first Duke of Marlborough, and perhaps earlier. Marlborough, who in the late seventeenth century was the greatest of English generals, as Winston Churchill himself observed, was "Sometimes overdaring and sometimes overprudent . . . separate states of mind, and he changed from one to the other in quite definite phases." Among other things, Marlborough was responsible for the building of Blenheim Palace. Situated in Woodstock, ten miles from Oxford, it is one of the most grandiose constructions to be found in the English countryside. The Churchill family depression, which Winston called his "Black Dog," seized the English statesman early in life. For Winston it was interspersed with periods of extraordinary vision and productivity, including the leadership of Britain during the Second World War. See A. L. Rowse, *The Early Churchills* (London: Macmillan, 1956), pp. 227–228; Anthony Storr, *Churchill's Black Dog and Kafka's Mice* (New York: Ballantine Books, 1990); and John Pearson, *The Private Lives of Winston Churchill* (New York: Simon & Schuster, 1991), pp. 24–30.

Page 246, JOHN BOWLBY: See John Bowlby, *Attachment and Loss*, 3 volumes (New York: Basic Books, 1980).

Page 246, INTERPERSONAL PSYCHOTHERAPY: Interpersonal Psychotherapy (IPT) is a time-limited psychotherapy designed specifically for the treatment of depression. IPT makes no assumption about the historical origin of depressive complaints, but ties the symptoms of depression to current interpersonal difficulties occurring in a patient's life situation. The program is structured through a manual and lasts usually 6–8 weeks. IPT began as a research program in the psychotherapy of depression and was the intervention used in the three-year Pittsburgh study by Frank and Kupfer (see note, below).

For details of IPT, see G. L. Klerman, M. M. Weissman, B. J. Rounsaville, E. S. Chevron, *Interpersonal Psychotherapy of Depression* (New York: Basic Books, 1984), and for a review, see M. M. Weissman, J. C. Markowitz, "Interpersonal psychotherapy: Current status," *Archives of General Psychiatry* 51 (1994): 599–606.

Page 246, AARON BECK: See earlier note on psychotherapy (page 335), and also "A psychiatrist who wouldn't take no for an answer," by Joel Greenberg (Science Times, *New York Times,* August 11, 1981). The *Times* article is interesting for its description of the research which led Dr. Beck to the development of cognitive therapy, and for its account of his struggles with the psychiatric and psychoanalytic establishment.

Page 248, ANTIDEPRESSANTS CAN . . . DESTABILIZE LIMBIC PATHWAYS,

EVEN IN PATIENTS WITHOUT BIPOLAR ILLNESS: Some psychiatrists, notably in Europe, have argued that the antidepressant drugs are being overused, especially in minor depressions, and through sensitization are causing recurrent episodes of illness that otherwise would not exist. For an interesting editorial, see Giovanni Fava, "Holding on: Depression sensitization by antidepressant drugs, and the prodigal experts," *Psychotherapy and Psychosomatics* 64 (1995): 57–61.

Page 248, PROFESSORS ELLEN FRANK AND DAVID KUPFER: For details of this landmark study, see E. Frank, D. J. Kupfer, J. M. Perel, C. Cornes, D. B. Jarrett, A. G. Mallinger, M. E. Thase, A. B. McEachran, V. J. Grochocinski, "Three-year outcomes for maintenance therapies in recurrent depression," *Archives of General Psychiatry* 47 (1990): 1093–1099; and G. L. Klerman, "Treatment of recurrent unipolar major depressive disorder: Commentary on the Pittsburgh Study," *Archives of General Psychiatry* 47 (1990): 1158–1162. The importance of maintenance antidepressant therapy in the treatment of recurrent depression was recognized in the early 1980s (see R. F. Prien, D. J. Kupfer, P. A. Mansky, J. G. Small, V. B. Tuason, C. B. Voss, W. E. Johnson, "Drug therapy in the prevention of recurrences in unipolar and bipolar affective disorders: report of the NIMH Collaborative Study Group," *Archives of General Psychiatry* 41 [1984]: 1096–1104) and confirmed in later studies. (J. H. Kocsis, R. A. Friedman, J. C. Markowitz, A. C. Leon, N. L. Miller, L. Gniwesch, M. Parides, "Maintenance treatment of chronic depression," *Archives of General Psychiatry* 53 [1996]: 763–774.) Antidepressants, particularly the serotonin reuptake inhibiting drugs, are also valuable in the long-term relief of dysthymia. For details, see M. E. Thase, M. Fava, U. Halbriech, J. H. Kocsis, L. Koram, J. Davidson, J. Rosenbaum, W. Harrison, "A placebo-controlled trial comparing sertraline and imipramine for the treatment of dysthymia," *Archives of General Psychiatry* 53 (1996): 777–784.

Page 249, COMPARISON . . . WITH DIABETES: For those who doubt such comparisons, see E. G. Eakin, R. E. Glasgow, "Physician's role in diabetes self-management: Helping patients to help themselves," *The Endocrinologist* 6 (1996): 186–195.

Page 250, THINKING ABOUT HOW ONE THINKS: Manuals for the cognitive treatment of bipolar disorder have been slow in coming compared to books on cognitive therapy for depression, even though at the University of Pennsylvania (and other centers) we have been using cognitive therapy for some time to help manage bipolar disorder. However, this situation has changed dramatically with the publication of two excellent volumes, M. Ramirez-Basco, A. J. Rush, *Cognitive-Behavioral Therapy for Bipolar Disorder* (New York: The Guilford Press, 1996); and M. Bauer, L. McBride, *Structured Group Psychotherapy for Bipolar Disorder: The Life Goals Program* (New York: Springer, 1996).

Page 251, THE MANIC DEPRESSION FELLOWSHIP: The Manic Depression Fellowship runs self-help groups for those with bipolar illness, and for family members, in over one hundred locations throughout Britain. For more information contact Manic Depression Fellowship, 13 Rosslyn Road, Twickenham, Middlesex, TW1 2AR, England.

Epilogue

Page 254, THE SYMPTOMS OF MOOD DISORDER ARE WIDESPREAD: For a concise review of the epidemiology of depressive illness and manic depression, see Paul Bebbington, "The epidemiology of depressive illness," in *Psychopharmacology of Depression*, ed. S. A. Montgomery and T. H. Corn (Oxford: Oxford University Press, 1994), pp. 1–18.

Page 254, MANIA AND MELANCHOLIA ARE . . . FOUND THROUGHOUT HUMAN SOCIETY: For a recent cross-cultural study, see M. M. Weissman, R. C. Bland, G. J. Canino, and colleagues, "Cross-national epidemiology of major depression and bipolar disorder," *Journal of the American Medical Association* 276 (1996): 293–299. Using similar methods, the authors estimated the lifetime rates of depression and bipolar illness in ten countries, including the United States, Canada, Puerto Rico, France, Germany, Italy, Lebanon, Taiwan, Korea, and New Zealand. Depression ranged from 1.5 cases per 100 adults in Taiwan to 19 cases per 100 adults in Beirut. Bipolar illness was more consistent, ranging from 0.3 percent in Taiwan to 1.5 percent in New Zealand.

Page 254, THE EVOLUTIONARY SIGNIFICANCE OF EMOTION: Drs. Price and Nesse have both written extensively about behavioral states, distinguishing evolutionary explanations from explanations of "proximate cause," the immediate disturbances that dictate behavioral change in an individual. See R. M. Nesse, "What good is feeling bad? The evolutionary benefits of psychic pain," *The Sciences,* November/December (1991): 30–37; and A. Stevens, J. Price, *Evolutionary Psychiatry, A New Beginning* (New York: Routledge, 1996).

Pages 254–55, "WILL ANTIDEPRESSANTS ELIMINATE SADNESS": This has become a popular topic. See the cover story in *Newsweek,* February 7, 1994, "Beyond Prozac. How science will let you change your personality with a pill." In fact, antidepressants do little for people who are not depressed.

Page 255, BIPOLAR ILLNESS CLUSTERS MORE COMMONLY IN PROSPEROUS FAMILIES: The best discussion of this complicated subject will be found in the chapter entitled Epidemiology in F. K. Goodwin, K. R. Jamison, *Manic-Depressive Illness* (New York: Oxford University Press, 1990), pp. 169–173.

Page 256, WINSTON CHURCHILL'S FAMILY LINE: See earlier reference note, Chapter 10, page 337.

Page 256, THE GENETIC VULNERABILITY TO ALZHEIMER'S DISEASE: There is increasing evidence that familial Alzheimer's disease is inherited through gene mutations on three different chromosomes (14, 1, and 21) which together induce the characteristic metabolic changes in amyloid proteins in the brain. "Sporadic" cases of Alzheimer's disease also may involve contributions from other genes. A technical, but comprehensive, review is by R. Sandbrink, T. Hartmann, C. L. Masters, K. Beyreuther, "Genes contributing to Alzheimer's disease," *Molecular Psychiatry* 1 (1996): 27–40.

Page 258, RISK-TAKING . . . WALL STREET TRADERS: See *The Wall Street Journal,* Tuesday, October 8, 1996, "The latest management craze: Crazy management."

Page 258, "GENOTYPING–COMPUTER CHIP": This is not science fiction. See "Chipping away at the human genome," a news article that appeared in *Science* magazine, June 21, 1996, page 1,737, and which describes the potential development of DNA genomic scanning using microchip technology.

Page 258, RESEARCH GROUPS . . . ARE . . . COMBING THE GENOME: Several chromosomes have been identified as candidates for carrying genes associated with manic depression but rarely have they been confirmed across different research groups. See "Manic-depression findings spark polarized debate," a news article in *Science* magazine, April 5, 1996, pp. 31–32. A promising chromosome is 18, loci being reported by Berrettini and colleagues (*Proceedings of the National Academy of Science,* USA, 91 [1994]: 5,918–5,921), and by Freimer and colleagues (*Nature Genetics* 12 [1996]: 436–441). Other reports suggest candidate loci on chromosomes 6, 13, 15, and 4, and an older finding, now withdrawn, implicated chromosome 11. It is now generally agreed that vulnerability to the illness likely involves an interaction among several genes.

As molecular techniques have improved, the search for genetic determinants of mammalian behavior has broadened. Those interested in the various methods being employed will find an excellent review in L. A. McInnes, N. B. Freimer, "Mapping genes for psychiatric disorders and behavioral traits," *Current Opinion in Genetics and Development,* 5 (1995): 376–381.

Page 258, GENES ARE NOT DESTINY: Some of the thoughts outlined in this epilogue were developed during my participation in the workshop *Manic-Depressive Illness: Evolutionary and Ethical Issues,* organized by Dr. Kay Jamison and held in October 1996 at the Banbury Center, Cold Spring Harbor Laboratory, New York, and funded by the Charles A. Dana Foundation.

Name Index

Abrams, R., 327n
Abramson, L. Y., 284n–85n
Adolphs, R., 304n
Aggleton, J. P., 303n
Akiskal, Hagop S., 108, 110–11, 284n, 293n–94n, 296n, 305n, 333n, 334n
Alavi, A., 303n
Alltop, L. B., 331n
Altman, P. M. E., 287n
Alvarez, A., 63, 282n
Ambelas, A., 280n
Amsterdam, J. D., 303n
Ashby, C., 287n
Aston-Jones, G., 313n
Auerbach, A., 336n
Augustine, 97, 289n
Avery, D., 296n, 307n, 311n
Bahr, Robert, 307n
Bailey, J. E., 315n
Ball, W., 303n
Ballenger, J. C., 281n, 312n, 329n
Baraban, J. M., 330n

Barondes, S. H., 286n, 323n
Bauer, M. S., 313n–14n, 328n, 338n
Bearn, J. A., 326n
Bebbington, Paul, 339n
Beck, Aaron T., 80, 242, 246, 247, 279n, 282n–83n, 285n, 335n, 336n, 337n
Bennett, E. L., 297n
Berger, Diane, 334n
Berger, Lisa, 334n
Bergman, Ingmar, 45, 279n
Bernard, Claude, xvii, 149, 151, 270n, 272n, 304n
Berrettini, Wade, 340n
Bertelsen, A., 296n
Beyreuther, K., 339n–40n
Bland, R. C., 339n
Bloom, Floyd E., 297n, 301n, 313n, 322n
Bloom, R. H., 322n
Bouchard, Thomas, 83–84, 286n
Bowden, C. L., 329n
Bowlby, John, 246

341

Subject Index